T0313755

Developmental Neuroscience

Developmental Neuroscience

A CONCISE INTRODUCTION

Susan E. Fahrbach

PRINCETON UNIVERSITY PRESS
Princeton and Oxford

Published by Princeton University Press, 41 William Street, Princeton, New Jersey 08540
In the United Kingdom: Princeton University Press, 6 Oxford Street, Woodstock,
Oxfordshire OX20 1TW

press.princeton.edu

ISBN 978-0-691-15098-7
Library of Congress Control Number: 2013936865

British Library Cataloging-in-Publication Data is available

This book has been composed in ITC Stone Serif with Whitney display
by Princeton Editorial Associates Inc., Scottsdale, Arizona.

Printed on acid-free paper. ∞

Printed in the United States of America

10 9 8 7 6 5 4 3 2 1

To Jon, for reminding me that it's an adventure

CONTENTS

ILLUSTRATIONS

PREFACE

The best investigators recognize interesting questions that don't fit into a predefined paradigm and follow the biology for its own sake. These curiosity-driven experiments are the ones that lead to truly surprising discoveries. We can expect that studies of seemingly exotic developmental events will continue to provide new perspectives on evolution and human biology.

Anderson and Ingham (2003), 285

It is long-established tradition that beginning biology students learn to identify the four categories of animal tissue: epithelial, connective, muscle, and nervous. The student learns that nervous tissue contains two cell types: neurons and glial cells. The glial cells are then immediately set aside as the focus shifts to the stars of the nervous system, the neurons. It will be noted that neurons are electrically excitable cells that integrate information and transmit it to other cells (most often, to other neurons), primarily by chemical signals; that neurons are unlike other cells in that they possess long, thin extensions of the cytoplasm called axons and dendrites; and that neurons form polarized cell junctions called synapses. Following this reductionist line of thought helps students understand that the study of nervous system development is the story of how newly born cells differentiate a neuronal phenotype: how they come to express voltage-gated ion channels, assemble an extended cytoskeleton, and position synaptic proteins in just the right locations. In the twenty-first century, the story of how cells acquire a neuronal phenotype can be told in terms of molecular signals and the cellular receptors for those signals. An introductory account of these signals fills major parts of the chapters in this book. The molecular story of neuronal differentiation can in fact be told over and over again, with subtle variations and surprising plot twists, because there are so many different types of neurons. It's been estimated that as many as 100 billion neurons make up the human brain, collectively representing thousands, maybe even tens of thousands, of different ways to be a neuron.

But molecular signaling is only part of our story. The ability of the nervous system to integrate environmental cues and internal signals such as hormones with remembered experience to produce thoughts and behavior depends on its wiring diagram. Information flows through the nervous system via polarized neural circuits (by *polarized* I mean simply that there are distinct input and output sides to the circuit). These sophisticated circuits have built-in feedbacks, delays, and convergences that collectively enable a single circuit to produce multiple outputs. The second part of our story therefore involves understanding how connections within neural circuits are formed and sustained. If the diversity of neuronal phenotypes in the human brain is surprising, the targeting of the estimated 100 trillion connections (synapses) in the brain to form circuits is absolutely astonishing.

The third part of our story is the plasticity of the nervous system. In a sense, the development of the nervous system is a never-ending story. Across the life span, nervous systems respond to internal and external signals by altering neuronal phenotype and refining neural circuits. Familiar examples of neural plasticity are the seasonal behaviors of temperate-zone animals, acquisition of a skill such as playing the violin or a new video game, and formation of long-lasting memories of life events such as our first day of school. Nervous systems also have the capacity to recover from many (but not all) injuries. Does lifelong plasticity reflect reengagement of the mechanisms that supported the formation of the embryonic nervous system? Until we have a fuller understanding of both development and plasticity, this fascinating question is impossible to answer. This book does not avoid topics related to plasticity, but its primary goal is to give the reader a thorough grounding in the earliest stages of development.

While some readers of this book will be interested in learning about the nervous system so that they can better understand brain evolution and animal behavior, others will want this information so that they can be better physicians and educators and, eventually, parents. The latter category of readers may be disappointed that so many chapters focus on species other than humans. These species—some of which play such an important role in studies of development that they are referred to as *model organisms*—have contributed so much to our understanding of development that it would be impossible to write a meaningful book without reference to them. But readers primarily interested in humans can take heart, because advances in our knowledge of the human genome and proteome paired with new techniques of noninvasive brain imaging mean that direct studies of the development of the human nervous system are increasingly informative. For example, studies of teenagers using noninvasive brain imaging have revealed surprising and useful information about brain development during adolescence (Chapter 9).

Some readers of this book may be considering careers in neuroscience research. In the early 1980s, I chose to investigate the changes that occur in insect nervous systems during metamorphosis because I did not foresee the rapidity with which exciting studies of the developing mammalian nervous system would become possible. I love insects and have never been unhappy with my choice, but students interested in research on development of the nervous system can now choose from a longer menu of enticing options.

Many of these new models and areas of research are described in this book. Is it a good time to choose a career in neuroscience research? I think that most neuroscientists would agree with me when I say that the answer to this question is always *yes*.

It is my hope that this introductory account of nervous system development inspires all readers, but it is particularly dedicated to undergraduates encountering the subject of development for the first time. I assume that most such readers will have completed an introductory biology course (or courses) covering the basics of physiology, cell biology, genetics, and molecular biology. Students ready for more information can consult the notes and source lists for each chapter. Many of the references cited are review articles. A well-written review is often fun to read because it provides a concise summary of an interesting topic, but the savvy student appreciates that every such article is also a database. The opinions expressed in a review eventually become dated, but the curated list of references at the end of the article is timeless. In other words, use review articles (the secondary literature) as your portal to the primary literature (research reports published as journal articles).

By the way, my surname is easier to pronounce than to spell: just say *farbock*, and you've got it.

ACKNOWLEDGMENTS

Thanks to my teachers, mentors, and colleagues, especially (in order of acquaintance) Paul Rozin, Jane Mellanby, Joan Morrell, Donald Pfaff, Toni Wolinsky, Robert Meisel, Jim Truman, Karen Mesce, Stanley Friedman, Hugh Robertson, Gene Robinson, Sarah Farris, Rodrigo Velarde, and Scott Dobrin. Thanks to Bill Greenough for introducing me to Illinois and fragile X syndrome and to Linda Restifo for her insights into intellectual disability. Thanks to the Z. Smith Reynolds Library of Wake Forest University for easy access to the resources needed to complete this book. Thanks to Jonathan Christman, Maurice Robinson, Nathaniel Robinson, and Kenneth Fahrbach for encouragement, never overdone but always supplied when needed. Thanks to Maggie Christman for the daiquiris by the pool.

WHAT ARE
INVESTIGATIVE READING QUESTIONS?

Investigative reading questions are found at the end of each chapter. These questions offer students the chance to test their understanding by thinking about experiments based on material presented in the chapter. The most difficult of these questions are highlighted as *Challenge Questions*. Answers can be checked by carefully reading the recommended article that is provided, although you should be open to the possibility that you will come up with a better answer than the original investigators did. Full citations for each article are given at the end of the book. All of the articles used as the basis of investigative reading questions can be accessed free of charge by anyone with an Internet connection, regardless of institutional affiliation. Some of the articles are accessible because the journal publishers (in particular, scientific societies) make their archives freely available. Many articles are archived in PubMed Central (PMC), a free, full-text online library of over 2 million biomedical and life sciences articles maintained by the U.S. National Institutes of Health's National Library of Medicine (http://www.ncbi.nlm.nih.gov/pmc/). All research supported by the National Institutes of Health, the major public funder of research in the United States, is required to be made available to the public via PMC no later than 12 months after initial publication. Other articles that serve as the basis for investigative reading questions were originally published by choice of the authors in an Open Access format. That is, some publishers routinely restrict access to newly published material to subscription holders but give authors the option of paying an Open Access fee to make their articles immediately available to all.

TEACHING USING THE PRIMARY LITERATURE AND INVESTIGATIVE READING QUESTIONS TO COMPLEMENT THE TEXT

Many neuroscience instructors introduce primary literature into their undergraduate classes, but the fact that this practice is common does not mean that it is easy for either instructor or student. My own teaching of the primary literature has been heavily influenced by the C.R.E.A.T.E. method developed by Sally Hoskins and colleagues. C.R.E.A.T.E. stands for **C**onsider, **R**ead, **E**lucidate the hypotheses, **A**nalyze and interpret the data, and **T**hink of the next **E**xperiment. This method has been described in several journal articles, and useful sample teaching modules are available on the C.R.E.A.T.E. Web site (www.teachcreate.org). Another effective approach modifies the familiar journal club format to teach undergraduates a systematic method for reading primary literature. A method to accomplish this, described by Katherine Robertson (2012), can be incorporated into existing courses and takes about four class sessions to complete.

Teaching References

Hoskins, Sally G. 2008. "Using a paradigm shift to teach neurobiology and the nature of science—a C.R.E.A.T.E.–based approach." *Journal of Undergraduate Neuroscience Education* 6: A40–52.

Hoskins, Sally G., Leslie H. Stevens, and Ross H. Nehm. 2007. "Selective use of primary literature transforms the classroom into a virtual laboratory." *Genetics* 176: 1381–89.

Robertson, Katherine. 2012. "A journal club workshop that teaches undergraduates a systematic method for reading, interpreting, and presenting primary literature." *Journal of College Science Teaching* 41: 25–31.

Developmental Neuroscience

ONE

Introduction

What Do We Mean When We Say "Neural Development"?

Development unfolds smoothly over time but can be divided for experimental analysis into successive stages, each with its own defining events. Some of these events have clear beginnings and endings, although others may be protracted, sometimes unexpectedly so. For example, myelination of axons in the human brain, a key event that supports behavioral development by increasing the rate of action potential transmission, begins approximately 24 weeks after conception and then continues for decades. In general, the earliest events are most easily categorized as discrete stages shared by almost all members of a species, whereas later events are best described as ongoing processes, the exact details of which are unique to each brain. This is particularly true in long-lived species such as humans, but the inherent ability of nervous systems to refine neural circuitry across the life span is evident even in short-lived invertebrates. The neuroscientist Martin Heisenberg and colleagues were reflecting on data obtained from neuroanatomical studies of fruit fly brains, not human brains, when they were inspired to write, "An individual's life experience can . . . be encoded in the volume of selected neuropil regions."[1]

What Is in This Book and How to Use It

After a brief presentation of methods (this chapter), an overview of human development (Chapter 2), and an introduction to animal models (Chapter 3), the subsequent chapters consider the molecular mechanisms of selected earlier and later events (Chapters 4 and 6), neurogenesis (Chapter 5), and formation of synapses (Chapter 7). Glial cells are the focus of Chapter 8. Chapter 9 describes the postembryonic maturation of the nervous system via metamorphosis in some species and adolescence in others. In Chapter 10 the focus shifts to human intellectual disabilities. This chapter attempts to build a case that at least some forms of human intellectual disability reflect reversible differences in developmental processes rather than permanent deficits. This chapter was inspired by my personal connections with two outstanding neuroscientists—William T. Greenough at the University of Illinois at Urbana-Champaign and Linda L. Restifo of the University of Arizona, a researcher who is also a physician. Many other outstanding investigators work in this field, but it was Greenough's studies of the fragile X protein in the context of his life's work on experience-driven brain plasticity and Restifo's studies of mental retardation genes in *Drosophila* that forced me to rethink my views on human intellectual disability.

Each chapter has notes. Some provide additional background information on the topic being discussed. This information may be useful and/or interesting, but it has been placed in the notes because I believe it is not essential for an understanding of neural development. The notes are probably most helpful if they are consulted the first time you read the chapter. Some notes link specific results recounted in the text to specific references. The full references can be found in the chapter-by-chapter reference lists that appear at the end of the book. These reference lists also include pertinent reviews and commentaries that provide additional context if you are interested in the history of developmental neuroscience. Short chapter-by-chapter lists of trustworthy online resources can also be found at the end of the book. These are intended to provide additional graphic material and technical details as well as links to selected patient information web sites. This material is also nonessential. It is included to allow you to follow up a specific interest, either as you read or in the future.

Students who want to go further will find that there are numerous points of entry into the research literature. The student can begin with the end-of-chapter suggestions for *Investigative Reading*. Each of these readings is introduced by a short question based on the chapter. The answer to the question (one answer; the student may well come up with a superior alternative) is contained in the recommended reading. Students are encouraged to try to answer the questions on their own before going online to retrieve the article. Note that only partial citations are provided at the end of each chapter. This is because the titles of the articles often give the answers away! The journal articles listed in the Investigative Reading sections are freely accessible online, and full citations are provided at the end of the book.

This text is designed to provide a concise introduction to nervous system development. This goal will be achieved, in part, by a nearly exclusive focus on the central nervous system (the brain and spinal cord in vertebrates, the brain and nerve cord in invertebrates). We'll venture into the peripheral nervous system primarily in Chapter 7, where I use the neuromuscular junction to describe how synapses form. Topics intentionally shortchanged for the sake of brevity include the history of embryology, the neural crest, development of vertebrate sense organs, and the emerging story of microRNAs (miRNAs) as posttranscriptional regulators of development.[2] In addition, many of the signal transduction pathways described in this book have been pruned for clarity. Even in the garden, pruning is a tricky business. I apologize in advance if I inadvertently clipped your favorite branches.

Methods for Studying Development of the Nervous System

The modern neuroscientist's tool kit is stocked with powerful tools for studying the structure and function of the nervous system. While the tools of the electrophysiologist (intra- and extracellular recordings of neuronal electrical activity) and the neuroanatomist (many variants of microscopy) are still in heavy use, many developmental neuroscientists routinely incorporate measures of gene expression and functional brain imaging into their studies. Others just as routinely use genetically engineered (transgenic) animals.

The following sections introduce key techniques used to study nervous system development. Some researchers specialize in a particular technique, but many investigators work in teams and combine multiple approaches to answer research questions. Students who delve into the primary research literature are often amazed at the number of techniques required to generate the data contained in a single paper. This is one of the reasons that many modern research papers feature lengthy author lists.

Birthdating

All cells, including neurons, are produced by division of other cells. The time at which the cell division occurs that produces a particular neuron is referred to as that neuron's birthdate. Knowing neuronal birthdates is important for understanding the sequence of events that builds a neural circuit or a brain. Birthdating is also important for exploring the capacity of mature brains to add new neurons. Neurons themselves do not divide—part of becoming a neuron involves saying farewell to the cell cycle—so the challenge to the developmental neuroscientist wishing to determine a birthdate is to catch the neuron in the act of being produced by a progenitor cell that by definition is not itself a neuron.

If an animal is small and transparent, the process of cell division can be observed directly using a microscope. Otherwise, developing tissues may be fixed (preserved by chemical treatment), sectioned into thin slices (section thickness is typically measured in micrometers, μm), and attached to glass slides for viewing with a microscope. Stains may be applied to the sections to enhance detection of dividing cells. A combination of hematoxylin and eosin reveals key features of many tissues, including nervous tissue, because hematoxylin stains nuclei blue and eosin stains most other structures red or pink. DNA stains aid the identification of mitotic profiles by making condensed metaphase chromosomes readily visible. The Feulgen stain is traditionally used to mark DNA for viewing with a standard bright-field microscope. Modern biologists with access to a fluorescence microscope can choose from an array of colorful dyes that bind to DNA.

A drawback to searching for mitotic profiles in tissue is that the window for detecting these profiles is often so brief that the likelihood of catching a neuron in the act of being born is small. An alternative approach also relies on detection of DNA, but instead of staining all of the nuclear DNA present in a tissue, the investigator labels only new DNA. This is accomplished by providing special DNA precursors to cells as they copy their nuclear DNA prior to cell division. These precursors do not occur naturally in cells. Because the precursor provided is incorporated into new DNA, any neurons born during the time the precursor was present contain labeled DNA and can therefore be distinguished from cells born when the precursor was not present.

In classic studies, living animals were injected with the nucleoside thymidine linked to a radioactive atom (a nucleoside is a purine or pyrimidine base attached to a ribose sugar molecule; a radioisotope commonly used to label nucleosides is tritium, a radioactive isotope of hydrogen). The distribution of radioactivity in a tissue section prepared from the treated animal was subsequently detected by applying the section to a photographic emulsion. The radioactive decay particles emitted from the radioisotope exposed the film. At the end of an exposure period typically measured in months, the location of nuclei with radiolabeled DNA was revealed by developing the emulsion using darkroom chemicals. This method of detecting the distribution of a radioisotope in tissue is known as autoradiography.

Tritiated thymidine (^{3}H-thymidine) was used in neuronal birthdating studies through the 1970s. Its use has been superseded by a method based on detection of bromodeoxyuridine, a synthetic nucleoside that is an analog of thymidine. Antibodies can be purchased that bind specifically to bromodeoxyuridine. Labels attached to these antibodies make the position of bromodeoxyuridine within a tissue section readily evident using standard techniques of light microscopy (fig. 1.1). Bromodeoxyuridine is commonly referred to by its nickname, BrdU, pronounced bee-are-dee-you. Oval spots representing BrdU-labeled nuclei flash before the mind's eye of a neuroscientist who hears the term *neuronal birthdating.*

Birthdating methods that rely on incorporated nucleosides work only when the investigator can introduce the marker at the appropriate stage without perturbing normal development. Depending on the species, this may be accomplished by injecting or feeding or by immersing the entire animal in a solution containing BrdU. An alternative approach relies on immunodetection of endogenous molecules expressed by dividing cells. This circumvents the need to introduce a marker. Antibodies are available that recognize proteins expressed during the cell cycle. These include antibodies that bind to proliferating cell nuclear antigen (PCNA) and a nuclear protein

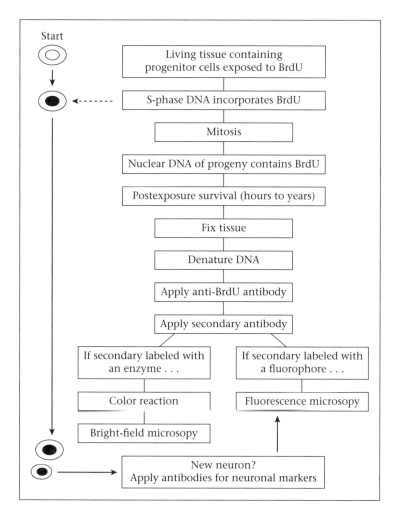

Figure 1.1. Determining neuronal birthdates by immunodetection of BrdU. DNA synthesized in the presence of the synthetic nucleoside BrdU can be detected in progenitor cells and their progeny. In the event depicted here on the left, a stem cell has completed a round of mitosis that regenerated the stem cell (the larger cell at the lower left) and produced a neural progenitor cell (the smaller cell at the lower left). An example of such an event is the production of a ganglion mother cell by a neuroblast in the developing ventral nerve cord of the fruit fly *Drosophila melanogaster* (Chapter 5). If the postexposure survival had been longer, three labeled nuclei would have been detected: the neuroblast and the two neuronal progeny of the ganglion mother cell.

called Ki-67. These proteins are not expressed by mature, postmitotic neurons, but they are good markers for progenitor cells and for newborn neurons, as they persist for several hours after mitosis before being metabolized.

One disadvantage of relying on the immunolabeling of endogenous proteins as markers for mitosis is that the antibodies used in these studies may not recognize proteins from a broad range of species. For example, antibodies raised against a fragment of human Ki-67 nuclear protein will likely cross-react with similar proteins expressed during the cell cycle in other mammals but are unlikely to bind even to related proteins in fish, birds, or insects. The challenges imposed by the need to introduce BrdU into developing tissues are often outweighed by the fact that this marker can be used to birthdate neurons (and other cells) in absolutely all animals, from hydras to humans.

Tissues exposed to BrdU can be chemically fixed for analysis shortly after introduction of the marker. This provides a snapshot of the cell divisions occurring at a specific point in development. Another approach is to allow the BrdU-treated animal to survive for some length of time after the marker is introduced. Because neurons do not divide, incorporated BrdU will persist in the nuclear DNA. The incorporated BrdU can be detected as long as the animal (or the neuron) lives.

Lineage Analysis

Lineage analysis refers to tracing the origin of a particular cell or cell type back through a series of successive cell divisions. This method differs from birthdating techniques in that the result is a family tree rather than a birth-date. Of course, the lineage of every cell in the body can be traced back to the fertilized egg (zygote), so it is not necessary to do experiments to prove this. It is the later portions of the lineage that are interesting to developmental neuroscientists, because these represent points at which developmental mechanisms act to limit a cell's fate.

The optical microscopes of the nineteenth century permitted biologists to observe living tissues directly to determine cell lineage. Direct observation remains a powerful tool for lineage analysis in small transparent or translucent embryos. In the 1970s and 1980s, direct observation of cell divisions was used to determine the lineage of every cell in *Caenorhabditis elegans*, a nematode worm. Researchers used the technique of differential interference contrast (DIC) microscopy to enhance the contrast of the un-stained living embryos and larvae they examined. But determination of cell lineage by direct observation is impossible in many animals, either because the embryo is not transparent or because development occurs inside an egg with an opaque shell or inside the mother's body. As a consequence, cell lineage determinations are often based on introduction of a marker into suspected progenitor cells. For example, a small amount of dye can be injected into a cell. If that cell subsequently divides, the resulting daughter cells will each contain some of the dye. It can be inferred that the injected cell was the parent of the pair of dyed cells. If one or both of the dyed daughter cells divides, the dye will also be found in the granddaughters of the injected cell.

A disadvantage of the direct injection method is dilution of the marker. Whatever dye was injected will be partitioned between pairs of progeny and hence diluted upon each successive division. Eventually the lineage marker will be so diluted that it will become undetectable. As a consequence, the simplest versions of this method cannot be used for analysis of long lineages.

But what if we could introduce a marker that replenished itself in the daughter cells after each division? Avoiding dilution would make it possible to trace long cell lineages. This approach is exploited by two powerful methods for determining cell lineage. The first is based on retroviruses that carry marker genes. The second produces marked cells by a process of gene recombination.

The genes of retroviruses are encoded in RNA instead of DNA. When a retrovirus infects a cell, the RNA-based genome of the virus is reverse transcribed and then integrated into the DNA of the infected cell. If the infected cell is a progenitor cell, its daughters inherit the viral genes along with the genes of the progenitor cell. Progenitor cells are typically infected by injecting the retrovirus into the extracellular fluid near the target cell or cells.

Natural retroviruses are altered for use in cell lineage tracing in the following ways. First, they are modified so that they cannot replicate. This modification prevents the virus from infecting neighboring cells, which may also be progenitors. Without this modification, it would be difficult to be certain that an inferred cell lineage was correct. Second, the retrovirus is modified so that it carries a reporter gene in addition to its own viral genes. A reporter gene produces a product that is easy to measure or see under a microscope. Two commonly used reporter genes code for the enzymes horseradish peroxidase (not surprisingly, a peroxidase enzyme produced by horseradish plants) and β-galactosidase (an enzyme encoded by the *lacZ* gene of *E. coli*). The presence of the retrovirus in a particular cell is revealed by supplying the tissue with the appropriate enzyme substrate; for example, the organic compound X-gal forms a blue product in the presence of β-galactosidase.

A second method for inserting a permanent marker into specific cell lineages is based on introduction of genes encoding site-specific DNA recombinases and their target sites into the genome of the species being studied. Because this method relies on transgenic animals with deliberately modified genomes, it is often used to study development of the nervous systems of genetic model organisms such as mice or fruit flies.

Site-specific DNA recombinases catalyze cleavage and rejoining of DNA strands at specific target sites. The recombinases used for experimental analysis of cell lineage have such large targets that they are unlikely to occur at random in an animal genome. This is the case for the Cre-*LoxP* system from the bacteriophage P1 (a bacteriophage is a virus that infects bacteria) and the FLP-*FRT* system from yeast. Both the Cre-*LoxP* and the FLP-*FRT* recombinase systems have been used to mark the progeny of specific neural progenitor cells (fig. 1.2). The introduced marker is permanent, so lineages marked in the embryo can be analyzed in the adult, even if the progeny have changed their morphology or migrated away from their place of birth. Because the

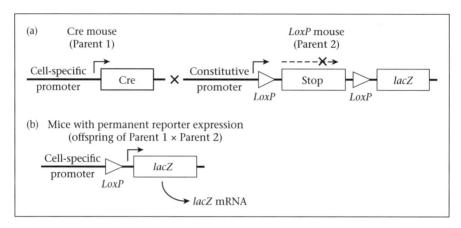

Figure 1.2. Use of the Cre-*LoxP* system for cell lineage tracing. One of the parent mice shown in (a) expresses the Cre recombinase enzyme under the control of a cell-type-specific promoter; the other parent is a reporter mouse in which a marker gene (*lacZ* in this example) is separated from a constitutively active promoter by a transcriptional stop sequence. The stop sequence is flanked by *LoxP* sites. Because *LoxP* is the target recognized by Cre recombinase, any cells in the offspring (b) that express Cre undergo removal of the stop sequences, which results in transcription of the reporter gene. The cells in which *LoxP* recombination occurs and any progeny they produce will be permanently marked. GFP is another popular Cre-dependent reporter gene. If the Cre recombinase gene is fused to a domain that binds a specific ligand (for example, a hormone), the Cre recombinase enzyme will be expressed only when the ligand is present. Arrows indicate the direction of transcription; triangles indicate the position of the *LoxP* sites.

logic of Cre-*LoxP* and FLP-*FRT* studies is similar, only the Cre-*LoxP* system is described here.

Cre, so named because it causes recombination, binds to a 34-base-pair target sequence called the locus of crossing over in bacteriophage P1, or simply *Lox*. If two directly repeated *Lox* targets are present in DNA, the Cre recombinase excises the intervening DNA, which is sometimes said to be floxed (flanked by *Lox*). If the excised DNA coded for stop of transcription, Cre action permanently turns on transcription of a previously silenced downstream gene. And if the previously silenced but now constitutively (continually) expressed downstream gene encodes a marker such as β-galactosidase, that marker becomes a lineage tracer.

In practice, use of DNA recombinase systems for lineage tracing requires extensive prior knowledge of the genes expressed in the progenitor cell populations being studied. This is because the expression of Cre recombinase must be limited to a particular progenitor cell or progenitor cell population if the results are to be interpretable. This is accomplished by placing Cre expression under control of promoter and enhancer elements from a gene with expression known from other studies to be restricted to the progenitor cells of interest.

You may have noted that use of the Cre-*LoxP* system for lineage analysis requires not one but two transgenes—in our example, the first transgene is the *Lox*-stop of transcription-*Lox-lacZ,* and the second is Cre recombinase driven in tandem with a developmental gene. If, for example, such a study is performed in mice, this means that two different lines of transgenic mice must be developed and then mated to produce double transgenic offspring. One quarter of the resulting offspring will have *lacZ* expression permanently turned on by Cre-mediated recombination. In practice, many different types of Cre transgenic mice, reflecting known patterns of developmental gene expression, can be interbred with a single strain of *Lox* mice.[3]

Microscopy

Microscopes are essential tools for developmental neuroscientists. Stereomicroscopes are commonly used in many laboratories for viewing whole animals and dissecting tissues of interest for subsequent analysis. Compound bright-field microscopes, which magnify light passed through sections of specimens using an objective and eyepieces, permit individual cells to be viewed directly. Eukaryotic cells in general and neurons in particular are often so large (relatively speaking) that it is easy to view not only the cells themselves but also large organelles through a compound microscope. The colored reaction products obtained when chromogenic (color-generating) substrates are supplied to cells labeled with enzyme markers are also easily seen with compound microscopes. A major limitation of bright-field microscopy, however, is the requirement that the sections be relatively thin—most studies are performed on sections thinner than 40 μm, sometimes much thinner. Tissues prepared for bright-field microscopy are therefore rather two-dimensional and, almost always, long dead. Development, by contrast, is three-dimensional and dynamic. Although generations of hardworking neuroscientists skillfully reconstructed key events in nervous system development by analyzing static, two-dimensional images, spectacular advances in microscopy and computer-based image acquisition and analysis now make it possible to visualize living cells as they make critical developmental decisions.

The development of confocal laser scanning microscopy (often referred to simply as confocal microscopy) for fluorescent markers is particularly noteworthy. Conventional fluorescence microscopes illuminate a tissue section with light of one wavelength that induces fluorescence in the sample. The resulting emitted light, which is of a longer wavelength than the illuminating light, is detected by the researcher through the microscope objective and eyepieces. Conventional fluorescence microscopy is valued for its sensitivity (a high signal-to-noise ratio) but shares with bright field microscopy the requirement for thin sections: unless the sections are thin, conventional fluorescent images are bright but blurry. Confocal microscopy also relies on fluorescent markers but uses point-by-point illumination with a focused laser beam paired with a moving (scanning) pinhole to eliminate out-of-focus emitted light.[4] The scanning is controlled by a computer, and images can be recorded in the z plane (depth of the tissue) as well as the x and y planes. The images acquired by laser scanning can be digitally stacked to produce a three-dimensional, high-resolution view of the specimen. Confocal images of cells and tissues are astonishingly crisp. Many well-characterized fluorescent molecules (fluorophores) are commercially available, permitting investigators to design sophisticated experiments using multiple markers.[5]

Near or at the top of many biologists' lists of favorite fluorophores is green fluorescent protein (GFP). GFP occurs naturally in the *Aequorea victoria* jellyfish as a component of this species' bioluminescent light organs. Note that many fluorophores are small molecules rather than proteins. This means that they are not encoded directly in the genome and therefore cannot be used to make fluorescent transgenes. But, as its name indicates, GFP is a protein. The cloning of the gene for GFP in 1992 allowed biologists to add the gene to other organisms in the form of transgenes coding for fusion proteins. A fusion protein is created by joining two genes that originally coded for separate proteins. If one of the fused genes is the gene for GFP, fluorescence can be detected whenever the cell produces the fusion protein. The GFP gene can also be attached to the regulatory region that drives expression of another gene so that a cell fluoresces whenever the other gene is being expressed. Today many variants of GFP (many of which are not green) are available as reporter molecules. For example, the gene encoding GFP can substitute for *lacZ* as the reporter gene in the Cre-*Lox* cell lineage analysis system described in the preceding section. The developers of GFP as a tool for studies of gene expression were Osamu Shimomura, Martin Chalfie, and Roger Tsien. They were recognized with the Nobel Prize in Chemistry in 2008.[6]

Gene Expression

The two most complex cell populations in your body are those of your nervous system and your immune system. It is true that many lymphocytes are genetically unique, but from the neuroscientist's perspective this diversity is accomplished by a sly bit of trickery involving DNA recombination during lymphocyte maturation. The nervous system achieves its diverse populations of neurons and glial cells not because these cells contain recombined genes but because at any given time they express different combinations of the genes encoded in the genome. This means that understanding nervous system development (and development in general) requires a clear accounting of transcriptional regulation in specific cell populations. At one time this task would have been conceptualized primarily in terms of identification of transcription factors, the protein signals that interact directly with nuclear

DNA to turn specific genes on and off. Transcription factors have not diminished in importance and are in fact a major focus of this book, but other factors such as noncoding RNAs (ncRNAs) and epigenetic modifications of chromatin are now recognized as essential to a complete understanding of cell differentiation.

Neuroscientists have just begun the hard work of listing the genes expressed at different stages of development of the nervous system. The challenge is great. First, multiple time points must be studied to capture all phases of development of a neuron's mature phenotype: neurogenesis, migration, growth of dendrites and axons, synaptogenesis, and establishment of neurotransmitter signaling systems. Second, the intrinsic plasticity of nervous systems means that even neurons that have differentiated a mature phenotype vary their gene expression profiles over multiple time scales. Third, cell populations within the nervous system are typically intermingled, so neurons from different lineages with different phenotypes can be neighbors. This mosaic quality of the brain means that tissue extract–based methods for assessing gene expression, such as DNA microarrays and whole-transcriptome shotgun sequencing (RNA-seq), tell only part of the story. Information flow in neural circuits depends on polarized synaptic connections rather than proximity. Studies of gene expression in the brain therefore require a merger of anatomical and molecular approaches.

A direct way of asking if a particular protein is expressed by a particular nerve cell or population of nerve cells at a particular point in time is to make the protein visible by tagging it with an antibody molecule. Different antibody molecules bind with specificity to different epitopes—typically fragments of proteins—and can themselves be tagged with fluorophores or enzyme markers. Identification of proteins in tissue samples or cell extracts using labeled antibodies is referred to as immunolabeling. Most techniques are based on chemically fixed tissues, because such fixation preserves the spatial distribution of proteins in the living cells. These techniques are referred to as immunohistochemistry or sometimes simply IHC (fig. 1.3). Both polyclonal antibodies (found in the serum of an animal exposed to the target protein) and monoclonal antibodies (found in the medium of hybridoma cells formed by fusing B lymphocytes from an animal exposed to the target protein with immortal tumor cells) are widely used in neuroscience research.

Antibodies can also be used to identify proteins present in extracts or homogenates of nervous tissue. A mixture of proteins can be separated by molecular mass using gel electrophoresis (passing an electrical current through a gel onto which the protein has been placed, or loaded). The separated proteins are in turn transferred to a membrane that is incubated in a solution containing the antibody; the antibody binds to any protein on the membrane that contains its epitope, and such binding (revealed by visualization of a marker) is evidence that the protein of interest was present in the tissue

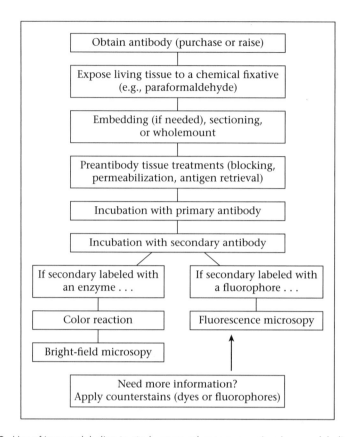

Figure 1.3. Use of immunolabeling to study neuronal gene expression. Immunolabeling provides a method for identifying cells that contain specific protein epitopes. The major steps are shown in sequence; actual experiments typically take days to complete, and each step in the procedure must be optimized to achieve meaningful results. The primary antibody, which may be obtained from any antibody-producing species other than the one being studied, binds to an epitope on the target protein; the secondary antibody binds to the primary antibody. The secondary antibody is modified by attachment of an enzyme or a fluorophore so that the presence of the target protein can be detected in the tissue by microscopy. For example, if the primary antibody was made in a mouse, the secondary antibody must be an anti-mouse antibody made in another species, such as a goat or a donkey. Note that a primary antibody made in a mouse cannot be confidently used to localize proteins in mouse tissue, because the anti-mouse secondary antibody would bind both to the primary antibody being used in the study and to endogenous mouse immunoglobulin molecules. Important control procedures include processing tissue with the primary antibody omitted and incubating the primary antibody with an excess of the target protein before applying it to tissue. This latter control is referred to as preabsorption. A preabsorbed antibody should not produce signal in tissue; if it does, the resulting immunolabeling will be difficult or impossible to interpret. The term *wholemount* refers to the processing of an entire embryo or, in some cases, the entire central nervous system, without sectioning.

from which the extract was prepared. This procedure is referred to as Western blotting or immunoblotting (the two terms mean the same thing).

Given that *DNA makes RNA makes protein* and *proteins get the job done*, students are sometimes surprised to learn that many studies of gene expression do not involve identification of proteins. Instead, the presence of messenger RNA (mRNA) in an extract prepared from a tissue sample is commonly taken as evidence that a particular gene is being expressed at that time in that cell population. This approach is popular in part because sequencing of nucleic acids is more easily and cheaply accomplished than sequencing of proteins. Probes for detection of specific mRNA sequences are easily synthesized, and the tiny amounts of mRNA present in individual cells or small pieces of tissue can be amplified (copied) for study by reverse transcribing the mRNA to its complementary DNA (cDNA). By contrast, proteins are more difficult (and costly) to sequence, are typically detected using antibodies produced by immunizing animals with proteins (which take weeks to months to produce), and cannot be readily copied for study. The number of predicted protein sequences available as a result of completed genome sequencing projects vastly outstrips the availability of antibodies targeted to specific proteins.

A standard method for detection of mRNA is Northern blotting. This procedure begins with isolation of total mRNA from a biological sample. Many kits are available that make the processes of tissue disruption and RNA stabilization efficient and reproducible, even for small samples. The extracted mRNA is loaded onto a gel and separated into fragments of different sizes by electrophoresis. The separated mRNAs are then transferred (blotted) to a filter. Specific mRNAs are identified by exposing the filter to a solution containing a radiolabeled single-stranded RNA or DNA probe complementary to the mRNA from the gene of interest. This radioactive probe can be used to expose X-ray film. Because only the target mRNA is visible on the film, the visible bands reveal how much of the target RNA was present in the sample. Standards can be electrophoresed and blotted side by side with the tissue-derived RNA so that the size of the visible band can be estimated.

Northern blotting is a tried and true method based on procedures many students will have an opportunity to perform in a laboratory class, but most researchers analyzing gene expression today prefer a fluorescence-based method called real-time quantitative reverse-transcription polymerase chain reaction, or real-time qRT-PCR (fig. 1.4). As its name indicates, real-time qRT-PCR is based on monitoring the progress of a PCR reaction in real time. Fluorescent reporter molecules fluoresce with increased intensity as the PCR product accumulates, allowing quantitation of the number of starting copies of mRNA. (The original mRNA, however, is no longer present because it was reverse transcribed into cDNA after being extracted from the sample to take advantage of the superior stability of DNA.) A single instrument runs the PCR reactions and monitors the fluorescent signal in multiwell plates. Many instruments run plates that contain 384 or even more wells, which means

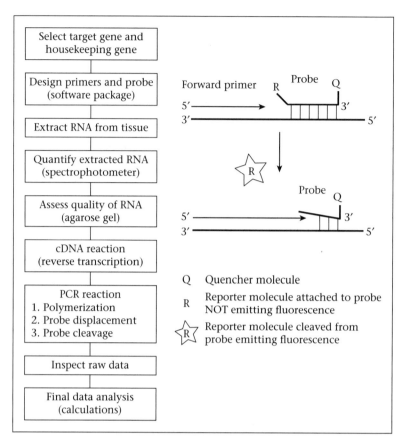

Figure 1.4. Use of real-time qRT-PCR to study neuronal gene expression. Several methods are widely used. The method depicted uses specific DNA probes that have a reporter fluorophore (R) attached at one end and a quencher fluorophore (Q) attached at the other end. The reporter fluorophore emits a detectable fluorescent signal only when it is cleaved from the probe. This occurs in the presence of an appropriate primer because DNA polymerase extends the forming DNA strand during the PCR reaction. The reaction takes place and the fluorescence is detected in an instrument typically referred to as a real-time PCR machine. The number of PCR reaction cycles required to produce detectable reporter fluorescence is related to the number of transcripts for that gene initially present in the tissue sample (more transcripts = fewer PCR cycles required to reach threshold). Gene expression is quantified by comparing the number of cycles required to reach the detection threshold for the gene of interest to the number of cycles required for a stably expressed housekeeping gene to be detected. Commonly used housekeeping genes are 18s rRNA, β-actin mRNA, and glyceraldehyde-3-phosphate dehydrogenase (GAPDH) mRNA. Choice of an appropriate housekeeping gene is particularly important for developmental studies, because a hallmark of development is changing gene expression as cells grow and differentiate.

that gene expression in hundreds of samples can be analyzed by a single researcher in a single afternoon.

The wonderful efficiency of real-time qRT-PCR may not compensate for the loss of spatial resolution inherent in methods based on analysis of isolated RNA, especially for researchers investigating the development of complex tissues such as those of the nervous system. Another tool for analysis of gene expression is in situ hybridization (fig. 1.5). In this method, target mRNA sequences are localized to specific cells in tissue sections, tissue pieces, or even whole embryos. This is accomplished by exposing the target sequence (in the tissue) to a complementary nucleotide probe (either cRNA or cDNA). Under the appropriate conditions, hybridization of the probe with the target occurs (in the context of molecular biology, the term *hybridization* means that a double-stranded nucleic acid forms as a result of complementary base pairing). The probe is tagged so that it can be recognized after hybridization. The most typical tags are radiolabeled nucleotides or digoxigenin, a steroid molecule found exclusively in parts of plants in the genus *Digitalis* (foxgloves). Because digoxigenin does not naturally occur in animal tissues and because it is highly immunogenic (which means that antibodies that bind strongly and selectively to digoxigenin are readily available), it provides an excellent nonradioactive label widely used by neuroscientists.

Tissue-based methods of studying gene expression leverage the information produced by more than a century of careful anatomical studies of development. It is therefore difficult to overstate the contribution of in situ hybridization to our knowledge of development. Although the tissue samples used for in situ hybridization are all fixed specimens (i.e., they are dead), analysis of carefully staged sequences of samples reveals the dynamic changes in gene expression that are the essence of development. This method has been partly superseded by the use of transgenic markers for gene expression in models such as the fruit fly and the mouse, but in situ hybridization remains in wide use by researchers working with other species.

In summary, developmental neuroscientists often study mRNA location and abundance because they can do so efficiently and quantitatively. The starting point for an RNA-based study of gene expression may be no more than a nucleotide sequence from a database. By contrast, detection of proteins typically requires production of specific antibodies. Making an antibody is not a particularly expensive endeavor (and it is typically done by contract with an academic or commercial facility), but it is time-consuming and unpredictable, two features guaranteed to make it unpopular with busy scientists.

RNA Interference

Once an investigator has a spatial and temporal map of a gene expressed in the nervous system during development, the next goal is often to test the

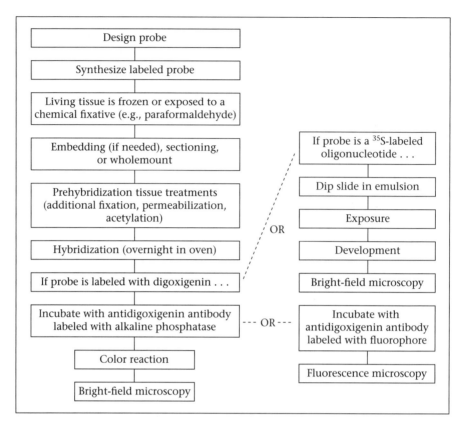

Figure 1.5. Use of in situ hybridization to study neuronal gene expression. In situ hybridization is a method for identifying cells that contain specific mRNA sequences. This method exploits the single-stranded state of mRNA within living cells. Probes must be antisense to the target mRNA that the investigator wishes to localize. Both radiolabeled oligonucleotide DNA probes and riboprobes (cRNA, complementary to the sequence of the target mRNA) are used for in situ hybridization studies. For riboprobes, RT-PCR is used to produce a DNA template that can be used to generate the riboprobe by in vitro transcription. A labeled ribonucleotide triphosphate (for example, digoxigenin-11-UTP or biotin-16-UTP) is added to the transcription reaction so that the synthesized RNA probe can be detected after it has been hybridized to the target mRNA. Different tissues require different protocols and, as in the case of immunolabeling, each step in the procedure must be optimized. An important control is to perform the procedure with the sense probe, which, in contrast to the antisense probe, should not yield specific signal.

hypothesis that the expression of the gene in that place at that time is required for normal development. A crude way to block gene expression is to destroy (lesion or ablate) the part of the tissue that expresses the gene of interest. Ablation was widely used by early embryologists to establish the source of critical developmental signals. By contrast, the modern developmental biologist, who often has the advantage of knowing the molecular identity of developmental signals, can substitute targeted gene ablations for tissue destruction. In genetic model organisms (several of which are described in Chapter 3), a normal gene can be replaced with an engineered mutant gene that does not express a functional version of the protein being studied. The ability to achieve targeted gene knockdowns accounts for much of the popularity of genetic model organisms. But targeted gene knockdowns can also be achieved in other species using the technique of RNA interference, commonly referred to as RNAi. RNAi methods are based on the surprising fact that double-stranded RNA (dsRNA) complementary to a target gene inhibits the activity of the target gene. The selective knockdown of gene expression initiated by introduction of artificial dsRNA into a cell or tissue is the result of the activation of a ribonuclease enzyme called Dicer, which cleaves the dsRNA into short fragments that in turn bind to the target mRNA and cause its cleavage so that it cannot be used for transcription. The gene-silencing action of endogenous dsRNA was first observed in plants and is now recognized as a regulatory mechanism shared by all eukaryotic cells. Gene knockdowns have been achieved in many different species of experimental animals using dsRNA-mediated interference. The discoverers of the mechanisms of RNAi, Andrew Fire and Craig Mello, were recognized with the Nobel Prize in Physiology or Medicine in 2006.[7]

Morpholino antisense oligonucleotides can also be used to reduce the expression of specific genes. Morpholinos are short, single-stranded oligonucleotide chains designed to be complementary to a target mRNA. Similar to DNA and RNA, they contain nitrogenous bases. But the bases in morpholinos are attached to a morpholine ring instead of a deoxyribose or ribose. The linkage between adjacent subunits in a morpholino is also different than in endogenous nucleic acids.

Morpholinos reduce gene expression by a different mechanism than dsRNAs. Instead of activating the Dicer ribonuclease, morpholinos bind to their target mRNAs in the cytoplasm and block translation. Morpholinos offer the advantages of being specific, relatively nontoxic, and extremely stable in living cells and tissues. For reasons that are not fully understood, morpholinos are most effective in frogs and zebrafish, and they have been used primarily in these animals. Many developmental neuroscientists now incorporate either RNAi- or morpholino-based gene knockdowns into their research. Examples will be found in many of the following chapters.

Human Brain Imaging

And now for something completely different.[8] The study of development at the cellular and molecular levels of analysis ultimately relies for meaning on our knowledge of neural circuit structure. The remarkable progress of recent decades in our understanding molecular and cellular mechanisms of development should not obscure the gaps in our knowledge of how the human brain develops, particularly during childhood and adolescence. The development of noninvasive pediatric (and even prenatal) neuroimaging is bridging these gaps and raising new questions.

Computer-assisted techniques that can be used to image the living human brain for the purpose of medical diagnosis can also be used to study brain development. The technique of computerized axial tomography uses multiple X-rays to map tissue density to produce maps of the brain referred to as CAT or CT scans. This technique is useful for detecting certain forms of brain damage or developmental abnormalities, but the resulting images are not detailed enough to track subtle longitudinal changes. Magnetic resonance imaging (MRI) techniques use powerful magnets to map brain density. Because MRI images are higher in resolution (the distinguishability of two adjacent points without blurring) than CAT scans, they reveal more meaningful neuroanatomical detail. MRI studies have revealed previously hidden dynamics of brain development, thereby generating new questions for cellular and molecular biologists to answer. For example, MRI studies of the cortices of a sample of more than 300 children and adolescents provocatively revealed that significant differences in rate of change of cortical thickness were correlated with performance on standardized intelligence quotient (IQ) tests.[9] The factors regulating these changes remain to be determined.

Several methods go beyond the descriptive neuroanatomical maps produced by MRI to provide glimpses of the brain at work. Positron emission tomography (PET) detects the distribution of injected radioactive tracers within the living brain, allowing neuroscientists to infer which brain regions have the highest levels of metabolic activity under particular test conditions. PET scans have provided new information about the brain, but the use of PET in children has been limited because of the requirement for intravenous injection of the radioactive tracer. Health concerns also limit the use of X-ray-based techniques in humans in general and children in particular, because exposure to X-rays is linked to death of cells undergoing mitosis at the time of exposure and the later development of cancers in exposed tissues. The development of functional MRI (fMRI) scanning in the 1990s solved this problem by using oscillating magnetic field gradients to monitor changes in the ratio of oxygenated hemoglobin to deoxyhemoglobin for the purpose of detecting localized changes in brain oxygen use. This type of fMRI is referred to as blood oxygenation level–dependent fMRI, or BOLD. The BOLD fMRI response is informative because increases in blood flow (hemodynamic response) and oxygen consumption are correlated with increased intensity of synaptic signaling. One of the first BOLD fMRI studies of children was published in 1995.[10] Three boys and three girls between the ages of 9 and 11 years had their brains imaged using BOLD fMRI while they performed a simple memory task requiring them to press a button whenever a particular sequence of letters appeared on a screen. The researchers specifically sought to understand the normal development of prefrontal circuitry. Other fMRI techniques continue to be developed, but BOLD fMRI is the leading technique currently used to study development of the human nervous system. BOLD fMRI offers the opportunity to determine the brain basis of age-related changes in behavior in both typically and atypically developing individuals. Note that studies of different age groups naturally present different challenges. Artifacts caused by head movements are most common in fMRI studies of preadolescents, but sleepiness sometimes interferes with the collection of reliable fMRI data from teenagers!

Longitudinal studies of human cortical development are described in Chapter 9. Note that noninvasive brain imaging is a rare example of a research tool that is more effectively applied to large brains (such as human brains) than to the smaller brains of popular animal models for the study of development. Only recently have advances in fMRI made it possible to study the brains of smaller species. For example, BOLD fMRI studies of zebra finches listening to recordings of their own song (the birds were anesthetized during the experiment) revealed a previously unknown right hemisphere lateralization of responses to a bird's own song.[11] These pioneering studies were performed using adult birds, but they raise the possibility that the brain events responsible for the development of singing, a behavior essential for avian social communication and reproduction, will one day be studied by making repeated BOLD fMRI measurements on individual birds as they learn to sing.

Another variant of MRI useful for developmental neuroscience is diffusion tensor imaging (DTI). This method exploits the tendency of water molecules to diffuse along nerve fiber tracts rather than crossing them and is therefore useful for visualizing the development of white matter (tracts of myelinated nerve fibers) in the brain.[12] For example, DTI has been used to monitor the postnatal development of the corpus callosum, the large bundle of nerve fibers that connects the right and left hemispheres of the human brain.

The Future

New methods continue to be developed to monitor and manipulate the function of neural circuits in the living brain. These novel approaches to classic questions, such as the changes that occur in synaptic structure as a result of experience (*learning*) can also be applied to problems in nervous system development. A particularly exciting new method introduces light-sensitive cation channels into neurons. This type of channel—an example is channelrhodopsin-2, a light-sensitive plasma membrane protein native to a species of single-celled algae—opens in response to blue light, with the result that any illuminated neuron containing the channel rapidly depolarizes. The ability to stimulate neurons without the need to penetrate them with traditional microelectrodes offers opportunities for minimally invasive manipulation of signaling in small embryos poorly suited for traditional electrophysiology. For example, rhythmic waves of electrical activity can be recorded from the spinal cords of chick and mouse embryos. Decreasing this spontaneous electrical activity by application of drugs that block the $GABA_A$ category of neurotransmitter receptors leads to disturbances in the position of motoneuron axons as they grow toward their muscle targets. Researchers asked if the axon pathfinding errors resulted specifically from the change in the pattern of electrical activity by observing axons growing in chicken embryos treated to express the channelrhodopsin-2 protein in the developing spinal cord.[13] Flashes of light were able to drive neural activity in the embryos even when the $GABA_A$ blocker picrotoxin was present. The result supported the view that normal patterns of spontaneous neural activity are required to guide axons to their embryonic muscle targets.

Notes

1. See Heisenberg et al. (1995), 1959.
2. Students interested in the development of nonneural tissues are encouraged to begin by consulting the beautifully illustrated text by Scott F. Gilbert, *Developmental Biology,* 9th edition (Sunderland, MA: Sinauer, 2010).
3. A list of mouse strains expressing Cre plus accessible explanations of Cre-recombinase systems is available on the web site of the Jackson Laboratory, an independent, nonprofit organization that has been supplying mice for research since 1933.
4. Free online tutorials and webinars in microscopy are offered by many microscope manufacturers.
5. The Molecular Probes Handbook, available via the Invitrogen web site, provides a comprehensive and frequently updated review of fluorophores developed for research in the life sciences.
6. See http://www.nobelprize.org/nobel_prizes/chemistry/laureates/2008/ for further information on the 2008 Nobel Prize in Chemistry.
7. See http://www.nobelprize.org/nobel_prizes/medicine/laureates/2006/ for further information on the 2006 Nobel Prize in Physiology or Medicine.
8. This useful phrase is borrowed from the title of a 1971 feature film released by Monty Python, a legendary British comedy group.
9. See Shaw et al. (2006).
10. See Casey et al. (1995).
11. See Poirier et al. (2009).
12. See Mukherjee and McKinstry (2006).
13. See Kastanenka and Landmesser (2010).

Investigative Reading

1. To study neurogenesis in the adult brain, you inject a group of mice with the thymidine analog, BrdU. You then prepare slides of brain sections and perform immunofluorescent colabeling with an antibody to BrdU and another to the cell cycle marker Ki-67. What percentage of your BrdU-labeled cells do you expect to coexpress the Ki-67 marker? Conversely, what percentage of Ki-67-labeled cells do you expect to contain incorporated

BrdU? What is the evidence that the newborn cells that you identify using antibodies to BrdU and Ki-67 are neurons?

Leuner, Benedetta, Erica R. Glasper, and Elizabeth Gould. 2009. *Journal of Comparative Neurology* 517: 123–33.

2. Huntington's disease is an incurable inherited disease of the nervous system. The disease typically appears between the ages of 30 and 50; it is progressive, meaning that the symptoms worsen with time. Patients suffer from uncontrollable movements, cognitive decline, and eventually dementia. At autopsy, brains from Huntington's patients have neurodegeneration in several regions. The most severe damage occurs in a part of the brain called the striatum, which controls behavior and coordinates movement. The mutation that causes Huntington's disease has been identified. The affected gene in humans, named the *Huntingtin* (*HTT*) gene, codes for an abnormally long version of the Huntingtin protein (HTT). It is not yet known how this mutant protein kills neurons. It is also not known why some areas of the brain are more vulnerable than others. One way to understand the mechanisms by which a mutant protein damages cells is to study the function of the normal protein in animal models. The genome of the fruit fly *Drosophila melanogaster* contains a gene that encodes a homologous protein called *dhtt*. Researchers used reverse transcription–PCR to show that the *dhtt* gene is expressed in adult fruit flies, but these studies were performed on RNA extracts of whole flies. You want to know if the *dhtt* gene is expressed in the fly nervous system; you also want to know if the protein is expressed in the cytoplasm or the nucleus. What methods do you use?

Zhang, Sheng, Mel B. Feany, Sudipta Saraswati, J. Troy Littleton, and Norbert Perrimon. 2009. *Disease Models and Mechanisms* 2: 247–66.

3. *Challenge Question:* You are using a mouse model to study a human disease called spinal muscular atrophy (SMA), an incurable inherited disease characterized by progressive muscle weakness caused by the deaths of spinal motoneurons. You find that motoneurons from your mutant mouse have reduced expression of β-actin, a cytoskeletal protein. β-actin is enriched in growth cones, the tips of growing axons. Your goal is to test the hypothesis that β-actin is required for the formation of the neuromuscular junction that normally occurs when growth cones reach their muscle targets. What technique do you use to delete β-actin from mouse motoneurons without affecting any other cell population?

Cheever, Thomas R., Emily A. Olson, and James M. Ervasti. 2011. *PLoS ONE* 6: e17768.

Overview of Nervous System Development In Humans

How Do We Know What We Know?

The goal of this chapter is to acquaint you with the major events that occur during development of the human nervous system. The device of starting our discussion of development with a large and complex mammalian nervous system that contains as many as 100 billion neurons—as opposed to starting with an invertebrate nervous system containing a few hundred or a few thousand neurons—is akin to making the human genome one of the first animal genomes to be sequenced.[1] In both cases there are good reasons beyond species chauvinism for putting humans at the front of the line. Detailed knowledge of human development—based on observation—focuses our experimental analyses of other species, guides our thinking about evolution, and helps the clinician detect patterns in human birth defects, the essential first step for devising therapeutic interventions. Practically speaking, starting with humans is also an efficient way to introduce the specialized vocabulary used by developmental biologists and neuroanatomists.

What is the source of our knowledge of human development? Study of the earliest stages of human development awaited the development of microscopes, because mammalian eggs are typically tiny. The fertilized human egg, for example, is no more than a tenth of a millimeter in diameter. Placental mammals also offer a challenge to the developmental biologist because embryonic and fetal development occur deep within the mother's body.

Early in the twentieth century, the Carnegie Institution of Washington, D.C., began to assemble what eventually became a massive collection of post mortem specimens of human embryos. Carnegie Institution embryologists devised a detailed timeline of early human development on the basis of careful analysis of this material. The 23 Carnegie stages cover the first 9 weeks (out of the 40 weeks total) of human gestation. Every student interested in human development should be familiar with the Carnegie stages, and they are reviewed from the perspective of nervous system development later in this chapter.

More recently, knowledge of the earliest stages of human development has been expanded as a consequence of assisted reproductive technologies

(ART). The development of in vitro fertilization as a treatment for human infertility was recognized by awarding the 2010 Nobel Prize in Physiology or Medicine to Robert G. Edwards.[2] The manipulation of oocytes and young embryos required to make ART routine has been used to advantage in human embryonic stem cell research and in procedures for preimplantation genetic diagnosis. However, the embryos created by in vitro fertilization are returned to the mother's (or surrogate's) uterus prior to the earliest signs of the nervous system, or indeed of any organs.

At the other end of prenatal life, anatomical MRI (as opposed to functional MRI) is used to study both normal and atypical development in the fetal brain (Chapter 1). Fetal MRI can be performed by placing the mother's abdomen inside the imaging coil. Most studies are performed after 18 weeks' gestation. Fetal MRI studies have provided new information about the later prenatal events in the development of the human nervous system, especially cortical development and the progress of myelination. An online MRI atlas of typical human fetal brain development has been published that covers the second and third trimesters.[3] In addition to improving our understanding of human brain development, such tools are aids to prenatal diagnosis of developmental disorders.

Start by Working Backward

How can we best organize information about the development of the human nervous system so that it is easy to retain? One approach is to start with an understanding of the structure of the mature brain, spinal cord, and peripheral nervous system. Working backward allows the sequence of events that links the fertilized egg to the brain to be immediately placed in meaningful context.

The brain and the spinal cord together constitute the central nervous system (fig. 2.1). If, when you think of the human brain, you immediately think of the large and convoluted cerebral cortex, you may be surprised to learn that the fundamental structure of the brain is a tube formed of layers of cells around a hollow, fluid-filled core. The other anatomical feature of the human central nervous system that may surprise you is that it develops as a segmented structure. Segments are serially reiterated cellular compartments that are initially similar in composition but are subsequently modified by segment-specific patterns of gene expression. Development by segmentation is a feature of the body plan of many different types of animals, and in many animals (think of an earthworm), segmental boundaries are visible externally, even in adults. By contrast, the fundamental segmentation of the human nervous system is most evident early in development. Later, the initially uniform segments become quite specialized.

The peripheral nervous system consists of neurons associated with sensory receptors that convey information to the brain and the spinal cord plus

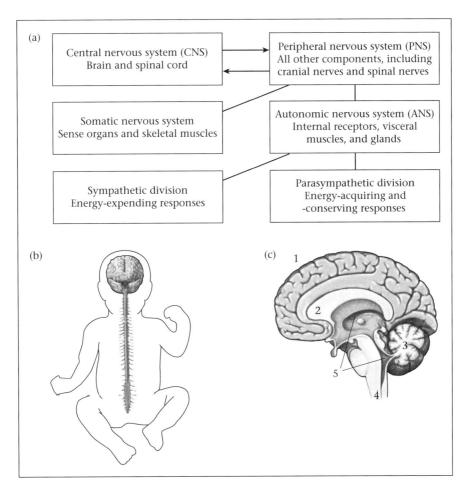

Figure 2.1. Organization of the human nervous system. The conventional names and abbreviations for the different divisions of the human nervous system are given in (a). The arrows indicate the many pathways that link the central nervous system and the peripheral nervous system. The central nervous system of a human infant is also depicted (b). As indicated, all of the major divisions of the nervous system develop before birth in humans. A cut-away side view of one hemisphere of the human brain is shown in (c). The numbers indicate five brain features that will be referred to in subsequent chapters: (1) cerebral cortex, (2) corpus callosum, (3) cere-bellum, (4) spinal cord, and (5) ventricles. Drawings for (b) and (c) based on images by Nucleus Media.

all axons (typically bundled into nerves) that leave the brain and spinal cord to control the skeletal muscles. In mammals, 12 pairs of cranial nerves connect the brain to the head, neck, and trunk; 31 pairs of spinal nerves innervate the trunk plus the arms and the legs. The peripheral nervous system also comprises the autonomic nervous system, which consists of clusters of neurons called ganglia and the axons of the neurons in these ganglia, which make synaptic connections with glands and smooth muscles in organs throughout the body. The autonomic nervous system has two divisions, the sympathetic and the parasympathetic. The sympathetic division contains a chain of ganglia that run alongside the vertebrae of the spinal cord; signaling in the sympathetic division is associated with states of arousal and energy expenditure. The parasympathetic division consists of ganglia located on or near the organ that is regulated (for example, the intestines or the bladder); signaling in the parasympathetic division is associated with states of relaxation and energy acquisition.

The human central nervous system is derived from an embryonic structure called the neural tube. The human peripheral nervous system is derived from embryonic structures called the neural crest. Both the neural tube and neural crest are derived from the embryonic tissue source layer called the ectoderm. The cells produced by successive divisions of embryonic cells (which is the meaning of the term *derived*) display changes in gene expression that simultaneously restrict their fate and promote adoption of a mature phenotype, a process referred to as differentiation. For example, embryonic tissue that has differentiated as ectoderm can produce skin or a brain but not muscles; later in development, cells derived from the ectoderm that have differentiated as the neural tube can produce a brain but not skin.

The Carnegie Stages of Embryonic Development

Carnegie Stages 1–23 cover the first 9 weeks of human development and are based on the appearance and internal structures of the embryo. The exact timing of events varies slightly from individual to individual, but the sequence of events is invariant. In postpubertal, premenopausal human females, a mature oocyte is released approximately every 28 days from one of the ovaries. The events of the ovarian cycle are controlled by hormones secreted by the brain, the pituitary, and the ovary itself. The oocyte enters the oviduct (also called the fallopian tube), a tube that connects each ovary to the uterus. Fertilization of the oocyte by a sperm cell occurs in the oviduct, and it is here that development begins. Prior to cell division, the fertilized oocyte is referred to as a zygote. The earliest cell divisions in a person's life are cleavage divisions. Cleavage divisions are cell divisions without growth, and the daughter cells produced by cleavage are called blastomeres. Several rounds of cell division occur while the forming human embryo is still in the oviduct, so the embryo enters the uterus as a compact little ball of 16 cells called the morula (fig. 2.2).

Carnegie Stage 1 is defined by the process of fertilization and lasts about a day. Carnegie Stage 2 covers the first cleavages and ends when the morula enters the uterus 3 or 4 days later. Cell divisions continue during Carnegie Stage 3, which is marked by the appearance of a new feature—an internal cavity. The formation of this cavity transforms the morula into the blasto-

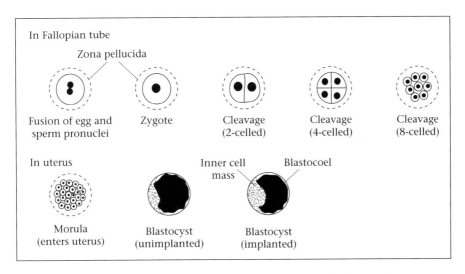

Figure 2.2. Early stages of development of the human embryo. After fertilization, the sperm and egg pronuclei fuse. Cleavage divisions begin shortly after formation of the zygote. At these stages, the zona pellucida, a protein-rich extracellular matrix secreted by the egg, is still present (dashed lines). The morula stage is a compact ball of cells that contains the inner cell mass and trophoblast cells. The morula enters the uterus. The stage that implants into the uterine endometrium on Day 6 after fertilization is called the blastocyst: at this time, the embryo consists of approximately 100 cells. The blastocyst is thin walled and hollow. The trophoblast layer of cells encloses the inner cell mass and lines the cavity, which is called the blastocoel. The inner cell mass will form the tissues of the body, including the central nervous system; the trophoblast cells will form the fetal membranes (chorion and amnion).

cyst (see fig. 2.2). The appearance of the blastocyst is an important transition, for it signals that there are now two distinct cell populations present: inner and outer. The cells on the outside of the blastocyst become extra-embryonic tissues, while some of the cells of the inner cell mass become the baby. At Carnegie Stage 4, the blastocyst sheds a coating called the zona pellucida and begins to implant into the uterine wall. Implantation begins between 5 and 6 days after fertilization and is complete by about 12 days after fertilization (Carnegie Stage 5). The tiny human embryo has just solved the problem of how to obtain the nutrients it needs to grow: it now shares the circulatory system of its mother.

At the time of implantation, the inner cell mass comprises two layers of cells: the epiblast and the hypoblast. The epiblast cells of the upper layer are the source of all of the embryonic tissues and the cells that will eventually line the fluid-filled amnionic cavity. The hypoblast cells of the lower layer become the yolk sac. This sorting of cells marks the beginning of a coordinated series of cell proliferation and migration processes collectively referred to as gastrulation. Gastrulation creates the distinct tissue source layers of the endoderm, mesoderm, and ectoderm (fig. 2.3). The appearance of the

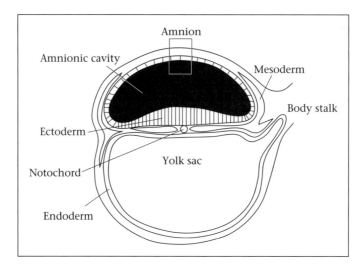

Figure 2.3. Origins of human embryonic tissues. A human embryo at Day 18 after fertilization is shown in cross section. *Gastrulation* is the term given to the cell movements that transform the two-layered embryonic disc of epiblast and hypoblast into a three-layered structure. Cells migrate from the surface of the epiblast, dive inside at the node and primitive streak (at the base of the amnionic cavity), and then continue to rearrange themselves once they are inside. It can be very difficult to understand this dynamic process by examining static photomicrographs or schematic diagrams, and relatively few biologists have ever seen an actual human embryo. To make things even more difficult, the gastrulating embryo is implanted into the uterine endometrium; the yolk sac and amnionic membranes further obscure the view. The embryo through which this particular cross section was imagined is roughly pear shaped and about 1 millimeter in length. Note the relative positions of the ectoderm, mesoderm, endoderm, and notochord, a temporary mesodermal structure required to induce the nervous system. The rectangle encloses a patch of amnion, also called the amnionic membrane. This membrane is formed by two extraembryonic cell layers. The inner layer of the amnion is ectoderm; the outer layer is mesoderm. Drawing loosely based on an *Encyclopedia Britannica* figure.

primitive streak at 13 days post fertilization signals the onset of gastrulation (Carnegie Stage 6). The primitive streak is a narrow line of cells that marks the midline and starts the formation of a third cell layer in what was previously a flattish two-layered embryo. The primitive streak forms as a result of migration of cells from the outer edge of the embryo to the midline. Eventually the innermost layer (endoderm) gives rise to the linings of many organs and forms the digestive tract; the middle layer (mesoderm) gives rise to muscles, bones, and the reproductive tract; and the upper layer (ectoderm) produces the integument, many sense organs, the teeth, and all components of the nervous system. At this stage of human development, however, the parts of the ectoderm that will become neural structures have not yet separated from general ectoderm. The embryo is now about 0.2 millimeters in length.

The cell migrations that defined gastrulation continue during Carnegie Stage 7. Carnegie Stage 8, which extends from Days 17–19 post fertilization, is characterized by the appearance of a thickened region of dorsal ectoderm called the neural plate (fig. 2.4). The thickening results from a change in shape of the individual epithelial cells that form the neural plate from cube-like to columnar. If one wanted to designate a particular stage in human embryonic development as the origin of the nervous system, a good case could be made for Carnegie Stage 8. During this stage a groove in the middle of the neural plate becomes evident (called, plainly enough, the neural groove). By the end of Carnegie Stage 8, the embryo is over 1 millimeter in length, and the head end is wider than the posterior end.

At Carnegie Stage 9, segmentally repeated mesodermal structures called somites appear on either side of the neural groove. The somites, which are precursors of dermal tissue, muscles, and vertebrae, are among the most obvious features of vertebrate embryos. At Carnegie Stage 10 (21–23 days after fertilization), the edges of the neural groove fuse to form the neural tube, and more somites are added on either side of the neural tube. Packets of cells at the sides of the neural tube begin to differentiate as neural crest cells, which will migrate away from the site of their origin and form the structures of the peripheral nervous system. Other organs, such as the heart, are now also forming.

During Carnegie Stage 11 (23–25 days post fertilization) more somites are added along the length of the embryo. Bulges on the side of the neural tube indicate that cell proliferation is beginning in the nervous system (see fig. 2.4). By Carnegie Stage 12 (25–27 days after fertilization), the neural tube closes, forming a cylindrical structure separate from other regions of the ectoderm. The open ends of the neural tube (the anterior neuropore and the posterior neuropore) also close. The sense organs of the head are now becoming evident. At Carnegie Stage 13 (through 30 days of development), the embryo is approximately 5 millimeters long and has more than 30 pairs of somites. Rapid brain growth begins during Carnegie Stage 14 (through about 35 days), and constrictions marking the division of the brain into forebrain, midbrain, and hindbrain are evident. By 38 days (Carnegie Stage 15), the cerebral hemispheres can be discerned. Rapid growth of the brain and spinal cord continues through the remaining Carnegie stages. By the end of 60 days of development, the embryo is 25 millimeters long and has formed its major organs. It has a brainy appearance, with the head appearing enlarged relative to the trunk and the small limbs. It is now approximately 10 weeks since the mother's last menstrual period and 8 weeks since fertilization, and the embryo is now referred to as a fetus.

To recap, the key embryonic events in the formation of the human nervous system are the differentiation of ectoderm, the formation of the neural plate, the invagination of the neural plate to form the neural groove, the closing of the neural tube, and the differentiation of the neural crest cells.

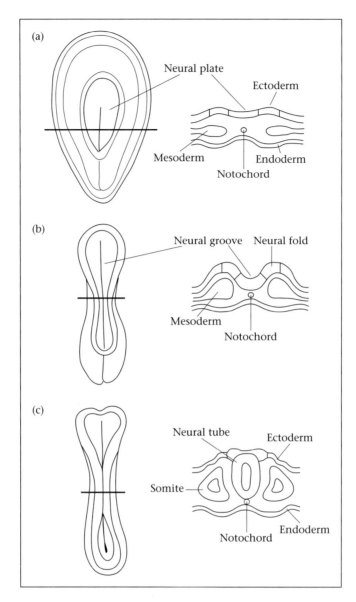

Figure 2.4. The neural tube of the human embryo. The neural plate (a) becomes the neural groove (b), which in turn becomes the neural tube (c). Note the position of the notochord under what will become the floorplate of the neural tube. The somites on either side of the neural tube are temporary mesodermal structures. They form as a series of regular, paired bumps on each side of the neural tube. The progeny of the somite cells will eventually form the cartilage of the vertebrate and ribs; the muscles of the ribs, limbs, and back; and the dermis of the dorsal skin. Drawing based on http://thebrain.mcgill.ca/flash/i/i_09/i_09_cr/i_09_cr_dev/i_09_cr_dev.html.

Once the neural tube closes, an intense process of cell proliferation results in the accumulation of layers of neurons and glial cells, and the walls of the neural tube become thicker and thicker. The hollow center of the tube is present throughout life; it forms the central canal of the spinal cord and the ventricular system of the brain. The term *neurulation* is sometimes used to describe the events that begin with formation of the neural plate and end with the closure of the neural tube. Each of these events is grist to the mill of modern developmental neuroscience because each reflects an intricate web of intracellular and intercellular molecular signaling events.[4] Some of these signals (and their receptors) are described in the subsequent chapters of this text.

Development of the Fetal Brain

It has been estimated that 250,000 neurons are added to the human brain per minute during the prenatal period. This seems an incredible number, but consider the following calculation, which starts with the assumption that the human brain contains 100 billion neurons. The human gestation period is 40 weeks or 280 days, which works out to be 6,720 hours or 403,200 minutes. Divide 100 billion by 403,200 to get 248,016 new neurons per minute. This means that the fetal period of brain development in humans is by and large a story of neurogenesis, the subject of Chapter 5. It is also, however, the story of how the neural tube subdivides into specialized regions. The boundaries of these functional regions are marked, often quite sharply, by changes in cellular gene expression. Examples of these gene expression domains and the molecular signals that define them are described in Chapter 6. Here I briefly review the major anatomical regions of the developing human central nervous system, document the transformation of the cerebral cortex from a smooth to a folded surface, and describe the clinical consequences that ensue if the neural tube fails to close.

The subdivision of the brain into three major regions is evident by the end of the embryonic stage of development. The anterior end of the neural tube swells to form three vesicles (cavities), which are divided from each other by slight constrictions in the neural tube. The walls of the anterior swelling define the forebrain, traditionally called the prosencephalon; the walls of the next swelling form the midbrain, or mesencephalon; and the walls around the third vesicle form the hindbrain, or rhombencephalon. As development continues, these three regions become more distinct. Each is also further subdivided. This is particularly noticeable in the hindbrain, which develops as a series of smaller units called rhombomeres (fig. 2.5).

The spinal cord is subdivided as it develops, both along its length and along the dorsal–ventral body axis. The result is a segmented structure subdivided longitudinally into consecutive cervical, thoracic, lumbar, and sacral

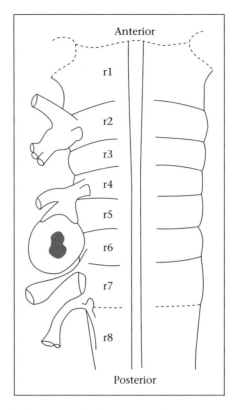

Figure 2.5. Vertebrate rhombomeres. This figure is based on the embryonic chicken hindbrain, Stage 21. Early in development, the hindbrain, or rhombencephalon, is subdivided into repeated segments called rhombomeres. The rhombomeres are first defined by patterns of gene expression (especially the homeodomain Hox genes, which are discussed at length in Chapter 6) but later become visible as alternating bulges and constrictions in the hindbrain. There are 8 rhombomeres (r1–r8). The cerebellum develops from dorsal r1, the pons develops from ventral r1, and cranial nerves IV–XII (shown only on the left side of the figure for clarity) have their origin in the rhombomeres. The Vth nerve (trigeminal) ganglion is associated with r2; the VIIth (geniculate) and VIIIth (auditory) nerve ganglia have their origins in r4; the IXth nerve (petrosal) ganglion is associated with the r6/r7 boundary; and the Xth nerve (jugular and nodose) ganglia grow from the r7/r8 boundary. (If you like this sort of thing, consider taking a neuroanatomy course!) Drawing adapted from Lumsden and Keynes (1989).

regions. The dorsal and ventral halves are separated by a longitudinal groove called the sulcus limitans.

The hemispheres of the cerebral cortex are visible as early as the fifth week of human gestation, but at this time they are smooth. In addition to becoming thicker, the surface of the cortex grows so much that later in gestation the cortices become folded. The tops of the folds are called gyri, and the valleys that separate gyri are called sulci. (*Gyri* and *sulci* are plurals: the singular forms are *gyrus* and *sulcus*.) Sulci appear in sequence in locations so predictable that the pattern of cortical sulcation can be used as an indicator of gestational age. The ultrasound examinations typically performed at an obstetrician's office can reveal gross abnormalities of brain development, but the images produced by sonography are not sufficiently detailed to catch subtle defects in cortical development. Comparison of sulcal development observed in post mortem specimens of known gestational age with sulcal development imaged using MRI has revealed close correspondence between the results produced by direct observation and MRI, although the sulci can be detected slightly earlier in tissue than in magnetic resonance images.[5] Abnormal patterns of sulci and gyri are associated with intellectual disability and seizures. In the extreme case of a birth defect called lissencephaly, the surface of the cortex is smooth or nearly smooth.

By the time fetal brain development is complete, the human brain weighs approximately 360–380 grams with baby boys having, on average, slightly heavier brains than baby girls. The high rate of neurogenesis characteristic of the prenatal period continues without slowing for two more years, by which time the brains of both boys and girls weigh more than 1,000 grams. The human brain continues to grow (more slowly) throughout childhood, and does not attain its final mature size until approximately 15 years after birth.[6]

Neural Tube Defects

Neural tube defects are common in human babies, occurring in approximately 1 in every 500 pregnancies (fig. 2.6). The failure of the neural tube to close leaves developing neural tissue exposed to amnionic fluid, and in this inappropriate extracellular environment, the neural tissue eventually dies. The site of the defect determines which part of the nervous system is damaged. Failure to close the anterior neuropore results in loss of most of the brain. This condition is called anencephaly, and affected babies die shortly after birth. Failure to close the middle section of the neural tube results in degeneration of the midbrain, hindbrain, and spinal cord. The resulting condition is called craniorachischisus, and it also results in early death. Failure to close the posterior neuropore results in damage to the posterior regions of the spinal cord. This condition is called spina bifida. Many babies affected by spina bifida survive and are of normal intelligence, although they often have conditions reflecting spinal cord damage, such as paralysis of the legs and poor control of bowel and bladder function.

The causes of neural tube defects are uncertain, despite the development of many mouse models of these disorders and evidence that increasing maternal intake of folic acid before conception and early in pregnancy reduces the incidence of babies born with these defects. The evidence that adequate intake of folic acid reduces neural tube defects is so strong, in fact, that mandatory fortification of enriched grain products with folic acid has been re-

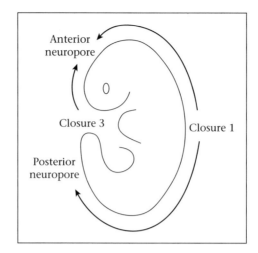

Figure 2.6. Origins of neural tube defects in humans. A highly schematized depiction of the origin of defects that arise from failure of the neural tube to initiate or complete closure at specific locations. The term *neuropore* is used to refer to the temporary openings in the neural tube that become apparent during the process of neurulation. Both ends of the neural tube (anterior and posterior neuropores) are initially open, and both close in humans by the 28th day after fertilization (Carnegie Stage 13). Arrows indicate the directions in which fusion proceeds from sites of fusion initiation called closures. Fusion moves rostrally and caudally from Closure 1 (located at the boundary between the hindbrain and the cervical spinal cord) and caudally from Closure 3 (located in the rostral forebrain). Failure to close the anterior neuropore results in anencephaly (lack of cerebral hemispheres). Failure to initiate closure properly at Closure 1 damages both the brain and the spinal cord, resulting in a condition called craniorachischisus. Failure to close the posterior neuropore results in spina bifida, a condition in which the brain develops normally but the spinal cord is damaged. What happened to Closure 2? Initiation of fusion in the region of the forebrain-midbrain boundary is referred to as Closure 2, but the occurrence and position of this closure in human embryos is variable, and hence this closure is not shown. Drawing adapted from Copp and Greene (2010).

quired in the United States since January 1998. The Centers for Disease Control and Prevention estimated in 2004 that a 26 percent reduction in the number of babies born with anencephaly or spina bifida has occurred since the initiation of folic acid fortification.[7] But at present there is no generally accepted model of how folic acid deficiency causes neural tube defects. Studies of neural tube defects in mice are frustrating in that they suggest that many, many factors, both environmental and genetic, cause these defects.

Notes

1. The Human Genome Project was formally initiated in 1990 and completed in 2003; an announcement of a complete first draft was made in 2000, and analysis of the sequence data continues (http://www.ornl.gov/sci/techresources/Human_Genome/

project/about.shtml). Sequencing of the much smaller genome of the nematode worm, *Caenorhabditis elegans,* was completed in 1999, and the genome of the fruit fly, *Drosophila melanogaster,* was published in 2000.

2. "The 2010 Nobel Prize in Physiology or Medicine—Advanced Information." Nobelprize.org, 3 Aug 2011. http://nobelprize.org/nobel_prizes/medicine/laureates/2010/adv.html.

3. See Chapman et al. (2010).

4. "Grist to the mill" is what the writer George Orwell referred to as a dying metaphor in his essay *Politics and the English Language.* Grist is the grain brought to the mill to be ground into flour, and hence represents the farmer's profit.

5. See Garel et al. (2001).

6. See Dekaban (1978).

7. See Centers for Disease Control and Prevention (2004).

Investigative Reading

1. You are a pediatric neuroradiologist. You are asked to evaluate a set of MRI brain images obtained from a human fetus. Gestational age is uncertain because the mother experienced extremely irregular menstrual cycles prior to conception. Her obstetrician wants to know, in your judgment, if the fetus was older or younger than gestational age 25 weeks at the time the set of brain images was acquired. You note that the following sulci are clearly visible on the scans: the interhemispheric fissure, the callosal sulcus, the cingular sulcus, the hippocampal fissure, and the superior frontal sulcus. You cannot see either the superior or the inferior temporal sulci. The central sulcus of the vertex (top) of the brain is just visible, but you cannot see either the precentral or the postcentral sulcus. Is this fetus older or younger than gestational age 25 weeks? Would your answer be different if you had been told that the fetus was a twin?

> Garel, Catherine, Emmanuel Chantrel, Hervé Brisse, Monique Elmaleh, Dominique Luton, Jean-Françoise Oury, Guy Sebag, and Max Hassan. 2001. *American Journal of Neuroradiology* 22: 184–89.

2. DNA microarrays permit a global analysis of embryonic gene expression at specific developmental time points. The results of such studies are lists of genes that display changes in the relative abundance of their mRNA transcripts over time. Subsequent analysis of gene expression is facilitated by clustering the genes on these often lengthy lists based on past studies of protein function and localization within the cell. A widely used system of categories, developed by the Gene Ontology (GO) Consortium (http://www.geneontology.org), categorizes genes based on biological process, molecular function, and cellular component. The results of a microarray study of human embryos spanning weeks 4–9 were analyzed in terms of GO categories related to the nervous system. Based on the Carnegie stages of development, during which of these weeks do you expect to find genes associated

with neural tube closure most abundantly expressed? Does this study confirm your prediction?

Yi, Hong, Lu Xue, Ming-Xiong Guo, Jian Ma, Yan Zeng, Wei Wang, Jin-Yang Cai, Hai-Ming Hu, Hong-Bing Shu, Yun-Bo Shi, and Wen-Xin Li. 2010. *FASEB Journal* 24: 3341–50.

3. In Greek mythology, a chimera is a monster formed from the head of a lion, the body of a goat, and a tail with a snake's head at the tip. In modern biology a chimera is an organism that contains cells of more than one genetic background. The cells may come from two individuals of the same species or of different species. Imagine that a human–mouse chimera could be created by transplanting human brain stem cells into a fetal mouse. What could be learned about brain development by studying such a chimera? What ethical issues would be raised by such a human neuron mouse?

Greely, Henry T., Mildred K. Cho, Linda F. Hogle, and Debra M. Satz. 2007. *American Journal of Bioethics* 7: 27–40.

THREE

Animal Models

Model Organisms

The term *model organism* is used to refer to nonhuman species studied by biologists because it is informative and convenient to do so. The history of life on earth is descent from a common ancestor and conservation of molecules that are the basis of cellular life, so model organisms can be studied with the strong presumption that what is learned from one species will apply to others. What defines convenience in a model organism? At a bare minimum, model organisms grow and reproduce in the laboratory, develop rapidly, and are inexpensive to maintain. They are not dangerous, rare, or endangered and are available for use anywhere in the world. A successful model organism is supported by a community of researchers, a substantial body of published literature, and a sequenced genome. Taken together, these traits make it possible for a researcher to focus his or her investigations on acquiring new knowledge rather than figuring out the basics of how a species works.

Although model organisms can be drawn from any domain or kingdom of life on Earth, models for nervous system development must of course be animals. Successful animal models for nervous system development share the traits of model organisms listed in the preceding paragraph, but each offers its own unusual and sometimes unique selling points to the neuroscientist. This chapter introduces as models for study of nervous system development a set of species: *Mus musculus, Danio rerio, Drosophila melanogaster,* and *Caenorhabditis elegans*. The advantages and disadvantages of each model will be described, followed by an overview of its life history and the basic structure of its nervous system. The chapters that follow draw primarily on experimental results from this restricted set of models. My intent is to keep the text short enough to enable the student to develop a clear, sturdy foundation for advanced courses and/or self-directed reading in the primary literature. In practice this means that many informative and historically significant species had to be omitted. In particular it is hard not to regret leaving out hydras (*Hydra vulgaris*), frogs (*Xenopus laevis*), rats (*Rattus norvegicus*), and chickens (*Gallus gallus*). But there is no doubt about the significance to developmental neuroscience of the four species that made the cut.

Some Helpful Concepts for Thinking about Animal Models

A specialized jargon is often used to describe and compare animal models. For example, you may hear someone say that mice are really good for reverse genetics, zebrafish are good for forward genetics, and fruit flies are good for both. Or you may be asked if you think a particular study would benefit from the use of an outbred stock of mice as opposed to an inbred strain. In your genetics class you may compare wild-type and mutant fruit flies, null mutations, and hypomorphs. It is helpful to define a few terms before we consider specific models.

Mutations are changes in genomic DNA sequence. They can result from errors during meiosis or DNA replication. They can also result from damage to DNA caused by radiation, viruses, and certain types of chemicals. Mobile DNA elements, called transposons, are sequences of DNA that can move from one position to another in the DNA of a single cell. Movement of transposons can alter genes at both the old and the new locations. If mutations occur in germ cells (sperm or oocyte precursors), a mutant gene can be passed from one generation to another. Some mutations have no effect on protein expression, but others result in a change in the protein that is expressed or in complete failure to express the protein. The latter is a null mutation. The term *hypomorph* is used to describe the situation in which the wild-type gene product is still expressed but at a significantly lower level than is typical.

If a change in phenotype occurs as a result of a mutation in a gene, analysis of the new phenotype may reveal the normal function of that gene. Analysis of mutants is a powerful tool widely used in the study of development. The term *wild type* is frequently used to refer to both the standard, unmutated phenotype and the unmutated gene that produces the standard phenotype. The concept of wild type can be a little fuzzy because different natural populations of an organism often have more than one allele at each genetic locus. *Wild type* can therefore be defined as the most prevalent allele of a gene in a population, but it can also be defined as the allele found in the strain in which a particular mutation has arisen or been induced. A strain is defined as a reproductively isolated group of animals. Strains are also referred to as lines or stocks and, in the wider world, breeds. For example, Jersey Giants, Leghorns, and Rhode Island Reds are among the many different breeds, or strains, of chickens available to poultry farmers. Strains provide genetically characterized populations for experiments. Some strains are the products of inbreeding, which is operationally defined as twenty or more consecutive generations of brother–sister matings. Inbreeding results in increased homozygosity, and properly inbred strains that have passed the twentieth-generation threshold can be assumed to have a 98 percent chance of being homozygous at any gene locus. The term *isogenic* is used to describe a strain in which every individual is genetically identical. Many different strains of mice and fruit flies are available for research from commercial breeders and nonprofit stock centers. Studies that use inbred strains should produce consistent results, down to the smallest details of development, in a laboratory anywhere in the world, but there is no guarantee of consistency if different strains are used.

The two major wet lab experimental approaches used to understand gene function are forward genetics and reverse genetics.[1] Scientists using forward genetics analyze mutant phenotypes to identify the gene or genes that are altered by the mutation; scientists using reverse genetics modify a gene and then analyze the phenotypic consequences. In forward genetics (sometimes also referred to as classical genetics), comparison of mutant and wild-type animals is the starting point for a project that, if successful, eventually results in identification of the gene or genes that, when altered, produce a particular mutant phenotype. Forward genetics studies can exploit spontaneously arising mutations (recall Thomas Hunt Morgan's famous white-eyed fruit fly), but it is also typical for forward genetics studies to be initiated by exposing individuals to a mutagen, mating them, and then screening the resulting offspring for mutant phenotypes (fig. 3.1, right side). An advantage of the forward genetics approach is that investigators are not biased by their prior knowledge of the process being studied. Genes not previously known to be associated with a particular developmental event can be discovered using a forward genetics screen. For example, a forward genetics screen used the offspring of male mice exposed to a mutagenic drug called N-ethyl-N-nitrosourea (ENU) mated to transgenic mice expressing GFP in their motoneuron axons to study the development of the spinal cord.[2] Abnormal axon outgrowth could be detected simply by examining the embryos resulting from this mating using confocal microscopy. A mutation of interest named Magellan by the investigators performing the screen was mapped to a region of a specific chromosome. Candidate genes known to be present in this region were then amplified and sequenced. The result was the discovery of a new function—stabilization of the cytoskeleton in growing axons—for a gene called *Phr1* that encodes a ubiquitin ligase enzyme. This screen therefore produced an interesting and absolutely novel finding. Is it significant? Until more is known, it is impossible to tell.

The counterpoint to forward genetics is reverse genetics (fig. 3.1, left side). Investigators taking a reverse genetics approach start with a known gene sequence and then alter or inactivate the gene with the goal of analyzing the phenotype of the resulting mutant. The sequences can be obtained by cloning and sequencing the gene of interest or, more typically, from a database. Knockout mice, described in the next section, provide a powerful example of reverse genetics. The power of this tool is such that an ongoing

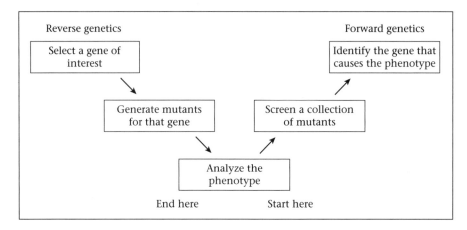

Figure 3.1. Comparison of forward and reverse genetics. Both approaches involve careful analysis of wild-type and mutant phenotypes. From the perspective of genetics, you are going in reverse as you move from the gene to the phenotype.

project at the U.S. National Institutes of Health called KOMP (the Knockout Mouse Project) has the goal of producing as a public resource a library of mouse embryonic stem (ES) cells containing a null mutation in every gene in the mouse genome.[3]

Students sometimes ask which is better: forward or reverse? The answer depends on both the extent of the researcher's prior knowledge of a subject and his or her patience. The genetic screens at the heart of forward genetics can yield exciting discoveries, but there are no guarantees that a specific screen will produce anything of interest. This is because mutant phenotypes may be so subtle that they are overlooked or because different genes have redundant functions, so a mutation in one gene is compensated for by the function of an unaltered gene.

Practical Considerations

Research on vertebrate animals is a regulated activity, even when the animals involved (for example, mice, and zebrafish) may be freely kept as household pets. In the United States, federal, state, and/or institutional laws and/or guidelines ensure that vertebrate animals used in the laboratory are well housed, well nourished, and handled with respect. Other countries impose comparable requirements. Researchers who use vertebrate animals may not enjoy the added costs, paperwork, and delays imposed by the requirement that research protocols be reviewed by institutional committees (often referred to as IACUCs—institutional animal care and use committees), but they welcome rigorous standards because careful handling of animal subjects is an essential component of quality research. Students should be aware that these laws and regulations also cover the use of vertebrate animals in the classroom and in undergraduate research projects.

Research on invertebrate animals (such as fruit flies and nematodes) is generally entirely unregulated, but new regulations adopted by the European Union has governed research on cephalopod mollusks (octopuses and cuttlefish) since January 2013, and it is not inconceivable that research on other invertebrates will eventually also be regulated.[4]

The Mouse, *Mus musculus*

The mouse tops our list of models for developmental neuroscience. The popularity of the mouse in all areas of biomedical research partly reflects that it is a mammal, and hence similar to humans in terms of genetics and physiology. Mice are also popular because their genome can be manipulated to produce *designer mice* customized to answer specific research questions (reverse genetics). This is achieved by exploiting homologous recombination in cultured embryonic stem cells. Knockout mice (sometimes referred to as KO mice) are produced by replacing genes with altered versions that produce nonfunctional products: analysis of the phenotype of mice with this type of genetic lesion can reveal the function of the normal version of those genes.[5] Mouse research is supported by decades of practical experience rearing mice in laboratories. Mouse embryonic stem cells are cultured more easily than those of other mammals. Reproductive technologies above and beyond those found in the human fertility clinic are available for maintenance of genetically defined lines of mice for research. As noted previously, many inbred strains of mice with predictable phenotypes and nearly identical genotypes are available for research. For example, the albino BALB/c strain of mice (what you likely think of when you think of a white laboratory mouse) was established through 15 years of brother–sister matings representing 26 generations.[6] Genetically modified mice produced for specific research projects are always based on a specific inbred strain, which can then be used as a control for phenotypic analysis.

Mice are members of order Rodentia (rodents). Taxonomists classify the species *Mus musculus* (common name, house mouse) as a member of the genus Mus and a subgenus also called Mus. The close relatives of *Mus musculus* are sometimes described as true Old World mice. The mouse lineage likely diverged from the lineage leading to humans approximately 75 million years ago. Humans have bred mice as pets for centuries, often prizing beautiful coat colors. The recognition of the relevance of Mendelian genetics to all organisms at the end of the nineteenth century led to the use of mice in experimental tests of inheritance. Often the phenotype used to test Mendel's laws involved those fancy coat colors. The sequence of the mouse genome was published in 2002, less than two years after the complete first draft of the human genome.

Mice reach sexual maturity at about 35 days of age. The female's ovarian cycle is 4–5 days in length, pregnancy lasts 19 days, and a female can produce 8 litters of approximately 8 pups per year. The typical mouse lifespan is between 500 and 1,000 days. This life history reflects a typical mammalian pattern, permitting all stages of nervous system development, from embryogenesis to senescence, to be studied in a convenient small package.

As in the case of human embryonic development (Chapter 2), the sequential events of mouse embryogenesis can be divided into stages defined by the appearance of specific morphological features. A widely used standard staging system is based on an atlas of mouse embryonic development compiled by the Swiss embryologist Karl Theiler.[7] The 26 Theiler stages of prenatal development are referenced to dpc, which can refer to days post conception or days post coitum. At Theiler Stage 1, the fertilized egg is approximately 50 micrometers in diameter. As in the human, cleavage results in the formation of a morula, which develops into a blastocyst containing an inner cell mass (fig. 3.2). At Theiler Stage 7, approximately 5 dpc, the blastocyst sheds its zona pellucida and implants in the uterine wall. At this point it does not look at all like the more or less flattened human embryo. Instead, the pregastrulation mouse embryo forms an elongated, cup-shaped structure called the egg cylinder, a configuration that persists through the end of Theiler Stage 11. Only rodent embryos form an egg cylinder, but many studies have shown that the tissues within the egg cylinder illustrate standard mammalian processes even though the early morphology is unusual.

The egg cylinder consists of three tissues: the epiblast, which becomes the embryo; the extraembryonic ectoderm; and the visceral endoderm, which wraps the epiblast and the extraembryonic ectoderm. The ectoplacental cone at the top of the egg cylinder attaches the embryo to the uterine lining during implantation. Layers of parietal endoderm and trophoblastic cells are separated from the visceral endoderm by the yolk sac cavity. Even at this very early stage of development, an amnionic cavity begins to form within the epiblast. As described in Chapter 4, at 5.5 dpc a subset of visceral endoderm cells from the distal tip of the egg cylinder migrates to the boundary between the epiblast and the extraembryonic ectoderm. This event is significant because it is an early—perhaps the earliest—indicator of the anterior–posterior axis of the mouse embryo. The destination of the migrating visceral endoderm cells becomes the site where the head and head-associated neural structures are induced in the underlying epiblast.

Key events in the formation of the mouse nervous system are as follows. The neural plate becomes defined anteriorly and the neural groove becomes visible at 7 dpc (Theiler Stages 11a–11d). The neural tube closes in the region of the fourth and fifth somites at 8 dpc (Theiler Stage 12). Neural tube closure continues, and the anterior neuropore closes during Theiler Stage 14 at 9 dpc. Later on Day 9 the posterior neuropore forms and the forebrain vesicle divides into the telencephalic and diencephalic vesicles (Theiler Stage 15).

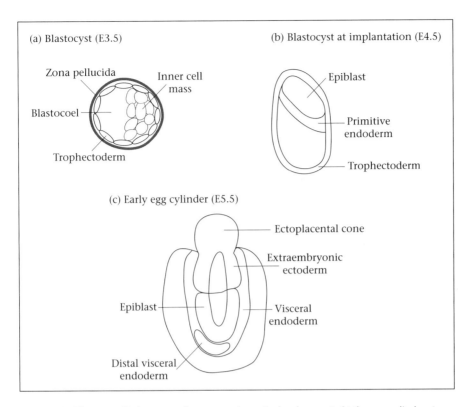

(a) Blastocyst (E3.5)

Zona pellucida
Inner cell mass
Blastocoel
Trophectoderm

(b) Blastocyst at implantation (E4.5)

Epiblast
Primitive endoderm
Trophectoderm

(c) Early egg cylinder (E5.5)

Ectoplacental cone
Extraembryonic ectoderm
Epiblast
Visceral endoderm
Distal visceral endoderm

Figure 3.2. The egg cylinder stage of mouse embryonic development. At the egg cylinder stage, the mouse embryo has a cup shape unusual among mammalian embryos. At 3.5 dpc (E3.5), the blastocyst (a) consists of the trophectoderm, the epithelium that surrounds the inner cell mass, and the blastocoel cavity. The noncellular zona pellucida that coats the oocyte is still present at this stage (compare with fig. 2.2). A day later (E4.5), at implantation (b), the epiblast is evident. This tissue will give rise to the ectoderm, mesoderm, and definitive endoderm during gastrulation. The ectoplacental cone is formed of proliferating cells that extend toward the uterine epithelium at the site of attachment. The visceral endoderm, derived from the primitive endoderm cells that lined the surface of the inner cell mass in the blastocyst, is evident on E5.5 in the egg cylinder stage (c). Changes in the position and thickness of the visceral endoderm are required to establish the anterior–posterior body axis. Drawing adapted from Rossant and Tam (2009).

At 10 dpc, the posterior neuropore closes (Theiler Stage 16). Rapid growth of the brain is evident from Day 11 (Theiler Stage 18) on. Equivalences between the days of human and mouse embryogenesis based on the Carnegie stages have been worked out, with Carnegie Stage 9 reached at 9 dpc in the mouse and on Day 20 in the human, Carnegie Stage 15 at 12 dpc in the mouse and on Day 36 in the human, and so on. It should be noted that, although many authors use either Theiler stages or dpc to describe the stage of development of the mouse embryos they are studying, others use the simpler designation of "E" followed by a number that indicates the age of the embryo in days.

For example, E11 indicates the eleventh day of embryonic life, as does 11 dpc or Theiler stage 18.

What are the *disadvantages* of using mice as a model? The major drawback, in addition to the cost of maintaining specialized rearing facilities, is that embryonic development occurs within the mother mouse's uterus, which blocks the view of embryonic development. Another drawback is that many interesting aspects of the later stages of human brain development—childhood, adolescence, and young adulthood—reflect postembryonic development of regions of the cortex that are significantly larger in primates than in rodents. A final concern is the many well-documented differences across mouse strains in almost every aspect of mouse physiology and behavior. Selection of the best strain for a given study is a critical aspect of experimental design, and it cannot be assumed that what is found in one strain will generalize to other strains.

In the second half of the twentieth century, neuroscientists tended to favor a larger, more robust rodent, the Norway rat (*Rattus norvegicus*), particularly for behavioral studies. This situation changed when the production of knockout mice using homologous recombination in cultured embryonic stem cells became routine. This new research tool was so compelling that many neuroscientists switched their investigations from rats to mice. Despite these animals' obvious similarities, many studies have revealed that mice are not just little rats. For example, a direct comparison of neurogenesis in the dentate gyrus of the hippocampus that used C57BL/6 mice and several different strains of rats (Sprague Dawley, CD1, and Long–Evans) revealed striking differences in the extent of adult neurogenesis between mice and rats. In particular, the number of new neurons added to the hippocampal dentate gyrus (Chapter 5) was significantly higher in rats than in mice.[8] What this difference means is not yet understood.

An alternative method of producing gene knockouts based on treatment of the rat embryos with zinc finger nucleases at the single-cell stage of development was developed in 2009 and has allowed production of a limited number of rats with specific gene deletions.[9] Reports of methods for culturing rat embryonic stem cells may mean that rats will continue to be used to study some aspects of neural development, although it seems unlikely that they will ever regain their former status as top rodent in the world of neuroscience.

The Zebrafish, *Danio rerio*

Relative to the mouse, the zebrafish is a newcomer to biomedical research. In the late 1960s and 1970s, phage geneticist George Streisinger promoted the idea that the zebrafish could serve as a useful model organism because it would permit the application of mutational analysis to vertebrate development. Streisinger, based at the University of Oregon, initiated a zebrafish research program that developed the methods needed to make his vision a reality. A paper published in 1981 introduced the zebrafish as a model and described basic procedures for studying and manipulating zebrafish genetics. Streisinger's colleagues at the University of Oregon quickly recognized the potential of the zebrafish to contribute to the understanding of nervous system development, primarily because of its transparent, relatively large, and free-living embryos. The Oregon group commenced the essential morphological studies that provide the foundation of contemporary genetic and molecular experiments on zebrafish.[10] In the early 1990s, these foundational studies supported a large forward genetics screen of more than a million zebrafish embryos. This screen was led by Nobel Laureate Christiane Nüsslein-Volhard, who will be mentioned again in this chapter in the section on *Drosophila melanogaster*. The embryos were the great-grandchildren of zebrafish males that had been exposed to a chemical mutagen; those with abnormal phenotypes constituted a library of mutations covering many aspects of development, from gastrulation to the formation of synaptic connections deep within the brain.

In addition, as befits a research community that grew in tandem with the World Wide Web, the zebrafish research community quickly developed an extraordinarily rich set of shared online resources. In fact, the very first web site that I ever visited (in June, 1994, at a workshop titled Establishing Identified Neuron Databases convened by the National Science Foundation in Arlington, Virginia) was a site containing zebrafish-rearing protocols. A zebrafish genome-sequencing project was initiated in 2001. The online Zebrafish Model Organism Database (http://zfin.org/) provides ready access to a zebrafish genome browser in addition to online versions of *The Zebrafish Book,* the essential guide to research on this species.[11]

Zebrafish are teleost fish within the vertebrate class Actinopterygii (teleosts are also referred to as Osteichthyes, or ray-finned fishes, in some texts). The last common ancestor of the lineages leading to zebrafish and humans probably lived 450 million years ago. Zebrafish are native to bodies of freshwater in the southeastern Himalayas but are found in pet stores and home fish tanks all over the world. They are a good example of the type of small fish referred to as minnows. Both males and females have dark horizontal stripes on the sides of their elongated bodies. With vigilant care, zebrafish can grow to a length of about 4 centimeters and have a lifespan of about two years. If males and females are housed in the same tank, adult females will ovulate and spawn (release their eggs into the water). A female can spawn hundreds of eggs every 2–3 days. Embryonic development begins at spawning but arrests unless fertilization occurs. One sign that fertilization has occurred is that fertilized eggs become transparent. Zebrafish eggs hatch approximately 48 hours after fertilization (the exact time depends on water temperature). A larval stage begins approximately 72 hours post fertilization. At this time the swim bladder inflates and the little fish begins to swim and look for food.

Kimmel and colleagues published a detailed description of zebrafish embryonic development in 1995. On the basis of observations of living embryos, they developed a morphological staging system of seven periods that links fertilization to hatching. The seven periods reflect events common to the embryonic development of all vertebrates. With the exception of the zygote (which contains only a single stage), each period can in turn be broken down into a series of stages recognizable in intact embryos (fig. 3.3).

The zebrafish oocyte has obvious polarity, even prior to fertilization: the yolky end is called the vegetal pole, and the end opposite the yolk is called the animal pole. The first rounds of cleavage result in formation of a cluster of cells the sit on top of a large yolk cell. This cluster of cells is called the blastoderm. After several more rounds of cell division, the blastoderm consists of two distinct cell populations. The outermost layer of the blastoderm is the enveloping layer (EVL). The cells directly underneath the EVL are called the deep cells. They are the equivalent of the epiblast cells of mammals, for they are the source of the cells that form the actual embryo. The cells at the vegetal edge of the blastoderm constitute a third layer called the yolk syncytial layer (YSL).

As in mammals, distinct tissue source layers of ectoderm, mesoderm, and endoderm form as part of the process of zebrafish gastrulation. Complex patterns of cell migration transform the blastoderm into a recognizably vertebrate embryo. The term *epiboly* is used to refer to the vegetal pole–directed movement of blastoderm cells. Eventually, the yolk is surrounded by blastoderm cells. Before this happens, however, the blastoderm thickens and becomes double layered. The outer layer, the epiblast, gives rise to the nervous system and the cells of the integument. The inner layer is the hypoblast. The fate of the cells of the hypoblast is to form the mesodermal and endodermal tissue layers.

The edge of the dorsal blastoderm becomes thicker as it spreads, and once the thickening is noticeable it is called the embryonic shield. The embryonic shield eventually becomes the notochord of the developing embryo. A subset of epiblast cells will be induced to produce neural structures as a result of interactions with the notochord-to-be. The neural-induced cells move to the dorsal midline of the embryo. In the zebrafish, epiboly is completed (100 percent) at approximately 10 hours post fertilization.

The next stage of zebrafish embryonic development is segmentation. During this stage, the mesodermal somites form and brain segments called neuromeres appear.[12] Early in the segmentation stage, the neural plate can be identified because its cells become columnar, making it the thickest part of the ectodermal epithelium. Like the neural plate of mammals, the neural plate of the zebrafish will form the neural tube. But the process by which the neural tube of zebrafish forms is so different from that seen in other vertebrates that it is given a different name: secondary neurulation. Primary neurulation in mammals forms the neural tube by *rolling* the neural plate into a

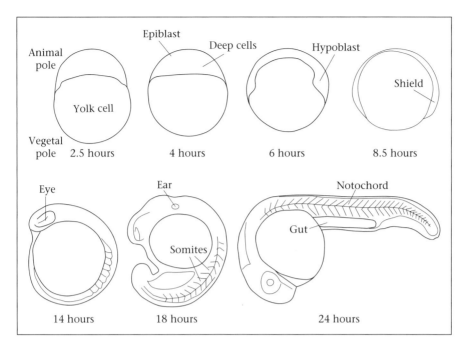

Figure 3.3. Early development of zebrafish embryos. The first 24 hours postfertilization are depicted schematically. The stages shown are early blastula (2.5 hours), midblastula (4 hours), early gastrula (6 hours), midgastrula (8.5 hours), early segmentation stage (14 hours), mid-segmentation stage (18 hours), and pharyngula stage (24 hours). Zebrafish eggs are yolky, and cell divisions take place only at the animal pole of the egg (the blastodisc). By 4 hours, the animal pole consists of a superficial monolayer of extraembryonic cells called the enveloping layer, and layers of deep cells that will form the embryo. By 6 hours, the population of deep cells has begun to form two layers: an outer layer called the epiblast and an inner layer called the hypoblast. Both epiblast and hypoblast cells will eventually form the slight thickening called the embryonic shield (visible at 8.5 hours). The head region (anterior) of the embryo is clearly defined by 14 hours, as shown. Drawing adapted from Amacher (2001).

hollow tube. Secondary neurulation, by contrast, is achieved by *forming an opening* in a preexisting rod of solid tissue. In zebrafish, the wedge of columnar epithelial cells that forms the neural plate is referred to as the neural keel. The neural keel cells form a rod that becomes the hollow neural tube by a process of directed cell movements called cavitation.

During the first half of the segmentation stage, the anterior end of the neural keel develops 10 swellings called neuromeres. The three most anterior neuromeres will form the forebrain and midbrain. The remaining seven will become the hindbrain. Because the hindbrain is also called the rhombencephalon, the posterior neuromeres in zebrafish are referred to as rhombomeres, as in other vertebrates (Chapter 6). The segmentation stage of development ends approximately 23 hours after fertilization, with all of the somites in place and neurogenesis well under way. During the next 24 hours, the major organs continue to develop as the body axis straightens and the fish begins to look like a fish. Hatching at 48 hours after fertilization is followed by a larval stage beginning at 72 hours. About 30 days later, the larva becomes a juvenile approximately 10 millimeters in length. Juveniles have adult pigmentation and fins but are not yet ready to breed. Reproductive maturity is reached at about 90 days. Neurogenesis in the central nervous system continues throughout adult life (Chapter 5).

The Fruit Fly, *Drosophila melanogaster*

Drosophila is the genus name for a large group of flies of the family Drosophilidae in the insect order Diptera, phylum Arthropoda. At the time I am writing, 1,579 species in this genus have been described, with several thousand likely remaining to be discovered. The common name for flies in this genus is fruit fly. One species of fruit fly, *Drosophila melanogaster,* was initially famed for its contributions to genetics research but now is also valued for the insights it has provided into early development. Here we use the genus name *Drosophila* as shorthand for the species *Drosophila melanogaster,* a practice that makes everyone else's life easier but drives entomologists crazy (they prefer *D. melanogaster*).

Drosophila are tiny flies. Adult *Drosophila* have tan bodies about 2.5 millimeters in length. Their eyes are red, and their abdomens are ringed with black bands. This cosmopolitan species is native to the Old World tropics but is now found wherever humans grow fruit. Adults feed on decaying plant material and are attracted to rotting fruit. Females lay their eggs in fruit that is just about to ripen so that newly hatched larvae do not have to look far for food. Embryonic development is completed prior to hatching, and fruit fly eggs hatch the day after they are laid. Three larval stages commonly referred to as instars follow. The first and second instars last one day each, while the third instar lasts two days. The instars are separated by the shedding of the cuticle of the previous instar, an event called a molt. Molts are a prerequisite for continuing growth. The tiny larvae have legless, segmented bodies and are referred to as maggots. At the end of the third instar, the larvae begin a roaming stage. This is very easy to detect in the laboratory because, prior to the onset of roaming, the larvae are found *in* whatever food they are offered. Once they begin to roam, they are instead found attached to the walls of their container. The roaming stage is terminated by pupariation, during which the final larval cuticle becomes a hardened case called the puparium. The pupa undergoes metamorphosis inside the puparium, concealed from view (Chapter 9). Pupariation occurs approximately 120 hours after an egg is laid (the exact time depends on the rearing temperature). The adult fly emerges from the puparium 10 or 11 days after hatching. Sexual maturity is attained in less than a day. The typical adult life span of a fruit fly is about two weeks. To summarize, embryogenesis is followed by hatching, larval life, metamorphosis, and adulthood. The pattern of postembryonic development displayed by *Drosophila* is referred to as complete metamorphosis or holometaboly, and this pattern is shared with many familiar insects such as butterflies, bees, and beetles, although only flies form a puparium. The nervous system of the adult fly consists of the brain, the ventral nerve cord, and sensory neurons associated with peripheral sense organs.

Modern biologists use *Drosophila* as a model species in part because the American geneticist Thomas Hunt Morgan did so, starting during the first decade of the twentieth century. Why did Morgan select this species? Statements he made when he won the Nobel Prize in Physiology or Medicine in 1933 indicate that Morgan credited American entomologist Charles W. Woodworth with the suggestion that *Drosophila* could be useful to geneticists, in large part because they can be reared at low cost in limited space.[13] Modern biologists still appreciate this feature of *Drosophila* but now also enjoy an extensive literature, a carefully annotated genome, and the ability to create targeted mutations. Even given these selling points, developmental neuroscientists might still eschew these creatures with tiny, yolky embryos if they did not have access to modern microscopy techniques. With these tools in hand, however, *Drosophila* researchers are among the most aggressive experimentalists on the planet, capable of performing sophisticated manipulations that remain the stuff of fantasy for neuroscientists using other models. The 1995 Nobel Prize in Physiology or Medicine was awarded jointly to Edward B. Lewis, Christiane Nüsslein-Volhard, and Eric F. Wieschaus for their discoveries concerning the genetic control of early embryonic development based on experimental studies of *Drosophila* embryos. Although these Nobel-winning studies tackled the broadest possible developmental questions (for example, how does an animal develop a body plan?), the forward genetics approach embodied in the work of these scientists has been successfully applied to all aspects of nervous system development.

The *Drosophila* egg is 0.5 millimeters long and 0.2 millimeters wide. It is protected by a chorion (eggshell) that can be removed by a bleach treatment to allow a better view of the developing embryo. *Drosophila* embryonic development has been divided into 17 stages based on the morphology of living dechorionated embryos. These stages are called Bownes stages after the creator of a photographic atlas of fruit fly embryonic development, Mary Bownes. It is now possible to relate patterns of gene expression to the visible events on which the Bownes stages are based.

Bownes Stages 1–3 reflect the unusual pattern of cleavage characteristic of insect embryos: division of nuclei without the formation of cellular membranes. The rounds of cleavage through cell cycle 13 result in the formation of a syncytial blastoderm (Bownes Stage 4, 90–130 minutes after fertilization). The nuclei produced by these early divisions then move to the periphery of the embryo, where folds of the oocyte plasma membrane finally wrap the new nuclei. The result is the cellular blastoderm (Bownes Stage 5, 130–180 minutes after fertilization) (fig. 3.4).

The process of gastrulation begins at Bownes Stage 6 (180–195 minutes after fertilization). A full description of the cell movements that result in the formation of *Drosophila* endoderm, mesoderm, and ectoderm is beyond the scope of this chapter, so a brief summary follows. The first visible event of gastrulation is the appearance of the ventral furrow, an inward fold at the ventral midline of the embryo. The cells that form this fold are future mesoderm. The cells at the surface are future ectoderm. The future mesoderm and future ectoderm at the ventral midline form the germ band, which in turn forms the trunk of the embryo. The germ band then elongates so much (Bownes Stage 7) that its posterior tip bends dorsally over the back end of the embryo. Eventually the germ band retracts (Bownes Stage 13) so that the posterior tip of the germ band is positioned at the posterior end of the embryo. Before the germ band retracts, however, the nervous system begins to form. The cells of the ectodermal epithelium that build the nervous system declare themselves by moving out of the epithe-

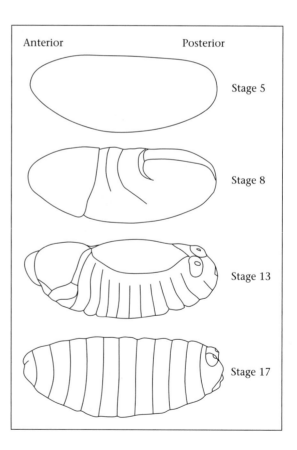

Figure 3.4. External view of development of *Drosophila* embryos. The selected stages depicted (Stage 5, Stage 8, Stage 13, Stage 17) cover cellular blastoderm formation through the end of embryonic life. Stage 8 shows the results of gastrulation, the segmentation of the epidermis is clearly evident at Stage 13, and, at the anterior end, the head structures have moved inward (involuted) by Stage 17. The posterior spiracles are evident at Stages 13 and 17. Drawings loosely adapted from Hartenstein (1993).

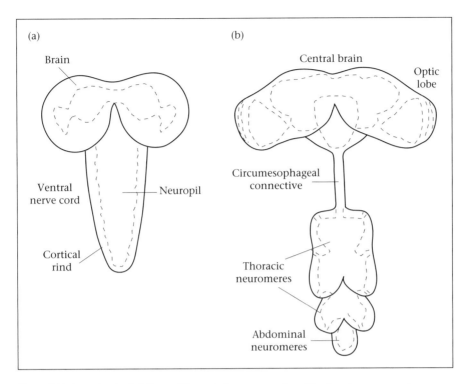

(a)

Brain

Ventral
nerve cord

Neuropil

Cortical
rind

(b)

Central brain

Optic
lobe

Circumesophageal
connective

Thoracic
neuromeres

Abdominal
neuromeres

Figure 3.5. Larval and adult *Drosophila* central nervous system. The larval nervous system (a) was dissected from a third-stage larva. It consists of bilaterally symmetric brain lobes and a ventral nerve cord formed from the thoracic and abdominal segmental ganglia. In the adult (b), the brain and ventral nerve cord are joined by the circumesophageal connectives. The optic lobes, which receive inputs from the compound eyes, are the most prominent features of the adult brain. Dashed lines indicate the boundaries of the central neuropils, which are surrounded by a cortical rind of neuronal somata.

lium in a process called delamination. These cells are called neuroblasts, and they begin to separate from the ectodermal cells that will remain at the surface to form the epidermis during Bownes Stage 9 (230–260 minutes after fertilization). The zones of the ectoderm where the neuroblasts form are called neurogenic regions (fig. 3.5). The neurogenic regions of the ectoderm are not a continuous cell population but instead form small neurogenic islands called proneural clusters. The ventral neurogenic regions generate the ventral nerve cord; the procephalic neurogenic region generates the brain. The delamination of neuroblasts continues through Bownes Stage 11 (320–440 minutes after fertilization). The body segments become visible during Bownes Stage 12 (440–580 minutes after fertilization). Both central and peripheral nervous system structures differentiate during Bownes Stage 13. The initially separate segmental neuromeres that form the central nervous system will fuse by hatching, which occurs 21–22 hours after fertilization. The boundaries of these neuromeres can still be recognized at later stages using appropriate molecular markers.

Neurogenesis continues during larval life and into the early part of the pupal stage, representing continuing mitosis by neuroblasts initially active during embryogenesis. Neurogenesis is one of the ways that the central nervous system adapts to the larval, pupal, and adult stages of postembryonic life (the nervous system is an exception to the rule that *Drosophila* larval tissues are replaced during metamorphosis). The other strategies that modify the embryonic nervous system to meet the needs of the postembryonic life stages are the programmed cell death of specific neuronal populations and the ability to reorganize the cytoskeleton of persisting neurons so that old synapses can be dismantled and new synapses formed (Chapter 9).

The Nematode Worm, *Caenorhabditis elegans*

"I would like to tame a small metazoan organism to study development directly," wrote Sydney Brenner in 1963 to Max Perutz, Nobel laureate and director of the MRC (Medical Research Council) Laboratory of Molecular Biology in Cambridge, England.[14] Brenner's switch from the classical molecular genetics of bacteriophages and bacteria to the worm is the origin of the last of our four models. Initial studies of the nematode *Caenorhabditis elegans* focused on development, including development of the nervous system. Today a broad range of neuroscientists, including those interested in learning, memory, and how neural circuits control behavior, turn their attention to this worm for the same reasons that Sydney Brenner did so in the 1960s.

Caenorhabditis elegans is commonly referred to as *C. elegans*, and that custom is followed in this text. *C. elegans* worms are members of the phylum Nematoda. This is a speciose and abundant phylum—it has been estimated that there may be as many as a million or more species in the phylum, most of them undescribed. Many nematodes are parasites, some with human hosts. Hookworms, pinworms, and filarial worms are examples of nematode worms that directly injure humans. Other nematodes harm humans indirectly because they are plant parasites that reduce the yield of food crops. *C. elegans* is, however, an example of a free-living, bacteria-feeding, non-parasitic nematode. Brenner obtained his *C. elegans* from a compost heap; in the laboratory, *C. elegans* live and feed on bacterial lawns growing on agar plates. Recent research has revealed that the natural habitat of *C. elegans* is not soil, as had long been thought, but rotting fruit. Relatively little is known about the ecology of *C. elegans* outside the laboratory.

By contrast, the list of what has been learned by studying *C. elegans* inside the laboratory is dazzling. Every cell of the worm's body has been cataloged, and the lineage of every cell in its body is known. *C. elegans* was the first multicellular organism to have its genome sequenced; it was the first animal that is not a jellyfish to express GFP; it can be genetically transformed by injection of DNA; and specific genes can be knocked down simply by feeding *C. elegans* bacteria engineered to express double-stranded RNA (dsRNA) targeting the gene of interest (see Chapter 1). Most *C. elegans* are self-fertilizing hermaphrodites, making it easy to maintain inbred stocks with low degrees of genetic variation. If their small size (about 1 millimeter in length when full grown), short life cycle (they are ready to reproduce in about 3 days), and transparent embryos were not advantages enough, young *C. elegans* larvae can be stored frozen in liquid nitrogen or a low-temperature freezer (–80° C) and thawed for study months or years later.[15] *C. elegans* stocks useful for research can be obtained for a small fee from the Caenorhabditis Genetics Center (CGC) based at the University of Minnesota.

The life cycle of *C. elegans* is straightforward (fig. 3.6). Fertilization occurs inside the body of the hermaphrodite, and cleavage initiates inside the parent, so the egg is laid at the gastrulation stage. The first-stage larva (L1) hatches 9 hours later; L1 molts to a second larval stage (L2) 12 hours later; the L2–L3 molt follows 8 hours later. The final molt (L3–L4) is completed in another 8 hours. After a further 10 hours, the L4 larva is considered a young adult; 8 hours later, the young adult is a mature hermaphrodite that initially produces sperm, then oocytes, and then up to 300 offspring. If L1 larvae are subjected to harsh conditions (overcrowding, starvation), instead of molting the larvae enter a survival stage during which they are called dauer larvae. The dauer form can be maintained for up to 4 months, after which time the larvae complete their development to the adult stage. A small percentage of males (an estimated 0.05 percent) are found in populations of *C. elegans*; they are slightly larger than the hermaphrodites, with which they mate.

C. elegans have an elongated body (see fig. 3.6). The most noticeable internal organs are those of the reproductive system. The mouth marks the anterior end, and there is a tapered postanal tail. A thin cuticle covers the hypodermis that forms the body wall; the cell bodies of neurons are found adjacent to the hypodermis, close to the layer of longitudinal muscles. The nervous system of a mature hermaphrodite comprises 302 neurons. The major part of the nervous system is called the somatic nervous system. The somatic nervous system controls all functions except ingestion, which is controlled by a small pharyngeal nervous system of 20 neurons. Most of the neurons of the somatic nervous system (the remainder of the 302 neurons) are clustered in ganglia in the head or the tail: nonganglionic neurons are found along the length of the body at the ventral midline. The ganglia contain nerve cell bodies but are not sites of information exchange, for they do not contain synapses. There is no brain, but a neuropil that wraps around the pharynx called the circumpharyngeal nerve ring serves some of the functions of a brain. The circumpharyngeal nerve ring is connected to the remainder of the body by peripheral nerve cords formed of longitudinal tracts of neuronal processes. Commissures connect the longitudinal tracts. Distinct populations of motoneurons, sensory neurons, and interneurons have been recognized, but some

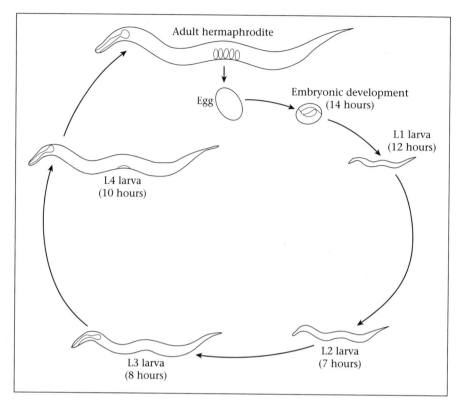

Figure 3.6. Life cycle of *C. elegans.* The adult hermaphrodite can self-fertilize. Development times are approximate and temperature dependent. The head end of each larva depicted is to the left (marked by the pharynx). Well-fed laboratory stocks of *C. elegans* proceed directly to the adult stage as shown. Under harsh environmental conditions (high population density, low food availability), late L1-stage larvae may enter the dauer stage of developmental arrest (not shown) instead of proceeding directly to the L2 stage. The dauer stage may last several months. Larvae leaving the dauer stage molt to the L4 larval stage. An adult hermaphrodite is typically slightly longer than a millimeter from nose to tail. Drawings based on Wormatlas.org.

neurons likely have multiple functions. The head contains a pair of sensory organs called amphids. Each amphid ends in a cuticular invagination at the anterior tip of the worm. Amphids consist of 12 sensory neurons with ciliated dendrites. One of these neurons is thermosensitive and is required for *C. elegans* to position itself appropriately on a thermal gradient. The other 11 sensory neurons of the amphids are chemosensory.

The neurons of *C. elegans* are located in predictable positions in the body, and each neuron bears a name that is a key to its lineage and its position relative to its siblings (anterior or posterior, right or left). The development of the nervous system can therefore be traced back to a small number of founder cells called blastomeres (fig. 3.7). As in other animals, the organization of

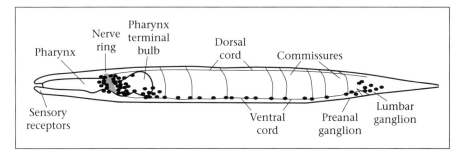

Figure 3.7. Nervous system of *C. elegans*. The major structures are depicted. Some axons run in longitudinal bundles, while others are circumferential. The major neuropils are the nerve ring associated with the pharynx and the ventral nerve cord, which extends posterior from the nerve ring in two tracts that follow the ventral midline. The nerve ring contains 100 axons produced by motoneurons, interneurons, and sensory neurons. Drawing adapted from Wadsworth and Hedgecock (1996).

cells into specific tissues and organs begins at gastrulation. The first event of gastrulation, which begins at the 24-cell stage, is the movement of gut precursor cells from the ventral side of the embryo into the center. This movement of gut precursors is followed by inward migration of the germ cells, followed, in turn, by inward movement of mesoderm. The nascent ectoderm maintains its superficial position, eventually giving rise to the hypodermis and the nervous system. Note that there is no dedicated neuronal lineage we can easily point to in *C. elegans,* as we can in our other models of nervous system development.

There are 959 somatic cells in the body of the adult hermaphrodite, but the newly hatched L1 larva contains only 558 somatic cells. The remaining 404 somatic cells are produced during larval life. Some of the larval-born cells include neurons. The three main periods of neuron production in *C. elegans* are during embryogenesis, late in the first larval stage, and throughout the second larval stage.

Typical Neurons

When you think of a typical neuron, you likely imagine something along the lines of figure 3.8a. This is a depiction of a vertebrate motoneuron. Branched dendrites extend directly from the neuronal soma (the term neuroscientists use to refer to the cell body of the neuron; the plural form of *soma* is *somata*), and a single long myelinated axon (fig. 3.8b) ends in branched axon terminals. The dendrites and the surface of the neuronal soma are postsynaptic, and the axon terminals are presynaptic. The flow of information through such a neuron is polarized: *in* through the dendrites

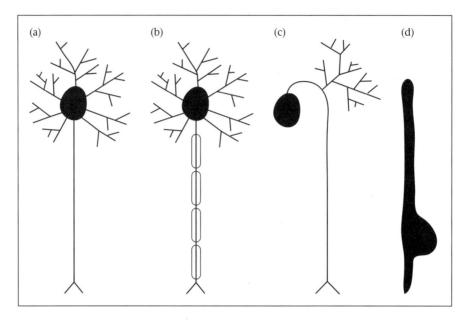

Figure 3.8. Typical neurons. Highly schematic depictions of neurons from different animals. Vertebrate neurons (a, b) are typically characterized by dendrites that extend directly from the neuronal soma and a single axon that also extends directly from the soma. Many vertebrate axons have a myelin sheath (b). Insect neurons (c) typically extend a single process called a neurite from the soma. One or more dendritic arborizations have their point of origin on the neurite, which then extends to form the axon. Nematode neurons, such as the *C. elegans* motoneuron depicted in (d), typically extend unbranched processes from the soma. Note that these depictions are grotesquely oversimplified for didactic purposes, that the axon terminals in vertebrates and insects are often highly branched, and that these neurons are not drawn to scale. Panel (d) is my interpretation of a VA- or VB-type motoneuron of *C. elegans,* as described by Schneider et al. (2012).

and surface of the soma, *out* through the axon terminals. Most students will already be aware that different populations of vertebrate neurons have more or less elaborately branched dendrites (see fig. 3.8a) and that not all vertebrate axons are myelinated (see fig. 3.8b for one that is). The vast majority of neurons on planet Earth, however, have a different phenotype because they do not reside in vertebrates. The typical invertebrate neuron does not extend dendrites directly from the surface of its soma. Instead, a single extension of the cytoplasm called a neurite grows out from the soma. After a short distance, the neurite branches to produce a dendritelike arborization and a separate axon. This arrangement is highly typical of arthropod neurons (fig. 3.8c). The dendritelike arborization, which may be profusely branched, constitutes the postsynaptic surface, for no synapses are formed directly onto the soma. No myelination is present. The flow of information through insect neurons is polarized but possibly less so than in vertebrate neurons, for bidirectional synapses have been observed in the complex dendritelike arborizations of some insects. The terms dendrite and axon are also difficult to apply to *C. elegans*. Nematode neurons typically extend a single neurite (fig. 3.8d). This process may be more or less branched. Synapses are typically formed en passant, between neurons and between neurons and muscles. One feature of the nematode nervous system worthy of further comment is the unusual (from the vertebrate and insect perspectives) relationship of motoneurons to their muscle targets. In *C. elegans,* the muscles come to the motoneurons rather than the other way around: muscles are innervated by neuromuscular junctions formed by motoneuron axon terminals onto unusual muscle arms extended by muscles toward their motoneurons. As in insects, no myelin is present in nematodes.

Gray Matter and White Matter

The terms *gray matter* and *white matter* are frequently used by neuroscientists to describe the two main compartments of vertebrate brains. The term *gray matter* is also used in informal speech to refer to the brain. These terms have their origin in the gray and white appearance of human brains preserved in formaldehyde. The gray matter of the brain comprises the neuronal somata and dendrites; the white matter consists of axons coated with myelin, which will be discussed in Chapter 8. Differences between white and gray matter are also evident in fresh brain tissue, although not to the same extent as in fixed tissue. Because myelin is almost the exclusive province of chordates, these terms are only applicable to the vertebrate model systems discussed in this book.[16] Many other types of animals, however, are also characterized by separate compartments for neuronal somata and neuropil (a neuropil is a region of synaptic interactions between dendrites and axon terminals). A common arrangement is to have a cortical rind of somata around a central neuropil. Such an arrangement is readily seen in the brain and ventral nerve cord of insects, including *Drosophila.*

Phylogenetic Relationships

Despite differences in cytoarchitecture and size of the nervous system, many of the early events important for the development of the nervous system and many of the signaling molecules that regulate development are shared across the animal kingdom. This reflects that fact that, taxonomically speaking, the animals are a very well-defined group. From the point of view of the proverbial scientist from Mars, all animals are much more similar to each other than to the other forms of life on Earth, and we scientists spend our time making much of small differences.[17] A number of traits that define the animals have already been noted: the blastula (also called blastocyst) embryo, gastrulation, neurons, and nervous systems. In subsequent chapters, signaling molecules will be introduced that have counterparts, as far as we know, in all animal embryos. These similarities reflect the common ancestor shared by humans and our four model species. But different types of animals can also be considered in the historical context of which groups arose when, and such information helps biologists know when to predict that two animal species will be very similar and when they will be less similar. The evolutionary relationships—the phylogeny—of different animal groups can be estimated using many forms of molecular evidence, including DNA sequence similarity, rare amino acid changes, intron locations, conservation of miRNAs, and mitochondrial gene order. A current view of animal phylogeny is shown in figure 3.9. You should regard this diagram as a hypothesis consistent with current evidence. All of the species mentioned in this book are represented in this diagram, but you will need to look hard to find them. Our three vertebrates—the zebrafish, mouse, and human—are members of a larger group of animals called deuterostomes that also includes the acorn worm (an example of a hemichordate) and starfish (an example of an echinoderm). Now look for the arthropods and nematodes. You may be surprised to see our arthropod (*Drosophila*) and our nematode (*C. elegans*) united in a multiphyla group named the Ecdysozoa that also includes the velvet worms (onychophorans) and the tiny water bears (tardigrades). This diagram is included to make two points. First, the coverage of nervous system development provided by our selected models omits all of a third large group of animals, the Lophotrochozoa. Second, most species simply don't have "the right stuff" to be a model organism.[18] The few that biologists have are treasures.

Figure 3.9. Animal phylogeny. Traditional views based primarily on morphology and embryology have been superseded by phylogenetic reconstructions based on DNA sequence comparisons. A well-supported example of such a phylogenetic reconstruction is shown here in the form of a branching diagram, which depicts the inferred relative evolutionary relationships among the groups included. The branch points, which are called nodes, represent the most recent common ancestor shared by the branches united at the nodes. Note that the animal models used in this book represent a very small portion of the animal kingdom: the nematode *C. elegans* and the arthropod *Drosophila melanogaster* are both ecdysozoans, whereas the vertebrates *Danio rerio, Mus musculus,* and *Homo sapiens* are all deuterostomes. The term *bilaterians* refers to all animals with a bilaterally symmetrical body plan. Adapted from Adouette et al. (2000).

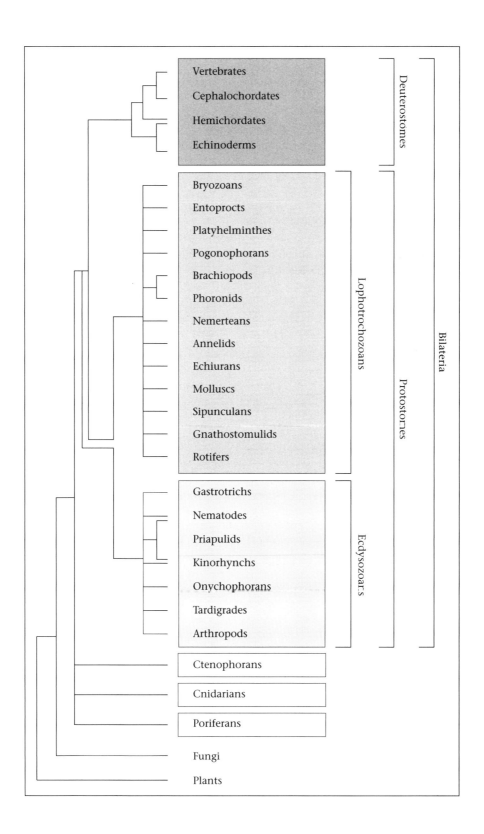

1. Why did I specify a wet lab in my introduction to experimental approaches for determination of gene function? Because *in silico* bioinformatics approaches can also be used to assess gene function.

2. See Lewcock et al. (2007).

3. See the U. S. National Institutes of Health Knockout Mouse Project web site: http://www.nih.gov/science/models/mouse/knockout/.

4. If you are interested in issues relating to the welfare of invertebrates used in research, a good place to start is by reading the article by Andrews (2011). This article is from a special issue of *ILAR Journal* titled "Spineless Wonders: Welfare and Use of Invertebrates in the Laboratory and Classroom."

5. The Nobel Prize in Physiology or Medicine for 2007 was awarded jointly to Mario R. Capecchi, Martin J. Evans, and Oliver Smithies for the discoveries that made the knockout mouse possible. See http://www.nobelprize.org/nobel_prizes/medicine/laureates/2007/.

6. A useful source of information about the history of laboratory mice is the online version of the classic *Biology of the Laboratory Mouse,* authored "by the staff of the Jackson Laboratory": http://www.informatics.jax.org/greenbook/index.shtml.

7. This atlas is now available online through The e-Mouse Atlas Project of the MRC Human Genetics Unit of the Western General Hospital, Edinburgh, UK: http://www.emouseatlas.org/CDROM_online/volume_01/macd/html_shdw_links/genex.html.

8. See Snyder *et al.* (2009).

9. See Geurts and Moreno (2010).

10. A valuable and interesting perspective on the origins of the zebrafish research field has been provided by researchers David Jonah Grunwald and Judith S. Eisen (2002).

11. Print versions of *The Zebrafish Book* are also available (see Westerfield 2007).

12. The term neuromere is used to indicate a segment of the developing nervous system and hence can be applied only to animals with a segmented body plan. In some cases the segments or neuromeres are obvious only during early development, as is the case in the vertebrate rhombencephalon. In other species, such as insects, some of the neuromeres may be plainly evident in adults as separate segmental ganglia. Even in many species of insect, however, embryonic neuromeres may fuse to form larger ganglia in later stages of development. Neuromeres can be defined on the basis of both morphology and the reiterated expression of molecular markers.

13. Information about Thomas Hunt Morgan is available from many sources. The official web site of the Nobel Prize, nobelprize.org, provides useful historical information about all Nobel laureates.

14. Recounted by Sydney Brenner in his foreword for the edited volume compiled by Wood (1988). Max Perutz, the recipient of the letter, was awarded (with John Kendrew) the Nobel Prize in Chemistry in 1962 for their studies of protein structure. Forty years later the Nobel Prize in Physiology or Medicine for 2002 was awarded jointly to Sydney Brenner, H. Robert Horvitz, and John E. Sulston for their research on *C. elegans.*

15. Very young mouse and human embryos can also complete development after freezing, as can the embryos of many domesticated animals such as sheep and cattle.

16. Chordates are members of Phylum Chordata, the animal phylum that includes the vertebrates and all other animals that develop a notochord, such as sea squirts and amphioxus. Myelin is abundantly expressed in chordate nervous systems, especially in vertebrates, but is also found in some species in two other phyla: the Annelida and Arthropoda. Current phylogenetic analyses suggest that myelin has evolved independently in each of these groups. See Roots (2008).

17. The proverbial man from Mars needs to have everything he sees on Earth carefully explained because he knows nothing about our planet or our lives. He also has a talent for spotting (and stating) the obvious. I morphed him into a gender-neutral proverbial scientist from Mars who has no time for details and just wants to see the big picture (in this case, all animals are pretty much the same). The phrase *proverbial man from Mars* is widely understood in the United States, but I do not know its origin or early uses.

18. The term *the right stuff* was used as the title of a widely read 1979 book about astronauts in the U.S. space program by American author Tom Wolfe. It means the qualities needed to succeed, particularly the qualities required to become part of an elite group.

Investigative Reading

1. You study the molecular mechanisms of vertebrate neurulation. You want to use the zebrafish model to test your hypothesis that regulation of cell–cell adhesion by the cell populations of the neural keel is the key event in neurulation. In particular you want to test the hypothesis that a protein encoded by a gene named *Lin7c* is essential for normal neurulation. There are no spontaneous zebrafish mutants that lack *Lin7c* expression (possibly because failure to complete neurulation is lethal). What research tool do you use to achieve a temporary and selective reduction of Lin7c protein in your zebrafish embryos? How do you assess the effects of your manipulations? Are your results potentially relevant to other vertebrates, including humans? To *Drosophila* and *C. elegans*?

Yang, Xiaojun, Jian Zou, David R. Hyde, Lance A. Davidson, and Xiangyun Wei. 2009. *Journal of Neuroscience* 29: 11426–40.

2. You read about an exciting new genetic screen that has discovered more than 150 regulators of molting in *C. elegans*. The screen was conducted by feeding young (early L1 Stage) *C. elegans* larvae bacteria expressing dsRNA. Different groups of larvae were fed different bacterial clones, each expressing a dsRNA corresponding to one worm gene. Problems with molting were detected by observation of defects such as failure to shed the old cuticle. You describe this study to a friend. Your friend asks, in return, if this is an example of a forward or a reverse genetic screen. What is your answer, and why?

Frand, Alison R., Sascha Russel, and Gary Ruvkun. 2005. *PLoS Biology* 3: e312.

3. *Challenge Question:* Whether strain differences in mice are a help or a hindrance to neuroscience research depends on the experimental question being asked. The rotarod task is a laboratory test of motor coordination in rodents. A mouse or a rat is placed on top of a rotating horizontal cylinder. The length of the time the animal is able to stay on the cylinder is taken as an indicator of motor coordination. Design a study using different inbred strains of mice to ask if differences in gene expression in a particular region of the brain are correlated with rotarod performance.

Nadler, Jessica J., Fei Zou, Hanwen Huang, Sheryl S. Moy, Jean Lauder, Jacqueline N. Crawley, David W. Threadgill, Fred A. Wright, and Terry R. Magnuson. 2006. *Genetics* 174: 1229–36.

FOUR

Early Events

Axis Determination and Neural Induction

The key early events covered in this chapter are the origin of the anterior–posterior body axis and the formation of the neural plate as a distinct region of the ectoderm, a process called neural induction. In actual embryos and in this chapter, axis determination and neural induction are inextricably intertwined: we recognize the anterior end of the body by its formation of head-typical structures such as the brain. As a consequence, it is difficult for an author to know how best to present the two topics. Neural induction is primarily a story of gastrulation, but the antecedents of the anterior–posterior axis can be glimpsed even earlier. Therefore the present chapter starts with a discussion of how various animal embryos develop their anterior end. But you should not be surprised to learn that overlapping molecular signals regulate body axis determination and neural induction.

Defining *Anterior* and Making a Head

We take for granted that brains are found in heads. Heads are also the body part that defines *anterior.* Animals with heads are referred to as cephalized, and cephalized animals are the rule, not the exception. Heads are a nearly ubiquitous trait in animals that have bilaterally symmetric body plans, a group that comprises almost all animals.[1]

Forming a head requires the embryo to have a working definition of *anterior* and *posterior* prior to the development of any body parts. The process of differentiating anterior and posterior can begin at the single-cell stage of development. Polarization may even be evident in the oocyte prior to fertilization. This early foreshadowing of the future anterior–posterior body axis relies on the partitioning of organelles and molecules into distinct cytoplasmic neighborhoods within the oocyte.

Signals from the parents provide cues for the definition of *anterior* and *posterior.* These cues are provided in the form of parental proteins and mRNAs that are active before the zygote (fertilized egg) is able to make its own. In most cases the parental contribution comes from the mother, but in *C. elegans* it is the position of the sperm pronucleus that initially defines the tail end. A sperm protein called CYK-4 sets in motion a series of events that polarizes the zygotic cytoskeleton.[2] This polarization is reflected in the accumulation of different sets of proteins called PARs at the opposite ends of the fertilized egg (fig. 4.1). The *Par* genes that encode the PAR proteins are named for the partitioning defects evident in mutants: when the *Par* genes are not correctly expressed, the resulting failure to develop anterior–posterior polarity is lethal. The specific biochemical targets of individual PAR proteins of *C. elegans* are not yet defined. This uncertainty reflects the complexity of the interactions between PAR proteins and other PAR proteins and the many other regulators of the actomyosin- and microtubule-based components of the cytoskeleton.

Par genes were first identified in a screen for polarity defects in *C. elegans*. A gene with homology[3] to the nematode *Par-3* gene was independently isolated in a screen for developmental defects in *Drosophila* embryos. This gene was initially named *bazooka* because the mutant phenotype is characterized by holes in the embryonic cuticle. PAR proteins are now known to function as regulators of cell polarization in many different cell types in a wide variety of animals, including mammals.

In contrast to the paternal contribution in *C. elegans,* the parental contribution to the specification of the anterior–posterior axis in *Drosophila* comes from the mother. Axis determination begins prior to fertilization, during oogenesis. Each oocyte matures in a tubelike compartment of the ovary called an ovariole. The oocyte is enclosed within a follicular epithelium, along with a set of 15 nurse cells. The nurse cells are the products of four divisions of a single germline precursor cell. Cytokinesis (division of the cytoplasm) is incomplete in all of these divisions. The result is a set of cells—one oocyte and 15 nurse cells—connected by cytoplasmic bridges. Elements of the cytoplasm, including organelles, proteins, and mRNAs, pass from nurse cell to nurse cell and into the oocyte. The oocyte is thereby stocked with the supplies it needs to build an embryo once fertilization occurs.

The nurse cells eventually undergo a form of cellular suicide called apoptosis, but before they die they will transfer mRNA encoding a protein called bicoid into the *Drosophila* oocyte (fig. 4.2). This protein is absolutely essential for the development of the head of the fly larva, but note that the *bicoid* mRNA from the mother's nurse cells enters an oocyte in which the cytoskeleton has already been polarized by signals from the follicular epithelium enclosing the egg chamber. The oocyte is further patterned by microtubule-dependent directed transport of specific mRNAs. *bicoid* mRNA is transported to the pole of the egg that will develop as the head, where a complex of regulatory proteins, RNA-binding proteins, and cytoskeletal components tether *bicoid* mRNA molecules via their 3′ untranslated regions.[4] The translation of *bicoid* mRNA into protein is activated when the egg is laid. The resulting

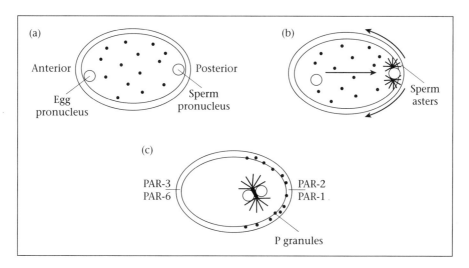

Figure 4.1. Development of asymmetry in the *C. elegans* oocyte. In the newly fertilized egg (a), the sperm pronucleus marks the future *posterior* end and the egg pronucleus marks the future *anterior* end of the worm. P granules (ribonucleoprotein complexes, depicted as black dots in this figure) are initially broadly distributed but become asymmetrically localized prior to the first cell division. Members of the PAR family of protein (PAR-1, PAR-2, PAR-3, PAR-6) generate movement of the cortex away from the point of sperm entry, which results in P granule movement and redistribution of the cytoplasm (see arrows in b). In addition to directing cytoplasmic flow, the PAR-1 protein also stabilizes the position of the P granules that reach the posterior cortex (c). The loss of cytoplasmic flow and/or the loss of P granule stabilization is reflected in the phenotypes of PAR mutants, which are characterized by inappropriate distribution of the P granules. A sperm aster is a centrosome with astral rays in the cytoplasm of a fertilized egg. The sperm aster serves as a microtubule organizing center and is essential for the fusion of the sperm and egg pronuclei. Drawings adapted from Cheeks et al. (2004).

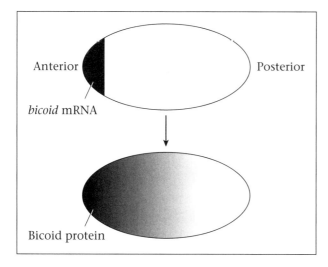

Figure 4.2. Polarization of the *Drosophila* oocyte. *bicoid* mRNA is tethered to the anterior pole of the oocyte by interactions of its 3′ untranslated region with microtubule-associated proteins. The untethered Bicoid protein, by contrast, forms an anterior-to-posterior gradient that regulates gene expression along the anterior–posterior axis.

bicoid protein moves away from its source as the fertilized egg undergoes nuclear divisions to form the syncytial blastoderm described in Chapter 3.

Bicoid protein has several known head-promoting functions, but its primary role is to serve as a transcription factor that activates expression of bicoid-target genes. Because the concentration of bicoid protein is highest near its mRNA source, the tethering of the *bicoid* mRNA molecules to one end of the oocyte results in a bicoid protein gradient. The bicoid gradient takes approximately 90 minutes to develop after the oocyte is activated. The gradient is significant because responses to high concentrations of bicoid protein are different from responses to lower concentrations of bicoid protein—the difference, for example, between the development of anterior head structures (in response to higher concentrations) and the development of a thorax (in response to lower concentrations).

Developmental biologists define *morphogens* as substances produced by embryos that determine different cell fates at different concentrations. Sources of morphogens are in turn defined as signaling centers. Bicoid is an example of a morphogen, and the anterior pole of the *Drosophila* oocyte is an example of a signaling center. Bicoid is such a powerful morphogen that injection of *bicoid* mRNA into the tail end of a wild-type embryo results in a short-lived embryo with heads at both ends.[5]

The development of the anterior–posterior axis in *Drosophila* has been intensively studied, and the contributions maternal genes other than *bicoid* make to the process have been well documented. The mRNA for one of these genes, *nanos,* is tethered to the opposite end of the oocyte from the *bicoid* mRNA. As in the case of *bicoid, nanos* mRNA is kept in its appropriate position by a localization signal in its 3′ untranslated region. Translation of the *nanos* mRNA results in a posterior-to-anterior gradient of nanos protein that opposes the anterior-to-posterior gradient of bicoid protein. The two gradients together define and refine patterns of gene expression that result in anterior structures at one end and posterior structures at the other.

Although the broad outlines of how bicoid controls head formation are understood, many questions remain unanswered. These include questions about the physical process by which the bicoid protein gradient is established. Another question concerns the relationship of the relatively smooth gradient of morphogen to the activation of bicoid-target genes, which are often expressed in patterns characterized by sharp boundaries.

Our quick look at anterior–posterior axis determination in the nematode and the fruit fly provides some of the basic concepts needed to understand the formation of the head in zebrafish. On the basis of the preceding sections, you might expect to find the following in the zebrafish: a patterned oocyte with an organized cytoskeleton, parental determinants, and secreted morphogens. None of your expectations would be disappointed, although the introduction of additional signaling pathways is required to place each of these factors into its full context.

Our current understanding of the body plan formation in vertebrate embryos has its foundation in experimental studies of amphibian embryos pioneered by Spemann and his students in the early twentieth century.[6] Now, in the twenty-first century, in this text, we will take this instructive historical body of research for granted, but it should be noted that the molecular signals described here are the embodiment of factors presumed to exist to account for the results of manipulations of frog and newt embryos. Two major concepts of development emerged from these studies, one general and one specific: inducers and the organizer. An inducer is defined as a population of cells that evokes a cellular response from another population of cells. The organizer is a specific population of cells—the cells forming the dorsal lip of the blastopore—with the power to determine the longitudinal body axis and induce the formation of the neural tube. Once the organizer cell population had been defined, developmental biologists began to search for the chemical basis of its inductive powers. The application of the tools of molecular biology, first to amphibian embryos and now to a whole zoo, including the model species discussed here, has led to the identification of molecules that can induce the development of specific aspects of the animal body plan.

The main body axes of the zebrafish are established during the first 10 hours of development. During this time, the development of dorsal and anterior (as opposed to ventral and posterior) are so closely intertwined—that is, dorsalizing factors are so important for the development of the head and the brain—that it is most appropriate to think in terms of development of a combined dorsoanterior compartment.

As described in Chapter 3, the zebrafish oocyte is polarized, with an easily recognizable animal pole and vegetal pole. Zebrafish embryos are ventralized by experimental removal of the vegetal region of the yolk prior to the first cell division. (The term *ventralized* indicates that the head and the upper half of the body do not begin to develop.) Ventralization can also be achieved via disruption of the cytoskeleton by treatment with drugs that affect microtubule assembly, such as colchicine. These observations of ventralization demonstrate the relevance of oocyte patterning to development of the zebrafish embryo, although the identity of the key vegetal signal or signals has not been determined.

Cell division is rapidly initiated by the zebrafish zygote, and the 128-cell blastula stage is achieved at 2.25 hours post fertilization. At this time, an asymmetric accumulation of a maternal protein called β-catenin occurs in the nuclei of dorsal blastomeres. This is just prior to the time at which zygotic gene expression is initiated. The β-catenin protein is a transcription factor that activates expression of zygotic genes that serve as essential dorsalizing factors. To streamline our account we focus on just two of these genes: *chordin* and *dickkopf*. The proteins encoded by these genes are among those expressed by the amphibian organizer and its counterparts in other species.

Experimental manipulations of β-catenin in amphibian and zebrafish embryos result in under- or overdorsalization, depending on whether β-catenin activity is inhibited or increased. More recently, analysis of gene expression in gastrulating *Xenopus* embryos using cDNA macroarrays has provided support for a model in which maternal β-catenin directly regulates zygotic expression of chordin.[7]

Chordin is an antagonist of the Bmp signaling pathway (read on to learn more about Bmp signaling). The term *antagonist* indicates that chordin blocks this signaling pathway. Mutants that lack functional chordin have a ventralized phenotype. This result indicates that chordin promotes dorsal development by blocking ventralizing pathways rather than by actively triggering dorsal-determining pathways. Such experimental results have led to a model in which dorsoanterior cell fate is a default pathway counteracted by signaling in ventralizing pathways. β-catenin-activated inhibitors such as chordin, in turn, counteract signaling that promotes ventralization.

The dickkopf protein also dorsalizes by blocking a ventralizing signaling pathway.[8] It does so by blocking activation of the Wnt signaling pathway. Mutants in which the dickkopf protein is expressed at abnormally high levels have an expanded forebrain, indicating that the balance between ventral-promoting and ventral-inhibiting signals has been disrupted. Taken together, results from studies of chordin and dickkopf suggest a model for the early stages of body axis determination in the zebrafish. First, as zygotic transcription begins, ventralizing genes are broadly expressed in the embryo, but maternal β-catenin (present only in a small region of the embryo) is locally dorsalizing. This β-catenin-rich region is the embryonic shield described in Chapter 3. The embryonic shield is therefore the functional equivalent of the dorsal lip of the amphibian blastopore, and in the sense intended by Spemann, can be called the zebrafish organizer. The embryonic shield is the source of many signals, but chordin and dickkopf are among the most important.

The Bmp and Wnt signaling pathways are highly conserved pathways that regulate many aspects of development throughout the animal kingdom (fig. 4.3). The terms *Bmp signaling pathway* and *Wnt signaling pathway* refer to specific intracellular signal transduction pathways activated by binding secreted ligands to their receptors. Note that, in vertebrates, proteins are often encoded in the genome in the form of families of proteins with similar amino acid sequences. These proteins, related by their evolutionary history and shared protein structure, can have distinct and/or overlapping functions.[9] Both the ligands and the receptors in the Bmp and Wnt pathways are members of large families of proteins. Bmp and Wnt signaling are so important in development that they require a full introduction here before we complete the story of head formation in zebrafish.

Bmp stands for bone morphogenetic protein. The name refers to only one of the many known actions of proteins in this family. Bmp ligands indi-

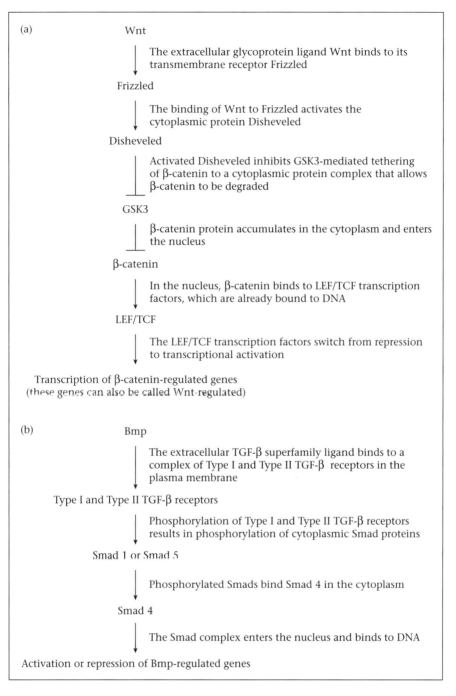

(a) Wnt

The extracellular glycoprotein ligand Wnt binds to its transmembrane receptor Frizzled

Frizzled

The binding of Wnt to Frizzled activates the cytoplasmic protein Disheveled

Disheveled

Activated Disheveled inhibits GSK3-mediated tethering of β-catenin to a cytoplasmic protein complex that allows β-catenin to be degraded

GSK3

β-catenin protein accumulates in the cytoplasm and enters the nucleus

β-catenin

In the nucleus, β-catenin binds to LEF/TCF transcription factors, which are already bound to DNA

LEF/TCF

The LEF/TCF transcription factors switch from repression to transcriptional activation

Transcription of β-catenin-regulated genes
(these genes can also be called Wnt-regulated)

(b) Bmp

The extracellular TGF-β superfamily ligand binds to a complex of Type I and Type II TGF-β receptors in the plasma membrane

Type I and Type II TGF-β receptors

Phosphorylation of Type I and Type II TGF-β receptors results in phosphorylation of cytoplasmic Smad proteins

Smad 1 or Smad 5

Phosphorylated Smads bind Smad 4 in the cytoplasm

Smad 4

The Smad complex enters the nucleus and binds to DNA

Activation or repression of Bmp-regulated genes

Figure 4.3. Wnt, β-catenin, and Bmp signaling pathways. Binding of Wnt to the Frizzled receptor (a) can result in the transcription of β-catenin-regulated genes. Binding of Bmp to Type I and Type II TGF-β receptors (b) can result in activation or repression of Bmp-regulated genes. (a) Based on Moon et al. (2004); (b) based on Gilbert (2010).

rectly activate transcription in their target cells. They accomplish this by binding to plasma membrane receptors that, when activated by binding a Bmp ligand, phosphorylate (and thereby activate) members of the Smad family of transcription factors. Phosphorylated Smads enter the nucleus, where they regulate expression of specific Smad-sensitive target genes. The Bmp family of proteins is, in turn, a member of an even larger family (so large that it is sometimes called a superfamily), the transforming growth factor-β (TGF-β) protein family. The receptors for TGF-β proteins, called TGF-β receptors, are transmembrane proteins with intracellular serine-threonine kinase domains. The TGF-β family member ligand binds to a complex formed by dimers of Type I and Type II TGF-β receptors.

Wnt combines the names of two of the first genes encoding members of this family to be discovered: *wingless* and *integrated*. Wnt is typically pronounced "wint." Wnt proteins bind to a family of receptors in the Frizzled family of proteins. The binding of a Wnt ligand to a Frizzled receptor triggers the following events. First, an intracellular protein called Disheveled is activated; Disheveled, in turn, activates an enzyme called glycogen synthase kinase (GSK3), which, when activated, allows β-catenin to enter the nucleus. In the nucleus, β-catenin binds to another transcription factor, which is thereby converted from a repressor of transcription to an activator. Although the mechanism is about as indirect as one can imagine, it is wholly appropriate to refer to genes regulated by this pathway as Wnt-responsive genes.

To recap: by the 128-cell stage of zebrafish development, maternal β-catenin accumulates in the nuclei of a small subset of dorsal margin blastomeres associated with the shield. (The factors that localize β-catenin to the nucleus remain to be defined.) As the switch from maternal to zygotic control proceeds, this nuclear β-catenin activates expression of the zygotic genes *chordin* and the gene that encodes the zebrafish dickkopf, *dkk1*. The products of both the *chordin* and *dkk1* genes are secreted proteins. Prior to the secretion of the chordin and dkk1 proteins, ventralizing genes are widely expressed throughout the embryo. By antagonizing, respectively, Bmp and Wnt signal transduction pathways, chordin and dkk1 inhibit ventralizing signaling in the dorsoanterior region of the fish body. As this simplified account predicts, blocking nuclear accumulation of β-catenin yields an embryo that is completely ventralized; reduction of chordin and dkk1 activities expands the region of ventral fates and reduces the dorsal region. (Details omitted from this account include multiple cross-regulatory interactions and the role of the genes *vox* and *vent* in indirectly promoting ventral development by repressing the expression of the *chordin* and *dkk1* genes).

The process of specifying the body axes of the zebrafish initiated by *chordin* and *dkk1* (and other genes downstream of β-catenin) continues into the gastrula stage. At this time, a different mechanism involving a morphogen becomes important. Recall the definition of a morphogen given when bicoid was introduced: a morphogen is a substance produced by embryos that determines different cell fates at different concentrations. The morphogen relevant to the development of the anterior–posterior axis in zebrafish and other vertebrates is retinoic acid. Because retinoic acid and related compounds are drugs used for the treatment of human skin disorders such as severe acne, this stage of vertebrate development, described here for the zebrafish, is relevant to women who may opt for this treatment.

Retinoic acid is naturally formed in the body as a derivative of Vitamin A. Its action in this context is to promote posterior development during the gastrula stage in zebrafish, particularly in the developing nervous system (fig. 4.4). The synthesis of retinoic acid depends on the enzyme retinaldehyde dehydrogenase (RALDH). Retinoic acid is broken down by the product of the *Cyp26* gene (P450RA1).[10] RALDH is expressed in posterior mesoderm cells during the gastrula stages. The resulting retinoic acid produced by this source binds to its receptors, which in turn bind to specific response elements in nuclear DNA to activate the transcription of genes required for the development of the posterior hindbrain and, eventually, the spinal cord. Concurrent expression of the *Cyp26* gene product in anterior structures protects the forming head and forebrain from being *posteriorized.*

The ability of retinoic acid to suppress anterior head and forebrain development makes it a powerful teratogen, a name given to any compound that causes birth defects. Exposure of vertebrates (not just zebrafish) to retinoic acid at early stages of embryonic development results in craniofacial and brain malformations. For this reason, women who are pregnant or who have even a slight chance of becoming pregnant should not take oral Vitamin A derivatives to treat skin disorders.[11] The period of greatest sensitivity to the teratogenic effects of retinoic acid extends from Week 3 to Week 8 of human

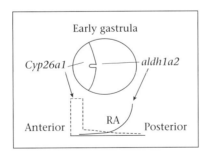

Figure 4.4. Retinoic acid action in zebrafish embryos. This simplified depiction shows a dorsal view of a gastrulating embryo (anterior to the left) and the associated retinoic acid gradient. Retinoic acid is produced as a result of expression of *aldh1a2,* the gene that encodes the enzyme retinaldehyde dehydrogenase in mesoderm. Once synthesized, retinoic acid diffuses away from its site of synthesis throughout the ectoderm. The expression of the *Cyp26a1* gene in the anterior ectoderm prevents the activation of posterior genes in the head region; in the posterior ectoderm, retinoic acid acts as a morphogen, exerting concentration-dependent posteriorizing effects. Drawing adapted from White et al. (2007).

gestation, indicating that this is the developmental stage in humans when endogenous retinoic acid is acting as a morphogen. This period covers Carnegie Stages 7–20 of human embryonic development.

Analysis of the mechanisms that establish the anterior–posterior axis is mammals is challenging because the developing embryo is hidden from view. Most of what is known is based on studies in mice. These studies have been informed by a substantial body of research on avian embryos. In birds, a midline structure formed by migrating cells called the primitive streak becomes visible at gastrulation. The primitive streak extends anterior to posterior and marks the longitudinal body axis. A basic understanding of the network of molecular signals that promotes formation of the primitive streak has been developed. This network includes Wnt and Bmp signaling regulated, respectively (and unsurprisingly), by dkk-1, the avian dickkopf, and chordin. Signaling in fibroblastic growth factor (FGF) pathways is also important.[12] Many of these signals converge on another signal transduction pathway called the nodal pathway. Like Bmps, nodal proteins are members of the TGF-β superfamily; like other TGF-β superfamily members, nodal proteins regulate transcription by controlling the phosphorylation of Smads.

Mouse embryos, like avian embryos, form a primitive streak. The appearance of this structure at approximately 6.5 dpc (the start of gastrulation, Theiler Stages 9a and 9b) was previously assumed to represent the origin of the body's anterior–posterior axis. Development was thought to proceed in a radial fashion prior to the formation of the primitive streak, a mode of development consistent with the roughly spherical shape of the young embryo. Recent studies, however, have found that indicators of the anterior–posterior axis are present in the mouse embryo more than a day prior to the appearance of a visible primitive streak. Somewhat surprisingly, the signals arise not in the embryo itself but from extraembryonic tissues. Recall (from Chapter 3) that the inner cell mass of the mouse blastocyst gives rise to two layers of tissue, the epiblast and the hypoblast. The epiblast, in turn, gives rise to all of the embryonic tissues, while the hypoblast is the source of extraembryonic tissues such as the placenta. A small group of hypoblast-derived cells called the anterior visceral endoderm (AVE) is initially positioned at the distal tip of the egg cylinder. The AVE cells migrate to a more anterior position on the side of the epiblast a full day before the appearance of the primitive streak. This asymmetrical migration, which has been witnessed in living embryos by videomicroscopy, is critical for the subsequent development of the anterior–posterior axis.[13]

The AVE cells exert their effect though the secretion of inhibitors of proteins that promote formation of posterior structures. Two of the secreted proteins, Lefty-1 and Cerberus, block nodal signaling. Once they arrive at their final, anterior location, the AVE cells also secrete the Wnt blocker dickkopf. The AVE cells, however, do not express any Bmp antagonists, and therefore they must work with another source of signals to complete formation of the head. The essential source of Bmp antagonists in the early mouse embryo is a tissue called the node, which secretes chordin. Many experiments have demonstrated that the inhibition of Bmp action by chordin is essential and that head structures, including the forebrain, do not develop in embryonic mice in which Bmp signaling is unopposed.[14] The confinement of retinoic acid to posterior compartments is also essential for normal development of mouse embryos; treatment of mother mice with retinoic acid early in their pregnancy disrupts axis formation by blocking the migration of the AVE cells.

What about humans? The head begins to develop during the fourth week of human gestation. Can any human developmental disorders be related to disruption of signaling in the molecular pathways discovered to be responsible for anterior–posterior axis formation using animal models? Analysis of the defects present in humans at birth likely leads to underestimates of the incidence of head development defects, given that severe defects are associated with prenatal mortality and spontaneous abortion (miscarriage). Some defects are associated with chromosomal abnormalities (for example, trisomy 13 and trisomy 18); some reflect exposure to teratogens such as retinoic acid at sensitive stages in development. Others, however, are linked to genetic mutations in the signaling pathways described in the preceding paragraphs. One approach to understanding human disorders is to compare details of human malformations to mutant phenotypes observed in specific knockout mice. Each of the signaling pathways described in this chapter is a set of hypotheses that can be experimentally tested by creating knockout mice that lack specific ligands or receptors. This experimental strategy is well illustrated by studies of the effects on mouse embryos of knocking out Bmp antagonists.

One of the most common malformations of the human forebrain is holoprosencephaly. Both the forebrain and facial features are affected in individuals with this condition. Holoprosencephaly occurs in approximately 1 in 10,000–15,000 human live births and in 1 in 250 first-trimester spontaneous abortions. Even mild cases that are not fatal are often associated with significant intellectual disability. The initial clinical diagnosis is based on visible malformations ranging in severity from a small head and closely set eyes to a single midline eye, a condition referred to as cyclopia. Abnormalities of the nose and a cleft palate are also often present. Many elements of human holoprosencephaly are strikingly reproduced by elimination or reduction of the Bmp antagonists chordin and noggin in the mouse embryo. (Chordin has already been mentioned; the pattern of expression and actions of the noggin protein overlap with those of chordin.) Severely affected humans and mice are both characterized by partial or total fusion of the eyes, a single nostril, cleft palate, and a single telencephalic lobe (instead of left and right cerebral hemispheres). Knockout mice permit us to link signal transduction pathways defined in animal models to human birth defects.

To summarize, the mechanisms of body axis determination described here simultaneously reflect both specific patterns of early embryonic development displayed by our selected animal models and general principles. In *C. elegans*, we noted the importance of information from the parents and the contributions of the oocyte cytoskeleton to the organization of the embryo. These features—parental mRNAs and proteins, interaction of signaling molecules with the cytoskeleton—are also critical in the fruit fly, but the *Drosophila* embryo also relies upon a morphogen called bicoid to get the job done. Strongly conserved vertebrate pathways for head formation, in particular a requirement for Bmp- and Wnt-signaling, are clearly evident in the zebrafish embryo, which also relies upon a morphogen (in this case, retinoic acid) to produce a head and a tail (and all the structures in between). Despite conservation of the molecular signaling pathways across the vertebrates, the context for early development is different in mammals than in fish and amphibians. Mammals make an earlier switch from maternal to zygotic control (as early as the 2-cell stage in mice and the 4-cell stage in humans) and rely upon a sustained connection to the mother via extraembryonic tissues rather than yolk for nutrients. Some of the extraembryonic tissues, such as the AVE, are sources of early signals that help determine the anterior–posterior body axis.

Defining an early signal for the positioning of the anterior–posterior axis provokes the logical follow-up question: What, then, is the earlier signal? The earliest signal? In the case of the nematode we know that the earliest signal is the site of sperm entry into the oocyte. In the case of the fruit fly, the earliest signal is provided by the orientation of the maternal egg tubes. In the case of the zebrafish, the animal-vegetal polarity of the oocyte cytoplasm, established during oogenesis, is the cue. The apparently unpolarized mammalian oocyte and the resulting spherical blastula embryo have not yet revealed the secret of their earliest signal.

Neural Induction

Defining the head end of an animal is not precisely the same as growing its brain, although in practice the two developmental processes can be hard to separate. The same molecular signals often promote development of both neural and non-neural anterior structures. As noted, we very often rely upon the emergence of the proto-brain (assessed by detecting the expression of neural markers) to confirm that we are indeed looking at the head end of a young embryo! Yet a simple equation of head and neural ignores the fact that the nervous system of most animals extends though the trunk from head to tail.

To induce is to use influence on, to persuade. In biology, the term induction is used to distinguish between intrinsic, cell autonomous regulation of development and processes that require cell-extrinsic signals. It is the latter that are said to be induced. Neural induction is the specific term used to refer to the process by which a subset of ectodermal cells commits to forming nervous tissues rather than other derivatives of embryonic ectoderm—in other words, a decision to produce progenitor cells to build brains and nerve cords rather than progenitor cells to make skin. The term neural induction applies to the process throughout the entire body (that is, it covers not just the brain but also the nerve cord in invertebrates and the spinal cord in vertebrates, plus assorted sense organs).

The outcome of neural induction is not neurons, but cells that will divide to produce the progenitor cells that will make neurons. This means that the process of neural induction can also be considered an early component of neurogenesis, the production of new neurons by mitosis. Neurogenesis is the topic of Chapter 5, and it is in that chapter that the origins of neural progenitor cells are described for *C. elegans* and *Drosophila*. The later stages of neurogenesis in vertebrates are also covered in Chapter 5. Here in Chapter 4 vertebrate embryos take center stage as we review current understanding of signaling events that induce the formation of the main neurogenic region of the vertebrate ectoderm, the neural plate described in Chapter 2.

The surprising result of decades of research is that all of the ectodermal cells in the early embryo have a strong proclivity to develop a neural fate unless inhibited from doing so by signals from adjacent cells. The strongest of these inhibitory signals are Bmps. In fact, so strong is this effect that one might appropriately think of Bmps as epidermal inducers, given that ectoderm becomes epidermal through the binding of Bmps to Bmp receptors. Neural tissue in general—and in the young embryo, the neural plate in particular—forms where and when it normally forms because Bmp inhibitors are present in that region of the embryo.

This view of neural induction in vertebrates is referred to as the default model, and is similar conceptually to the model of body axis determination by chordin and dkk1, which dorsalize by blocking ventralizing factors rather than by activating genes that trigger dorsal development. This model predicts that successful knockdown of Bmp expression (sometimes difficult to achieve experimentally because multiple Bmp family members must be knocked down, not just one) will lead to overdevelopment of neural structures such as the brain, and that knockdown of Bmp antagonists will block formation of the nervous system (fig. 4.5). These predictions have been experimentally confirmed.[15]

The signaling events of neural induction occur at the gastrula stage of embryonic development. The complex tissue rearrangements of gastrulation result in the formation of the ectodermal, mesodermal, and endodermal tis-

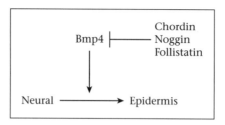

Figure 4.5. Default model of neural induction in vertebrates. This model assumes that ectodermal cells *want* to become neurons but are prevented from doing so by Bmps, particularly Bmp4. Any factor that blocks Bmp signaling results in neural induction. Known Bmp antagonists include the proteins chordin, noggin, and follistatin.

sue source layers. A specific region of the ectoderm then becomes the neural plate, the flat but thickened epithelium (thickened because the cells have assumed the classic columnar epithelia shape of bricks standing on end) that will eventually form the neural tube. The neural tube will in turn form the brain and spinal cord. Classic experiments on amphibian embryos defined the dorsal lip of the blastopore, a mesodermal tissue, as the source of the signals that induce the overlying ectoderm to become the neural plate. The dorsal lip of the blastopore (or its equivalent in a non-amphibian species) therefore functions as the organizer described earlier in this chapter. The first signaling molecule produced by the dorsal lip of the blastopore to be identified as a neural inducer was noggin. Subsequent studies, also performed in amphibian embryos, identified two more proteins that function as neural inducers: chordin and follistatin. All three of these molecules are secreted proteins that interfere with Bmp signaling and, as one would expect, their concentrations are highest nearest their sources. Although expression of these neural inducers eventually extends to other tissues such as the notochord, at the earliest stages of gastrulation their mRNAs are expressed exclusively in the dorsal blastopore lip or its equivalent. The mechanisms of neural induction defined initially in newts and frogs are conserved in other vertebrates.

As the reappearance in our account of previously described signaling molecules such as chordin implies, the processes of body-axis determination and neural induction share common signaling pathways and common sources of secreted signals. One difference is that the key events of neural induction occur later in developmental time than does body-axis determination. Neural induction is a phenomenon of the increasingly elaborate patterning of embryonic tissues that develops as gastrulation proceeds.

In zebrafish, the embryonic shield is the early source of two Bmp antagonists, noggin and chordin. The actions of chordin in this regard have already been described. In mammalian embryos, the formation of the neural plate also depends on inductive signaling from a bit of mesodermal tissue called the gastrula organizer, the primary source of Bmp antagonists during gastrulation. Mutant mice that do not form a gastrula organizer (often referred to as the node) do not experience neural induction and form no neural ectodermal derivatives.

Many details have been omitted from this simplified account of neural induction to make the major steps evident. In particular, it should be noted that neural induction in vertebrates is not a single, one-time-only event but instead plays out over time. For example, posterior neural tissues are induced not by signals from the organizer but rather by sequential derivatives of the original gastrula organizer. Many additional signaling molecules act at slightly different times and in slightly different locations (and in different concentrations, if they are morphogens) to reinforce, stabilize, and refine the events set in motion by the inhibition of Bmps. The fibroblastic growth factors (FGFs) also support neural induction, and genetic disruption of FGF signaling leads to defects in posterior neural tissues. At present, it is certain that Bmp inhibition is absolutely necessary for neural induction but that other signaling pathways also contribute to the formation of the neural plate.

The clinical impact of mutations in genes implicated in signaling pathways for head development and neural induction was covered in our earlier discussion of holoprosencephaly. Greater knowledge of the molecular signaling pathways will lead to a more complete understanding of the causes of many birth defects. It is also possible that this knowledge will lead to new therapies for brain disorders. For example, if development as neural tissue is truly the default pathway, mammalian embryonic stem cells grown in cultures without the inhibitory signals produced by other cells should differentiate as neural and glial progenitor cells. This is exactly what happens when mouse embryonic stem cells are grown at low density in a minimal culture medium that contains no serum (and hence no endogenous signaling molecules).[16] Neural stem cells produced by allowing the default pathway to run its course in vitro might one day be used to repair damaged or diseased nervous systems.

Notes

1. Bilateral animals—the bilaterians—are animals that display bilateral symmetry in the larval and/or adult stage. In addition to having a shared body plan, all bilaterians form three layers of tissue early in embryonic development (ectoderm, mesoderm, and endoderm), and all have a head. Analysis of DNA sequences has led to the definition of three bilaterian superphyla: Deuterostomia, Ecdysozoa, and Lophotrochozoa. Humans, mice, and zebrafish are representative deuterostomes; fruit flies and nematodes are ecdysozoans. Lophotrochozoans are unrepresented in these chapters. The phylogenetic relationship of bilaterians to the nonbilaterian animals (poriferans, placozoans, cnidarians, and ctenophores) remains an area of active research. You might want to refer back to figure 3.9.

2. *C. elegans* hermaphrodites self-fertilize. They first produce sperm, which are stored in an organ called the spermatheca, and then oocytes. During the first day after the L4/adult molt, hermaphrodites accumulate fertilized eggs in an organ called the uterus. Egg laying occurs when specialized muscles contract, the vulva opens, and eggs are expelled. When males mate with hermaphrodites, they respond to contact with the hermaphrodite by inserting organs called spicules into the hermaphrodite uterus. Ejaculation follows.

3. The term *homolog* is used to refer to two or more similar DNA or amino acid sequences. It is relatively easy to identify homologs using modern bioinformatics tools. Homologs may arise as a result of descent from a common ancestor or gene duplication. The term *ortholog* is used to refer to genes in separate species deriving directly from a common ancestor without duplication. Strictly speaking, orthologs can be defined only by phylogenetic analysis, not solely on the basis of sequence similarity.

4. A 3′ untranslated region (UTR) of a messenger RNA molecule (mRNA) is the section of the mRNA between the coding region and the polyA tail, the sequence of repeating adenylate residues added to the 3′ end of a newly transcribed mRNA before it exits the nucleus. Although the 3′ UTR does not code for protein, it has important functions based on the binding of regulatory proteins and miRNAs; this sequence also contains the signal for polyadenylation.

5. See Driever et al. (1990).

6. Hans Spemann was awarded the Nobel Prize in Physiology or Medicine in 1935 for the discovery of the organizer effect in embryonic development. At the Nobel web site you can read his Nobel Lecture (in English) and his banquet speech (in German). See the list of online resources.

7. What is the difference between DNA macroarrays and microarrays? Both methods are used for the study of gene expression. DNA macroarrays are created by immobilizing pieces of DNA (polymerase chain reaction products or oligonucleotides) on nylon membranes. The macroarray is hybridized with labeled complementary DNAs reverse transcribed from RNAs extracted from the tissue being studied. Because the DNA spots on the membrane are relatively large, only a limited number of genes (in the low thousands) can be included in a single macroarray study. In microarrays, tiny DNA spots are created by a robot on a piece of glass similar to a microscope slide. The slide is then hybridized to labeled cDNAs. The number of spots per slide can be extremely large (in the hundreds of thousands, with each gene represented multiple times).

8. The gene name *dickkopf* is derived from a German word meaning stubborn (literally, thick head). *dickkopf* was identified in a screen for mRNAs with head-inducing activity in *Xenopus* embryos. See Glinka et al. (1998).

9. The multiple members of many vertebrate protein families often have overlapping functions. This redundancy complicates experimental analysis. One of the surprising lessons learned from knockout mice is how many single-gene knockouts have no phenotype.

10. If retinoids are known teratogens, why are they prescribed for treatment of acne, a skin disorder that primarily affects persons of reproductive age? The simple answer is that they work, and work well, when other treatments are ineffective. They are relatively safe when taken under appropriate medical supervision unless you are an embryo. It has been reported that daily oral intake of the retinoid isotretinoin for

a period of 4–6 months cures or significantly improves acne in a majority of patients. Oral isotretinoin ingestion by women during very early pregnancy has been definitively linked to human birth defects. Topical treatments with another retinoid, tretinoin, have not been directly linked to birth defects, but women who may become pregnant are cautioned against the topical use of all retinoids, including tretinoin. See Lowenstein and Lowenstein (2011) for a discussion from the dermatological perspective.

11. The large cytochrome P450 family of proteins comprises enzymes catalyzing the oxidation of many different types of organic molecules. The genes encoding P450 proteins are named the *Cyp* genes. The term P450 refers to the observation that these enzymes have maximal absorbance to wavelengths of 450 nm (UV light) when exposed to carbon monoxide.

12. The large fibroblast growth factor (FGF) family of chemical signals regulates many aspects of development. FGFs bind to tyrosine kinase receptors in the plasma membrane of target cells. The culmination of the signal transduction cascade activated by the binding of an FGF to a tyrosine kinase receptor is regulation of transcription by extracellular signal regulated kinase (ERK)–mediated phosphorylation of intranuclear transcription factors.

13. See Srinivas et al. (2004).

14. See Anderson et al. (2002) and Klingensmith et al. (2010).

15. See Stern (2006) and Levine and Brivanlou (2007).

16. See Rowland et al. (2011) for an example comparing in vitro and in vivo studies of neural stem cells.

Investigative Reading

1. You are in the middle of a project requiring you to produce an inbred strain of zebrafish carrying a recessive mutation. As you sort through embryos you discover many with an unexpected phenotype. These unusual embryos have overdeveloped ventral tail fins but in the most extreme cases completely lack heads. You identify the female fish that are producing these embryos and mate them with wild-type males. Almost all (99–100 percent) of the embryos resulting from these crosses are abnormal. You then mate the siblings (males and females) of the original females and score their progeny for the headless phenotype. Your sibling matings produce mostly normal embryos, but some crosses give 100 percent headless phenotypes. Study of the morphology of the embryos produced by the latter crosses reveals that they do not form an embryonic shield. How do you explain the observed pattern of inheritance of this mutation? What mRNA or protein do you inject into your mutant embryos to try to rescue the development of a head?

Kelly, Christina, Alvin J. Chin, Judith L. Leatherman, David J. Kozlowski, and Eric S. Weinberg. 2000. *Development* 127: 3899–911.

2. Many experimental studies in zebrafish embryos have demonstrated the importance of Bmp signaling before and during gastrulation. It has

been observed, however, that Bmps are expressed after gastrulation, particularly in the ventral and posterior mesoderm. The critical role of Bmp during gastrulation, however, means that mutations in the Bmp pathway typically arrest development at this stage. Outline a method that could be used to study the function of Bmps at postgastrulation stages of development in zebrafish.

Pyati, Ujwal J., Ashley E. Webb, and David Kimelman. 2005. *Development* 132: 2333–43.

3. One goal of research with human embryonic stem cells is to identify signals that lead stem cells to produce neural progenitor cells. Knowledge of these signals could allow the production of neurons for therapy. Based on the signals shown to be vital for induction of anterior structures, including the brain, in animal models, propose a list of molecules that could be added to cultures of human embryonic stem cells to promote the production of neurons over other categories of cells.

Dhara, Sujoy K., and Steven L. Stice. 2008. *Journal of Cellular Biochemistry* 105: 633–40.

FIVE

Neurogenesis

Production of Neurons by Neural Progenitors

The term *neurogenesis* is defined as production of new neurons by the process of mitosis. Because differentiated neurons are in a state of cell cycle arrest (a term defined in the following section), the parent cell that divides to produce daughter neurons is not a neuron but is instead a mitotic neural progenitor cell. Neural progenitor cells have been given different names in different species and in different parts of the nervous system. They are sometimes referred to generically as neural stem cells. Neural stem cells are most abundant and most active during embryogenesis, when an enormous expansion of neuronal populations is needed to build the nervous system. In many species (possibly most), neurogenesis is also characteristic of part or all of the immediate postembryonic period. Neurogenesis also occurs in some adult animals, including many mammals.[1] Postembryonic neurogenesis necessarily implies persistence or replenishment of the embryonic population of mitotic neural progenitor cells.

The defining features of the cell cycle are shared by all eukaryotic cells, including neural progenitor cells (fig. 5.1). The cell cycle is divided into four successive phases: G1, S, G2, and M. Late in G1 (Gap or Growth Phase 1), molecular events occur that commit the cell to divide. Two of the most important regulatory events are the phosphorylation of the tumor suppressor retinoblastoma gene (*Rb*) and the activation of a transcription factor called E2F1. G1 is followed by S, during which DNA synthesis occurs. This is the only time when the marker BrdU (Chapter 1) can be incorporated into DNA. During the next phase, G2, the components required to complete cell division become organized. Finally, in M (mitosis) phase, the duplicated chromosomes and the cytoplasm are divided. At the end of mitosis, RB, the protein product of the *Rb* gene, is dephosphorylated. A family of proteins referred to as cyclins promotes progression through the cell cycle, in part through the activation of cyclin-dependent kinases (CDKs).

In many cell populations, the two resulting daughter cells enter G1, and the cell cycle begins anew. In other cell populations such as neurons, how-

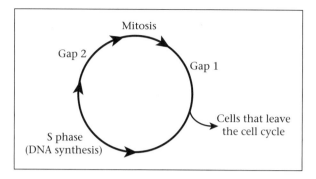

Figure 5.1. Eukaryotic cell cycle. The incorporation of markers for mitosis such as BrdU occurs only during S phase of the cell cycle. Mitosis is divided into four major stages—prophase, metaphase, anaphase, and telophase—and is concluded by cytokinesis, the division of the cytoplasm. Neurons provide an excellent example of cells that no longer divide.

ever, the probability of cell division is so low (possibly even nonexistent) that the cells are said to be in a state of cell cycle arrest (G0).

The simplest version of the cell cycle involves symmetrical partitioning of the contents of the parent cell to each of two daughters. The result is a doubling in the number of a particular type of daughter cell and the loss of the parent. But now imagine that a parent cell–to–be is not required to treat its future offspring fairly. If the cytoplasmic components of the parent cell are distributed unequally between its daughters, the result will be two new cells with dissimilar phenotypes. This asymmetrical partitioning is possible because the contents of the cytoplasm are not distributed uniformly throughout cells but are instead stabilized into subcellular neighborhoods by means of interactions with the cytoskeleton.

Asymmetrical cell divisions play a key role in diversification of cell types during development. For example, a single neural progenitor cell can produce a neuron and a glial cell (more typically, the direct progenitors thereof) through the process of asymmetrical cell division. But asymmetrical cell divisions are important in the developing nervous system because of the G0 status of neurons. If each neural progenitor produced only a single pair of neurons, the number of neural progenitor cells required to stock the nervous system would be huge (half the number of neurons). But if the offspring of a neural progenitor cell are not two postmitotic neurons but rather a neuron and a cell cycle–competent neural progenitor cell (like the parent, or, alternatively, one with a more restricted, intermediate potential), a relatively small number of neural progenitor cells can produce the neurons needed to build a nervous system. As we will see, reiterated use of a relatively small number of neural progenitor cells is the strategy of choice of many species. This chapter uses examples from each of our models to review the basic processes of neurogenesis. One topic that has received relatively little attention is the interesting question of how differentiated neurons so effectively block their reentry into the cell cycle.

Neurogenesis in *C. elegans*

bHLH Proteins

As described in Chapter 3, the nervous system of the *C. elegans* hermaphrodite contains 302 neurons, of which 118 are recognized as morphologically distinct. It is not at all difficult to look up the lineages of each of these neurons, tracing each back to a specific embryonic blastomere or to a particular postembryonic blast cell.[2]

The family trees of nematode neurons are somewhat more complicated than one might naïvely guess. There is no dedicated neuronal lineage in *C. elegans* in which neural progenitor cells produce one neuron after another. Instead, neurons are scattered across different lineages, and neurons with similar functions and/or phenotypes are often the products of different lineages. Fortunately, the stereotyped pattern of cell divisions that characterizes *C. elegans* makes forward genetics a powerful tool for the study of neurogenesis in this species. Many genes have been identified in this species that either promote (proneural) or inhibit (antineural) neuronal fate.

Instructive examples of proneural and antineural gene expression were discovered by analysis of the effects of mutations in two genes called *lin-22* and *lin-32*. These genes are important for neurogenesis in lineages produced during the postembryonic period (after hatching) in a set of progenitor cells called the V blast cells (fig. 5.2). The designation *lin* in the name of these genes refers to the observed phenotype of a *lin*eage defect.

The V blast cells are a set of six progenitor cells located on each side of the lateral hypodermis (skin). Each divides a fixed number of times to produce a mix of cell types: neurons, glial cells, muscles, and hypodermal cells. Here we compare the progeny produced by the V4 and V5 blast cells in hermaphrodites (the pattern is slightly different in the rarer males). The wild-type lineages produced by the V4 and V5 blast cells are shown in figure 5.2. The V4 blast cell lineage produces only hypodermal cells, whereas the V5 lineage produces hypodermal cells and neurons. Mutations of *lin-22* transform a subset of V4 progeny into neurons, in effect converting the V4 lineage into the V5 lineage. (In *lin-22* mutants the V5 lineage develops as in wild type.) Mutations of *lin-32* have the opposite effect, resulting in the loss of neuronal fate in the V5 lineage, whereas the V4 lineage contains the expected hypodermal cells. These results can be summarized as follows: loss of *lin-22* activity results in an excess of neurons in the mutant compared with wild-type cells, whereas loss of *lin-32* activity results in an excess of hypodermal cells compared with wild-type cells. The most straightforward interpretation of these results is that the product of the *lin-22* gene has antineural activity and the product of the *lin-32* gene has proneural activity.

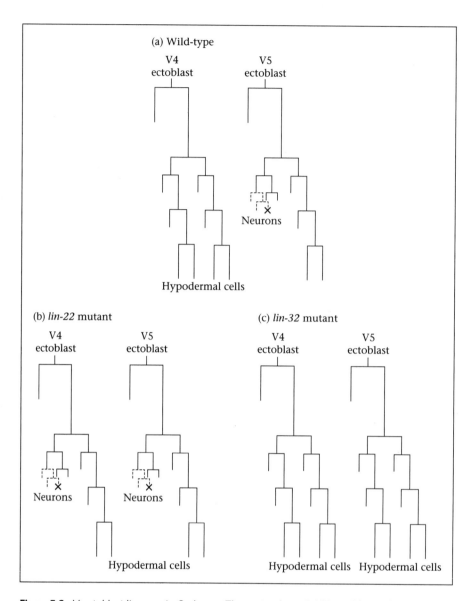

Figure 5.2. V ectoblast lineages in *C. elegans*. The postembryonic V4 ectoblasts of wild-type *C. elegans* produce hypodermal cells, while the V5 ectoblasts produce a mixture of neurons (indicated by dashed lines) and hypodermal cells (a). In *lin-22* mutants, some of the V4 progeny acquire a neuronal fate (b); in *lin-32* mutants, all V5 progeny acquire a hypodermal fate (c). Analysis of these mutations revealed the critical role of bHLH proteins in the regulation of cell fate in the nervous system. X indicates a cell that undergoes developmental programmed cell death; hyp indicates a cell that becomes part of the hypodermis. Drawing adapted from Hobert (2010).

The *lin-22* and *lin-32* genes each encode a basic helix-loop-helix (bHLH)-containing protein that functions as a transcription factor (fig. 5.3). A primary function of these proteins is to regulate expression of target genes by binding to a hexanucleotide (6-mer) sequence of DNA called an E-box. E-boxes are typically found upstream of promoters: binding of a bHLH protein to an E-box typically enhances transcription of the downstream gene. Proteins in this family are known for their regulation of cell proliferation and differentiation in many animals, so their role in *C. elegans* neurogenesis is not surprising. All bHLH proteins share a common structure of a DNA-binding basic region approximately 60 amino acids long (the source of the *b* in the name of these proteins) followed by two α-helices separated by a variable loop region (the source of the HLH in the name). The HLH domain functions to promote dimerization, allowing the formation of either homodimers (complexes of two identical proteins) or heterodimers (complexes between different members of the bHLH family). It is these dimeric complexes that bind to specific hexanucleotide sequences in DNA. Thirty-five

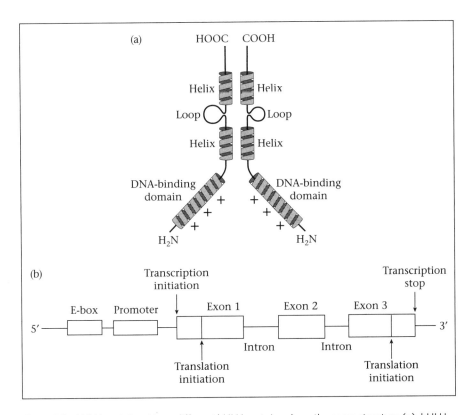

Figure 5.3. bHLH proteins. Many different bHLH proteins share the same structure (a). bHLH proteins bind as dimers to E-box elements in DNA (b). The E-box DNA sequence is CANNTG, in which N can be any nucleotide. The E-box is usually found upstream of a gene in the promoter region, as shown in (b), but may also be located inside target gene introns or exons.

bHLH genes are encoded in the genome of *C. elegans;* most have identified homologs in other animals. Some *lin-22* and *lin-32* homologs are described later in this chapter.

Because the products of the *lin-22* and *lin-32* genes are transcription factors, understanding their function requires identifying the genes that have their expression regulated by these proteins. Analysis of the cell lineages produced by the V4 and V5 blasts in double mutants lacking both LIN-22 and LIN-32 proteins revealed that the V4 lineage was wild type in these animals but that the V5 lineage was completely hypodermal. Taken together with the observations of the single mutants, these results suggest that LIN-22 inhibits *lin-32* expression in the V4 blast, consequently restricting neurogenesis in the V4 lineage.

These observations raise many additional questions. What factor(s) restricts *lin-22* expression to the anterior V blasts (V1–V4)? How general are the antineural effects of *lin-22* and the proneural effects of *lin-32*? Can they be observed in other lineages in *C. elegans* and at other stages of nervous system development? What is the function of homologous genes in other species? We might also ask what factor(s) triggered the proliferative activity of the V blasts in the first place.

Some of these questions can be answered by more extensive analysis of the phenotypes of additional mutations in these genes; others can be addressed through comparative studies of the function of homologous genes in other model animals. We will take one more look at *lin-32* in *C. elegans* before shifting our focus to the *atonal* gene of *Drosophila*, which encodes another important bHLH protein directly involved in neural development.

Rays

Our prior example focused on two postembryonic lineages in the lateral hypodermis of hermaphrodite *C. elegans*. The effects of mutations in *lin-32* have also been examined in males (fig. 5.4). Male *C. elegans* have sensory structures called rays on their tails. Rays provide tactile information used for positioning when males mate with hermaphrodites. Each of the nine rays found on each side of wild-type males contains two neurons and a sensory cell. The rays are produced by nine pairs of hypodermal ray precursor cells (Rn). A diagram of the lineage that produces the two neurons (RnA and RnB) and the ray structural cell (Rnst) that form each ray is shown in figure 5.4. Mutations in *lin-32* that disrupt the function of the lin-32 protein block the formation of rays because all three of the component cells are lacking. This rayless phenotype is in accord with the proneural function of the *lin-32* described for the hermaphrodite. Reporter gene–based studies of *lin-32* expression, however, revealed that *lin-32* is expressed not only by the hypodermal ray precursor cells (Rn) but also by subsequent cells in the lineage: Rn.a, Rn.aa, and Rn.ap. These observations suggest that *lin-32* likely functions not

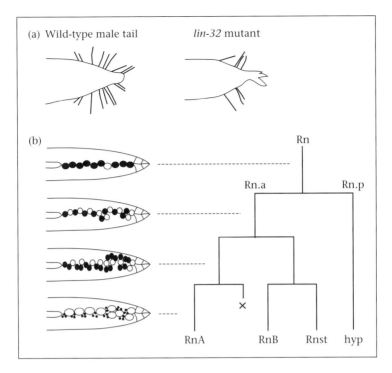

Figure 5.4. Neurogenesis in tail of *C. elegans* males. The wild-type male tail has nine rays on each side; different mutant alleles of *lin-32* produce different ray-loss phenotypes, one of which is shown (a). The positions of the ray precursor cells and their progeny are illustrated (in black) to the left of the ray sublineage (b). The ray precursor cells (Rn cells), the ray neuroblasts (Rn.a cells), and their descendants (the ray neurons RnA and RnB and the ray structural cell Rnst). X indicates a cell that undergoes developmental programmed cell death; hyp indicates a non-neural cell that becomes part of the hypodermis. Drawing adapted from Portman and Emmons (2000).

only to permit production of direct neuronal precursor cells (in this case, Rn.a) but also influences the more restricted fate of cells produced later in the lineage. A detailed study of ray development in mutant males carrying hypomorphic alleles of *lin-32* supported the hypothesis that *lin-32* is active at different steps in the ray lineage. This finding, of course, in turn raises the question of how a single protein can have multiple functions, which in this case must involve transcriptional regulation of different target genes.

These results, plus others based on experimental analysis of neurogenesis in *C. elegans*, can be summarized as follows. First, neurogenesis proceeds through the successive production of neuronal precursors, and several rounds of cell division often separate the initial expression of a proneural gene in a progenitor cell from the production of an actual neuron. Second, many of the genes for the regulation of neurogenesis encode transcription factors such as members of the bHLH family of proteins. Third, antineural factors such as the *lin-22* gene product can actively inhibit production of neural progenitors and neurons. The careful reader will note that the fundamental process of asymmetrical cell division tacitly lurks in the background even when it is not the focus of a particular study. An example is the coproduction of hypodermal cells and neurons in the V5 lineage of the hermaphrodite (see fig. 5.2).

Neurogenesis in *Drosophila*

Chordotonal Organs

Our initial example from *Drosophila* is also drawn from the postembryonic development of the peripheral nervous system. During metamorphosis, undifferentiated ectodermal cells differentiate as sense organ precursor cells, which divide to produce sensory neurons or sense organs that contain sensory neurons. Many sense organs, such as bristles or hairs, are expressed on the surface of the body, but stretch- and vibration-sensitive sense organs called chordotonal organs are located inside the body (fig. 5.5). A chordotonal organ comprises one or more basic chordotonal subunits. Each subunit, called a scolopidium, consists of one, two, or three bipolar neurons and three supporting cells. An example of a large chordotonal organ is Johnston's organ, which is located at the base of insect antennae. In *Drosophila*, Johnston's organ is formed of more than 200 scolopidia. The vibration sensitivity of Johnston's organ is the basis of antennal hearing in fruit flies. Other chordotonal organs are associated with the joints of the leg and the wall of the abdomen.

Two proneural genes, *achaete* and *scute,* had already been identified in the fruit fly prior to investigation of chordotonal organ formation. It had been established that development of almost the whole of the adult peripheral nervous system is dependent on the normal function of *achaete* and *scute*. Among the rare exceptions to this rule are the chordotonal organs, and it is here we begin our story.

As in the case of *lin-22* and *lin-32* of *C. elegans*, *achaete* and *scute* encode bHLH transcription factors in *Drosophila*. A search was mounted for additional bHLH genes with proneural function. At this time the *Drosophila* genome had not yet been sequenced, so candidate bHLH genes were sought using a polymerase chain reaction (PCR) performed with degenerate primers. (A PCR primer is considered degenerate if some of its positions have several possible bases.) This strategy permits the amplification of new, previously undescribed members of known gene families on the basis of conserved motifs in the protein sequence. This search resulted in the identification of *atonal*, a bHLH protein–encoding gene with a proneural function.

A strong case can be made that *atonal* is a chordotonal organ proneural gene. *Atonal* mRNA is expressed in every proneural cluster and sense organ precursor cell destined to produce a chordotonal organ. In fact, careful analysis of the pattern of *atonal* mRNA expression in developing fruit flies using

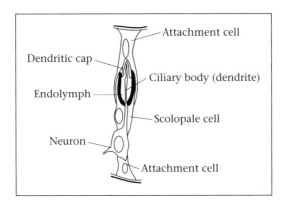

Figure 5.5. A chordotonal organ in *Drosophila*. Chordotonal organs are arthropod stretch receptors and auditory organs. They are diverse in structure and size, and the version illustrated here is quite simplified. Bipolar sensory neurons are the essential elements of chordotonal organs. These neurons produce specialized ciliated dendrites enveloped in a pool of receptor endolymph contained within a scolopale cell. The sensory neurons are sheathed by glial cells (not shown in this simplified diagram). The attachment cells located at both ends of the structure connect the chordotonal organ to the inner surface of the hypodermis and internal connective tissues.

in situ hybridization led to the discovery of a previously undescribed chordotonal organ in the tegula, a small patch of cuticle found at the base of the wings. Ectopic expression of a gene—that is, expression of a gene in a cell population in which it is normally not expressed—can be used to test hypotheses concerning the function of a gene. Ectopic expression of the *atonal* gene during metamorphosis led, as predicted, to the formation of extra chordotonal organs in unusual locations throughout the body.

Fly larvae do not develop any chordotonal organs as elaborate as the Johnston's organ of adult antennae, but they do have small chordotonal organs associated with the body wall. These chordotonal organs serve as stretch receptors. These stretch receptors help keep larvae from overeating prior to molting! As is the case during metamorphosis, expression of *atonal* mRNA is found at every site at which a chordotonal organ forms. Mutant embryos that lack *atonal* expression do not form the normal complement of larval chordotonal organs. Conversely, ectopic expression of *atonal* in embryos leads to the formation of supernumerary chordotonal organs.

Comparison of sequences of known bHLH genes revealed that *atonal* in *Drosophila* is a homolog of *lin-32* in *C. elegans*. The proneural function of this gene has been conserved along with its sequence.

Bristles

The significance of bHLH proteins as proneural factors in the development of the nervous system can hardly be overstated. Let us emphasize this point by returning to *achaete* and *scute*. These genes are two members of a group of four *Drosophila* genes found near the tip of the fly X chromosome referred to collectively as the *achaete–scute* complex (sometimes abbreviated as *AS*-C). The other members of the complex are *lethal of scute* and *asense*, but *achaete* and *scute* deservedly get most of the attention.

The term *complex* is used to refer to a group of genes with overlapping or redundant function that are located close together on the same chromosome. The genes in such complexes are evolutionarily related, and their expression is jointly regulated. All four of the genes in the *AS*-C encode bHLH transcription factors essential in the decision of certain epidermal cells to commit to a neural progenitor fate. (Reflect on the concept of neural induction referred to in Chapter 4: in many cases observations of neurogenesis form the basis of the inference that neural induction has occurred.) The proneural function of the *AS*-C was initially established in the context of bristle formation, in part because bristle pattern is a very easily studied phenotype. Here we linger on the topic of fruit fly bristles not only for the sake of *achaete* and *scute* but also so that we can meet the Notch signaling pathway.

Bristles are visible sense organs (mechanoreceptors or chemoreceptors) that project from the surface of adult fruit flies in stereotyped arrays. Tactile stimulation of mechanoreceptor bristles typically induces a behavioral response—for example, touching the bristles on the thorax triggers a cleaning reflex during which a leg is swept across the stimulated area. A typical bristle consists of four cells: a hair (also called the chaete), a socket cell, a sensory neuron, and a support cell (fig. 5.6). These four cells are the products of two rounds of cell division that begin with a cell called a sensillum mother cell. Depending on their location in the body, the sensillum mother cells differentiate from the epidermal epithelium or from an internal structure called an imaginal disc.[3] The process of becoming a sensillum mother cell depends on expression of the *AS*-C, because mutations that disrupt the *AS*-C suppress formation of bristles. Conversely, ectopic expression of the complex results in extra bristles.

Many interesting aspects of the proneural function of bHLH proteins have been revealed though study of the *AS*-C. For example, it has been shown that the *achaete* and *scute* gene products cross-induce each other's expression in a web of interactions that complicates the analysis of mutant phenotypes. Another striking aspect of *AS*-C function is that all of the *achaete-scute* protein products regulate transcription by binding to DNA E-boxes as heterodimers that they form with yet another bHLH protein, daughterless. Because *AS*-C proteins cannot function without daughterless, the *daughterless* gene is also described as proneural, although it is ubiquitously expressed and functions to promote neural identity only when partnered with *AS*-C proteins.

But—and this is extremely important—the story of bristle development in the fruit fly is not just another story of proneural bHLH signaling. Think back to the epithelial cells that became the sensillum mother cells. They

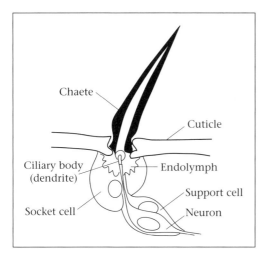

Figure 5.6. Bristles in *Drosophila*. The chaete is the visible structure; it is innervated by a sensory neuron that projects a ciliated dendrite into a pool of receptor endolymph contained within a socket cell. A nonneuronal support cell is associated with the sensory neuron.

were initially part of the small islands of ectoderm called proneural clusters described in Chapter 3. The proneural clusters that eventually produce a single sensillum mother cell (and hence a single bristle) each contain six to seven epidermal cells. This means that each of the epithelial cells that become the sensillum mother cells was initially one among many similar cells in a proneural cluster. A mutation in a *Drosophila* gene named *shaggy* results in the formation of a tuft of bristles where there should be only one. The phenotype of this mutation suggests that the cells that form the proneural cluster were initially equivalent in that at one time each had the potential to become a sensillum mother cell. Why don't all of the cells in the proneural clusters go on to produce sense organs? The answer to this question involves inhibitory interactions among the cells in the cluster mediated by a receptor called Notch (fig. 5.7). This signaling pathway is so widespread in multicellular organisms and plays so many roles in development that it is well worth the effort required to grasp an overview of its components.

The *Notch* gene is named for the dominant phenotype displayed by heterozygotes carrying one mutated copy of the gene—a slight notch, or little split, in the wing. Homozygotes lacking *Notch* expression die as embryos because so many epidermal cells differentiate as neurons that too little cuticle is produced. The overproduction of neurons in the absence of *Notch* expression implies that one of the functions of Notch protein is to suppress neurogenesis. Not surprisingly, this is achieved via interference with bHLH-based proneural signaling. But we need to enlarge our cast of genes, gene products, and pathways beyond bHLH transcription factors to see how this works.

The Notch protein is a transmembrane receptor that binds protein ligands located on the surface of adjacent cells. This signaling pathway allows two neighboring cells to have a private conversation. Notch binds several ligands,

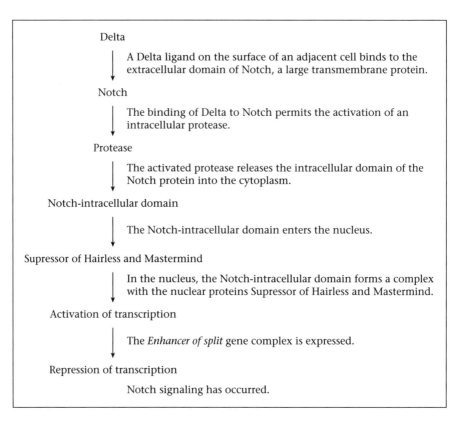

Delta

| A Delta ligand on the surface of an adjacent cell binds to the
↓ extracellular domain of Notch, a large transmembrane protein.

Notch

| The binding of Delta to Notch permits the activation of an
↓ intracellular protease.

Protease

| The activated protease releases the intracellular domain of the
↓ Notch protein into the cytoplasm.

Notch-intracellular domain

| The Notch-intracellular domain enters the nucleus.
↓

Supressor of Hairless and Mastermind

| In the nucleus, the Notch-intracellular domain forms a complex
↓ with the nuclear proteins Supressor of Hairless and Mastermind.

Activation of transcription

| The *Enhancer of split* gene complex is expressed.
↓

Repression of transcription

Notch signaling has occurred.

Figure 5.7. Notch signaling. Notch receptors are activated by binding to Delta or other Deltalike ligands on the surface of adjacent cells. In the canonical *Drosophila* Notch signaling pathway illustrated here, many of the known Notch target genes are located in the *Enhancer of split* gene complex, which contains at least 8 genes. Some of these genes encode bHLH proteins, which in turn act as transcriptional repressors. One of the genes regulated by the *Enhancer of split* complex is Delta. *split* (*spl*) is another name for the *Notch* gene; it refers to the characteristic wing abnormality that results when Notch is mutated.

but for now we focus on a ligand named Delta. When Notch binds Delta (as a result of the expression of these proteins on the surface of adjacent cells), the resulting change in Notch conformation releases a protein that, in the absence of the Delta ligand, binds to the cytoplasmic part of the Notch protein. The protein that binds to the cytoplasmic part of Notch is a transcription factor called Suppressor of Hairless. When released from Notch, Suppressor of Hairless enters the nucleus, binds to DNA, and activates transcription of a set of genes collectively referred to as the *Enhancer of split* complex. The *Enhancer of split* complex encodes seven bHLH proteins plus an additional three non-bHLH proteins. These proteins interfere in various ways with signaling by the proneural *AS*-C proteins, thereby preventing the Notch-bearing cell from becoming a sensillum mother cell. Be careful to note exactly what has happened here: by expressing the Delta protein on its surface, a cell has gained control of transcription in a neighboring cell! The products of the *achaete* and *scute* genes promote expression of the Delta ligand, so we can envision a feedback loop within a proneural cluster: any cell with even slightly higher expression of Delta, achaete, and/or scute proteins than found in its neighbors gradually but inexorably becomes the sole sensillum mother cell in the cluster. As a result, each proneural cluster forms but a single bristle sense organ.

Notch and Neuroblasts

Now that we have described a fundamental mechanism—lateral inhibition—that limits production of bristle sense organs in the periphery to one organ per cluster, can we say anything about neurogenesis in the central nervous system of the fly? The answer is definitely yes. Recall (from Chapter 3) that the neurons of the brain and ventral nerve cord are products of neural progenitor cells called neuroblasts that separate from the general neuroepithelium. As in the case of the sensillum mother cells, the neuroblasts initially resided in proneural clusters. The proneural clusters that are the source of the neuroblasts initially contain up to six cells. The proneural clusters can be identified prior to the emergence of the neuroblasts because, unlike the surrounding cells, they express *AS*-C genes. But only one of these cells becomes a neuroblast. How is that cell selected? Round up the usual suspects: bHLH proteins, Notch, and Delta.[4]

Mutations that eliminate or reduce expression of achaete, scute, or daughterless proteins lead to a reduction in the number of neuroblasts and an increase in the number of epidermal progenitor cells. Mutants with loss of *Notch* or *Delta* function have the opposite problem—too many neuroblasts and too few progenitor cells to form a complete integument. It is because of these mutant phenotypes that *Notch* and *Delta* are sometimes referred to as neurogenic genes.

Given these striking mutant phenotypes, it is surprising that separation of neuroblasts from the proneural clusters proceeds normally in flies manipulated to have uniform expression of Notch. The process is also normal or almost normal if Delta is ubiquitously expressed. Therefore the determination of the neuroblasts is not absolutely dependent on Notch–Delta signaling but is dependent on the expression of proneural genes. These results suggest the following scenario. First, imagine that all cells in the neuroepithelium mutually suppress their neighbors from selecting the fate of a neural progenitor (the neuroblast). Only a small number of cells with high levels of proneural activity—typically only one per cluster—can overcome this suppression. In this scenario, the primary functions of Notch–Delta signaling are to reinforce any initial, likely very slight, heterogeneity in level of proneural activity and, practically speaking, to ensure that enough epidermal cells are produced to cover the body surface. The source of the initial heterogeneity in proneural activity is undoubtedly regulated in a complicated fashion, given the long (but assuredly incomplete) list of proteins already known to interact with proneural genes and the Notch signaling pathway. It also remains formally possible that the initial heterogeneity in proneural activity arises by chance, an assertion difficult to prove or disprove. A memorable experiment depicted in fig. 5.8 (performed on grasshopper embryos because fly embryos are too small to permit such manipulations)

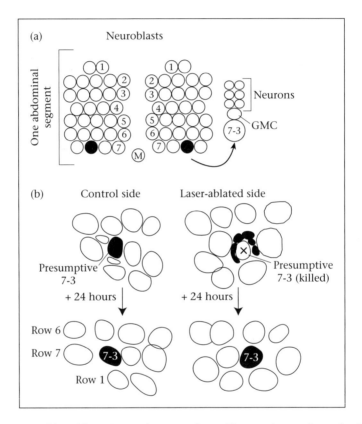

Figure 5.8. Neuroblast ablation in grasshopper embryos. The neural progenitor cells of the thoracic and abdominal segments form bilaterally symmetric arrays of 30 cells on each side of the embryo (plus one midline cell). The 30 neuroblasts are arranged in a bilaterally symmetrical array of seven rows, separated by one unpaired midline neuroblast (M). The right and left neuroblasts are numbered according to their row and position within the row (for example, 7-3, which is shown as a black dot, indicates the third neuroblast in the seventh row of the array (a). Each neuroblast undergoes a characteristic series of asymmetric divisions that produce ganglion mother cells (GMCs) that, in turn, each divide to produce two neurons. When a presumptive neuroblast is killed by laser ablation early in embryonic development (b), an adjacent cell takes its place, suggesting that all neuroepithelial cells have the potential to become neuroblasts and that cell–cell interactions within clusters regulate cell fate. Drawing adapted from Taghert et al. (1984).

revealed that other cells in a cluster can replace a neuroblast that meets an untimely fate (in this case, an unfortunate encounter in a laboratory with a laser microbeam).[5] This result provides support for a role for cell–cell signaling in the determination of neuroblasts in the insect nervous system.

Notch and Numb

Here we have just touched on the basics of the Notch signaling pathway in fruit flies. There will be more to say later in the chapter about the Notch pathway in zebrafish, mice, and humans. But familiarity with Notch will also stand you in good stead as we consider a final example drawn from studies of neurogenesis in *Drosophila.*

The structure of a bristle has already been described. Now consider bristle function. The adult fruit fly has a tough exoskeleton. As a result, a fruit fly's sense of touch depends on sense organs inserted into its cuticle, and bristles are one type of sense organ. A fruit fly without bristles would be relatively insensitive to touch. This is exactly the phenotype of *Numb,* a mutation that results in deletion of most peripheral sensory neurons. Visible sensory structures are present at the right locations in the cuticle of *Numb* mutants, but they are all for show because they lack sensory neurons to transduce the mechanical signal.

The presence of the nonneural parts of the bristle sense organ reveals that the precursor cells (the aforementioned sensillum mother cells) must have been present but that something has gone awry in the process of neurogenesis. Part of that something gone wrong is unrestrained Notch signaling in both of the daughters of the precursor cell. This out-of-control Notch signaling, in turn, instructs both daughters to produce nonneural progeny. The *Numb* gene is a major part of this story. In fact, the Numb protein encoded by this gene was the first molecule discovered to influence cell fate as a consequence of being asymmetrically distributed to the daughter cells during cell division.

In wild-type flies, each sensillum mother cell divides along a plane of division that yields nonequivalent anterior and posterior daughter cells. In the simplest case, the posterior daughter cell (typically referred to as secondary precursor cell pIIa) divides one more time to produce nonneural progeny: a socket cell and a hair cell. The anterior daughter cell (pIIb) divides one more time to produce neural progeny: a sensory neuron and a sheath (glial) cell. The production of neural progeny is dependent on the presence of Numb protein—if Numb protein is absent, two pIIa cells are produced, both of which are unable to produce neurons. Why does pIIb have Numb, whereas pIIa does not? The answer lies not in differential gene expression in the two cells after they are born. Instead, pIIb and pIIa are already distinct at their birth (fig. 5.9). The reason is that Numb protein has a very restricted distribution within the sensillum mother cell. Prior to cytokinesis, Numb is found

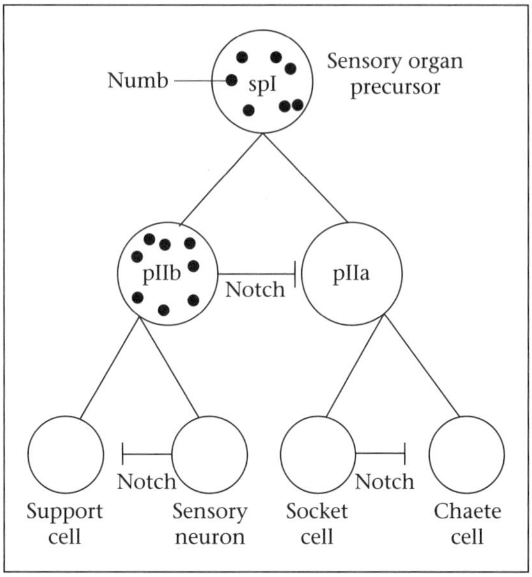

Figure 5.9. Asymmetrical partitioning of Numb in the lineage of *Drosophila* bristle sensory organs. The sensory organ precursor spl expresses Numb, but when spl divides only one of the daughter cells (pIIb) receives Numb, which is represented by the black dots. Mutants that lack Numb produce neither neurons nor support cells.

only in the anterior outer edge (also called the cortex or rind) of the sensillum mother cell. The combination of the polarized distribution of Numb with a specific alignment of the mitotic spindle results in an asymmetric distribution of Numb in the progeny: all to the anterior daughter (pIIb), none to the posterior daughter (pIIa)

Numb functions as a Notch pathway antagonist in pIIb. The Numb protein binds directly to the intracellular domain of Notch and to other proteins critical for receptor-mediated endocytosis, a process by which intracellular vesicles are formed from pinched-off pieces of plasma membrane. Collectively, these actions mean that, although both pIIa and pIIb express Delta and Notch, Delta on pIIb activates Notch on pIIa, but the reverse is not true.

Asymmetrical distribution of Numb protein to a pair of daughter cells is also a factor in the asymmetrical outcomes of neuroblast divisions in the *Drosophila* central nervous system. Parent neuroblasts produce two progeny: a renewed large neuroblast and a smaller ganglion mother cell. The daughter neuroblast will produce yet another neuroblast and another ganglion mother cell, whereas the ganglion mother cell will divide once to produce a pair of neurons. As you might predict, based on Numb's inhibition of Notch signaling, it is the ganglion mother cell that receives the Numb protein. The plane of division of neuroblasts in the central nervous system is apical–basal. The ganglion mother cell is the basal daughter. Therefore, prior to cell division, Numb protein is restricted to the basal cortex of the parent neuroblast.

In both examples (bristle and neuroblast), the polarized distribution of Numb is the result of a series of protein phosphorylations. The Numb protein of *Drosophila* has five sites that can be phosphorylated by a regulatory protein kinase. A version of Numb bearing mutations that block phosphorylation at all five of these sites results in an abnormally broad distribution of Numb protein in the parent cell. This mutation often results in a neural fate for both of the daughters instead of only one.

Neurogenesis in Zebrafish

Are the mechanisms of neurogenesis conserved across the invertebrate–vertebrate divide? The answer given to this question, as noted in the preceding chapter, is partly a matter of a scientist's preferred scientific style. As long as we are cognizant of the critical contributions that studies of invertebrates make to our understanding of development, it's more or less a personal choice whether to be surprised or unsurprised by the conservation of molecules and processes over the long span of evolutionary time. Our surprise (or lack thereof), however, is best paired with a keen eye for evolutionary novelties.

Large-scale screens for mutations affecting the development of zebrafish embryos recovered many mutants with obvious brain phenotypes. Some of the mutant brains were disorganized—thicker than usual in one place, thinner than usual in another—whereas others were characterized by collapsed vesicles and slow development. But several mutants displayed an increase in the number of neurons, a phenomenon sometimes referred to as neural hyperplasia (*hyperplasia* is a general term meaning an increase in the number of cells).

One of the neural hyperplasia mutants was named *white tail* (*wit*), because it lacks pigment cells on its posterior trunk and tail. During embryogenesis, the developing nervous system of *white tail* mutants displays subtle abnormalities. The neural keel is slightly misshapen, and the standard rhombomeric boundaries are indistinct. But the overall profile of development is more or less normal, and all major subdivisions of the brain are present. The effects of the mutation, however, become obvious when the nervous system is probed with antibodies that recognize proteins associated with neurons. For example, an antibody that recognizes a neurofilament epitope immuno-labels only one interneuron on each side of the fourth rhombomere in wild-type zebrafish but marks three or four interneurons in *white tail* embryos.[4] Many different populations of cells are affected: for example, the mutant also has more GABAergic interneurons in its developing hindbrain and spinal cord than are seen in normal zebrafish.[5]

The *white tail* phenotype is so striking in large part because it is reminiscent of the phenotype displayed by fruit flies lacking *Notch* or *Delta* function. Does the gene (or genes) at the *white tail* locus play a role in neurogenesis by mediating cell-to-cell interactions? Despite the fact that the *white tail* mutant was first described in 1996, the answer to this question is still maybe, because this mutation remains uncharacterized at the gene level.

Despite the incomplete analysis of *white tail,* there remains no doubt concerning the importance of Notch signaling in zebrafish development. By the early 1990s it was already known that the genomes of many vertebrates, including zebrafish, encode *Notch* homologs. Genes encoding Deltalike ligands were also identified in vertebrate genomes. Mutations in these genes resulted in a clear neurogenic phenotype.

The similarities between flies and zebrafish also extend to the proneural genes. Homologs of the *achaete–scute* and *atonal* genes of *Drosophila* are expressed in the developing zebrafish nervous system. Sequence comparisons reveal that zebrafish neurogenin1 in particular and vertebrate neurogenins in general are more similar to atonal than to any other *Drosophila* members of the bHLH transcription factor family. There is strong evidence that the zebrafish *neurogenin1* gene (*ngn1*) is expressed in the neural plate in neural progenitor cells prior to the expression of other neural markers (fig. 5.10). Injection of *neurogenin1* mRNA into very early embryos resulted in ectopic production of neurons. The resulting ectopic neurons are inevitably found scattered among nonneural cells in a pattern referred to as salt-and-pepper. This finding implies an interaction between proneural signaling and lateral inhibition comparable to that seen in the neurogenic regions of *Drosophila.* In addition, the supernumerary neurons generated by ectopic *neurogenin1* expression adopt fates appropriate to their location in the nervous system, providing a reminder that local patterning mechanisms interact with proneural signaling to determine the type of neuron produced once the number is set.

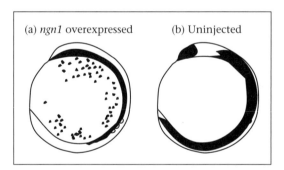

Figure 5.10. Misexpression of neurogenin1 (*ngn1*) in zebrafish embryos. Misexpression demonstrates the proneural actions of this bHLH protein in the developing nervous system. Injection of *ngn1* transcript into one- to two-cell zebrafish embryos resulted in overexpression of *ngn1* in the surface ectoderm overlying the body axis. One of the consequences of misexpression of *ngn1* (assessed using in situ hybridization for neuronal markers at the three-somite stage, approximately 11 hpf) was the development of ectopic neurons (represented by the scattered spots representing the distribution of neuronal markers) in the yolk sac ectoderm (a); no ectopic neurons were detected in uninjected control embryos (b). The embryos depicted were redrawn from photomicrographs in Blader et al. (1997).

All of the accounts of signaling pathways provided in this text are simplified for didactic purposes, but the process of simplification is more challenging for vertebrate pathways because of the expansion in gene number characteristic of vertebrate genomes. The greater number of genes found in vertebrate signaling pathways reflects gene duplication and divergence events that have occurred since the lineages split. For example, the Notch ligand Delta is encoded by a single gene (*Delta*) in Drosophila, but at least four *delta* genes—*deltaA, deltaB, deltaC,* and *deltaD*—are present in the zebrafish genome. Four different zebrafish Notch receptors—Notch 1a, Notch1b, Notch2, and Notch3—vie for these four *delta* ligands. Comparisons of the putative proteins encoded by these genes indicate that the extracellular (ligand-binding) domain is highly conserved, whereas the intracellular domains vary. In situ hybridization studies of the distribution of the *delta* mRNAs in the neural plate at the five-somite stage in the zebrafish revealed that each gene has its own map: *deltaA* and *deltaD* mRNAs were expressed diffusely in proneural patches, whereas *deltaB* mRNA was expressed strongly in scattered cells, and *deltaC* mRNA was not present in this tissue at this time. It's complicated.

Neurogenesis in the Mouse

More Neurons

In general, fish have smaller brains than birds or mammals of comparable size, and among the fish, zebrafish are not known to be particularly brainy. It has been estimated that the total number of neurons and glial cells in the adult zebrafish brain is less than half a million.[6] By contrast, estimates for the number of neurons in the mouse brain range as high as 75 million (the number is strain dependent, and this number probably represents the high end of the range for this species), whereas the standard number given for the human brain is 100 billion or more. These numbers imply that neurogenesis in mammalian embryos must proceed with an intensity far exceeding that in other taxa.

Despite the almost incomprehensibly enormous number of neurons involved, the basic features of neurogenesis in mammals are those already described for *C. elegans, Drosophila,* and the zebrafish. Neural progenitor cells are produced by divisions of epithelial cells; the progenitor cells either divide symmetrically to increase the population of neural progenitor cells or divide asymmetrically to produce neuronal and glial daughters while simultaneously maintaining the population of progenitor cells. As in all vertebrates, in mammals the epithelium that generates the neural progenitor cells is the monolayer of epithelial cells that forms the neural tube. The enlargement of the vesicles in the developing mammalian brain provides a large surface area from which the necessarily vast number of progenitor cells can arise.

Mash1

The proneural genes expressed in the mouse neural tube epithelium and neural progenitor cells include the homologs of the bHLH transcription factors that serve this function in *Drosophila*. Among the best characterized of the many mouse proneural genes are mammalian *achaete–scute homolog 1* (*Mash1*), the aforementioned *neurogenin1* (*ngn1*), and *neurogenin2* (*ngn2*). In situ hybridization studies have revealed that each of these proneural genes has a specific map of expression in the developing nervous system, as is the case for *atonal* and *achaete–scute* in *Drosophila*. For example, in the developing telencephalon, the distribution of *Mash1* mRNA revealed by in situ hybridization is associated with the cells that give rise to the basal ganglia, whereas the *ngn1* gene is expressed in the cells that produce the cerebral cortex. Double in situ hybridization studies using probes for both genes revealed that the patterns of *Mash1* and *ngn1* expression did not overlap. This result indicates a likely association of these proneural genes both with neurogenesis and with the specification of a brain region-specific neural fate. But be careful not to overgeneralize! In other regions of the mouse central nervous system, these same two genes were expressed in overlapping rather than mutually exclusive patterns. These regions included the olfactory epithelium, the midbrain, and the spinal cord. The appropriate conclusion is not that expression of individual bHLH genes determines specific neuronal identities but rather that individual bHLH genes act in combination with other restricted-distribution transcriptional regulators to determine neuronal identity. As I have noted before in this chapter, it's complicated.

Knockout mice homozygous for a null mutation of *Mash1* appear normal at birth but breathe with difficulty and do not nurse. They typically die before the end of postnatal Day 1. The widespread expression of *Mash1* in the developing central nervous system described in the previous paragraph predicts that the mutants will have defects in their brains and spinal cords, but these structures were normal both in overall appearance and in the expression of markers for neural tissue. Parts of the peripheral nervous system, however, were severely underdeveloped. Affected tissues were characterized by reductions in the number of neurons (compared with wild-type tissues) and included the olfactory epithelium, the autonomic nervous system, and ganglia associated with the gut (enteric nervous system). This pattern of defects is consistent with the known expression pattern for *Mash1* in the embryonic peripheral nervous system but at odds with the well-documented expression of *Mash1* in the embryonic central nervous system. Here the appropriate conclusion is not that *Mash1* plays no role in neurogenesis in the central nervous system but rather that other genes with overlapping patterns of expression and overlapping functions likely compensated for the absence of *Mash1* in the developing brain. It is also possible that reexamination of the *Mash1* knockouts using other molecular markers will reveal subtle brain defects. The absence of an obvious knockout phenotype in a particular tissue is often taken as evidence for fitness-enhancing genetic redundancy (a property sometimes referred to as genetic robustness), a reflection of the existence of multiple genes with similar sequences, gene products, and function in the mouse genome. A multiple knockout of two or more related genes may be required to unmask a phenotype undetectable in a single knockout.

Identification of proneural transcription factors such as *Mash1* and *ngn1* inevitably leads to questions about the identity of the genes they regulate. An essential next step is the construction of a gene regulatory network that links regulation of transcription with a particular cellular outcome. The targets of a particular transcription factor are defined by the response element (the specific short DNA sequence) to which the transcription factor binds. Direct methods used to find transcription factor targets include gel-shift assays (the rate at which a DNA fragment moves through a gel is studied both in the presence and in the absence of the putative transcription factor), cotransfection assays (two nucleic acid fragments are introduced into a host cell and gene expression is monitored), and chromatin-immunoprecipitation (ChIP). In a ChIP assay, DNA and proteins are mixed under conditions that allow the proteins (in our example, transcription factors) to bind at naturally occurring response elements. Whatever DNA–protein bonds form are then stabilized by exposure to formaldehyde prior to shearing the DNA into fragments. Antibodies to the transcription factors of interest can then be used to retrieve the DNA–protein complexes of interest for subsequent study. Note that this method can be used to identify previously unknown DNA binding sites but requires good supplies of a specific antibody for the transcription factor of interest.

Gene expression can also be studied in loss- or gain-of-function mutants. Such studies traditionally required considerable background information on the brain region to be studied, given that one needed to select in advance which gene products to measure. For example, a researcher studying proneural genes might hypothesize that genes involved in Notch signaling are regulated by *Mash1* binding to DNA and prepare qRT-PCR probes for these targets to compare wild-type and mutant tissues. A complementary postgenomic approach is to perform DNA microarray analyses of gene expression in tissues from different genotypes. Such a study, performed on mouse embryonic cortex, identified several known genes from the Notch pathway —for example, *Deltalike1* (*Dll1*) and *Hairy enhancer of split 5* (*Hes5*)—but also provided evidence for more than 40 additional *Mash1*-responsive genes.[7] It is simultaneously exciting and daunting to consider how these additional genes might be involved in cortical neurogenesis. As the cost of sequencing decreases, future projects of this type will often bypass the microarray, relying instead on direct sequencing of cDNAs as an index of gene expression in a particular cell population.[8]

As the topics covered in this chapter indicate, many studies on neurogenesis focus on the signaling pathways that define neuronal phenotypes. The larger brains of mammals, however, draw our attention back to the growth of cell populations. The layered structure of the cerebral cortex self-assembles to become the basis of perception, judgment, learning, and memory. Experimental studies in the mouse have contributed significantly to our understanding of cortical histogenesis, the coordinated cellular events required to build cortical tissue (fig. 5.11). Neuroanatomical analyses of fixed sections of mouse brain have been complemented by time-lapse observations of neurogenesis in living brain slices, in cultures of individual neural progenitor cells, and in dishes of embryonic stem cells dividing and differentiating in the laboratory. The story to date is described here. It comes with a spoiler alert: glial cells have their own chapter later in this book, but they are going to make an unexpected and significant appearance here in the role of neural progenitor cells.

The cerebral cortex has its origins in the single layer of epithelial cells lining the walls of the telencephalic vesicles. Neurogenesis converts this single layer into a stack of six layers, each defined by the shape, packing density, and connections of the neurons it contains. The upper layers (Layers 2 and 3) send their axons to other regions of the cortex, while the lower layers (Layers 5 and 6) project to subcortical targets. (Layers 1 and 4 receive inputs from other brain regions.)

Neurogenesis begins in most regions of the mouse central nervous system between embryonic dpc 10–12 (E10–E12). Approximately 11 cell cycles are required to build the cortex, and most cortical neurons have a birthdate between E14 and E17. The lack of earlier birthdates is explained by the observation that the earliest cell divisions in the cortex produce few daughters that leave the cell cycle. Instead, the earliest rounds of the cell cycle build the size of the progenitor population as stem cells make more stem cells.

These cell divisions take place in a layer of epithelial cells adjacent to the telencephalic vesicles, which will eventually become the brain's lateral ventricles. This layer is defined as the ventricular zone (VZ), and the cells that proliferate in this zone are referred to as apical progenitors. They are polarized cells: their apical surface faces the lumen of the ventricle, and their basal surface faces what will become the top of the cortex. Asymmetrical divisions of the apical progenitor cells often result in the generation of a single, postmitotic neuron and a restored apical progenitor cell. The newborn neuron then migrates in a basal direction (away from the ventricle) to become part of a layer of the cortex.

Some apical progenitor cells of the ventricular zone do not directly produce postmitotic neurons but instead generate another type of cell: the intermediate progenitor cell. Like newborn neurons, these cells also migrate

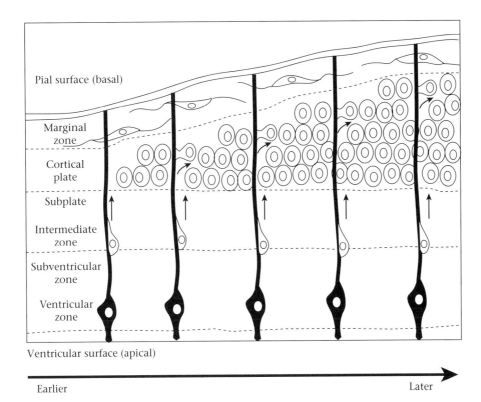

Figure 5.11. Histogenesis of the mouse cortex. In this simplified schematic diagram of cortical neurogenesis, neurons produced by radial glial cells (apical progenitor cells) in the ventricular zone migrate along the pial-directed long process of the radial glial cell to the cortical plate, passing successively through the subventricular zone, the intermediate zone, and the subplate. The earliest-born neurons are positioned deepest within the cortical plate; as shown, younger neurons move past them to a more superficial position. Elongated Cajal-Retzius cells are a distinctive feature of the immature cortex. They are born outside the cortex and migrate into the marginal zone, following a path roughly perpendicular to the orientation of the radial glia. See Chapter 6 for further discussion of the Cajal-Retzius cells. Drawing adapted from Bielas et al. (2004).

away from the ventricular zone, but they do not go into the cortex. Instead, they stop in a layer immediately basal to the ventricular zone called the subventricular zone (SVZ). Once in the subventricular zone, the intermediate progenitor cells often divide symmetrically to produce two neurons. These new neurons, in turn, migrate into the cortex.

Many events happen simultaneously as the cortex forms: apical progenitor cells proliferate, intermediate progenitor cells proliferate, and newborn neurons migrate. The oldest neurons start to differentiate their mature neuronal phenotypes as younger neurons begin their journeys. Careful birthdating studies have revealed that the neurons found in the layers of the cortex closest to the ventricle are born earlier than those in the basal, more superficial layers. In the mouse, cortical neurogenesis (birth of the cortical neurons) is completed by E18, but the process of cortical histogenesis—the construction of a functional cortex in terms of neuronal morphology and circuitry—continues into postnatal life.

For many years, apical progenitor cells were described in the scientific literature under another name: radial glial cells (RG). In fact, this name is still used by many neuroscientists when they refer to apical progenitor cells. It can be argued that to apply the term *glial cell* to a ventricular zone neural progenitor cell is certain to confuse, but the newer terminology can also be confusing. You will be on secure ground if you appreciate that it is only the nomenclature that is uncertain, not the biology. Regardless of whether they are referred to as apical progenitors or radial glial cells, these stem cells are the founders of almost all (if not all) neuronal lineages in the vertebrate central nervous system.

Apical progenitor cells can be recognized by the location of their somata in the ventricular zone and by their characteristic cytoplasmic extensions, referred to as processes (fig. 5.12). A short apical process contacts the lumen of the ventricle. A long basal process extends to the base of the pia mater (the innermost of the three meninges, or membranes, of the brain), where it forms branching terminals called end feet. The pial process (which gets longer as the growing cortex gets thicker) is retained through the cell cycle. When the progenitor cell divides, the pial process remains with the apical progenitor cell produced by the division. Apical progenitor cells are also characterized by a phenomenon called interkinetic nuclear migration: their nuclei move up and down within the ventricular zone at different stages of the cell cycle.

Markers for apical progenitor cells include vimentin and nestin, two intermediate filament proteins that are components of the cytoskeleton. Vimentin and nestin are also expressed in some populations of glial cells. The boundaries of progenitor cell or glial cell identity are further blurred by the fate of the apical progenitor cells when cortical neurogenesis is complete: they transform into astrocytic glial cells and migrate away from the ventricular zone.

The specialized morphological features of an apical progenitor cell contribute to the formation of the cortex. The short apical process anchors the epithelium through the formation of adherens junctional complexes and allows the cell to sample signals in the ventricular lumen.[9] Because several of the transmembrane proteins localized to the apical plasma membrane (for example, megalin and prominin-1) interact with cholesterol, it is possible that cholesterol-dependent signaling is involved.[10] The end feet of the long pial process can also take up extracellular signals. These could include signals secreted by the overlying meninges. One such signal is retinoic acid (previously mentioned in Chapter 4 in its role as a morphogen), which pro-

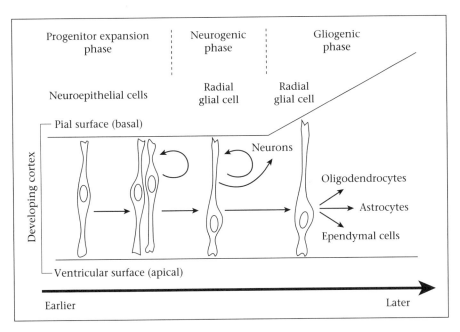

Figure 5.12. Radial glial cells in the developing mouse cortex. Radial glial cells (also called apical progenitor cells) extend a short process to the ventricular surface and a longer process to the pial surface. The population of neuroepithelial cells expands by repeated symmetric divisions. Some neuroepithelial cells elongate to become radial glial cells that span the entire cortex. Initially, radial glial cells divide repeatedly to produce neurons and intermediate progenitor cells (not shown). After a gliogenic phase, some or all radial glial cells differentiate as astrocytes. The cortex becomes thicker as neurogenesis proceeds, as indicated to the right in this diagram. Drawing adapted from Shimojo et al. (2011).

motes a switch from symmetrical to asymmetrical divisions of the apical progenitor cells.

The mysterious interkinetic nuclear migration of the apical progenitor cell nucleus during the cell cycle may reduce the exposure of nuclear DNA to Notch signaling. Recall that, based on the function of this pathway during neurogenesis in *Drosophila*, Notch signaling is predicted to inhibit neurogenesis. At present this scenario is speculative, but there is evidence that both intermediate progenitor cells and migrating neurons activate Notch in apical progenitor cells, thereby keeping the apical progenitors in a stem cell–like state. This is accomplished by expression of the gene *mind bomb-1*, which encodes a ubiquitin ligase enzyme essential for signaling by mouse homologs of the *Drosophila Delta* gene called *Deltalike* and *Jagged*. In mutants that lack the *mind bomb-1* gene product, all apical progenitor cell divisions result in either a pair of neurons or a pair of intermediate progenitor cells; that is, they are symmetrical divisions.

Neurogenesis in Humans

The Largest Cortex

Studies of the developing mouse brain are a primary source of our understanding of mammalian neurogenesis. Because rodent and primate brains share a common architecture, it is expected that the mechanisms of neurogenesis share many, if not most, features. But the greatly enlarged brains of primates imply that there are likely differences between rodents and primates in the regulation of neurogenesis. The spectacularly enlarged cerebral cortex of humans (large even compared with those of other primates) suggests that there may be unusual features associated with cortical histogenesis in humans. This means that we are interested not only in how human development differs from that of mammals in general but also in how humans differ from closely related nonhuman primates. Delineation of these differences will provide insights not only for evolutionary biology but also for clinical neuroscience, given that many devastating developmental disorders affect cortical brain regions.

How can neuroscientists study neurogenesis in the developing human brain? In part, the answer lies in very careful use of tissues obtained from spontaneous miscarriages and elective pregnancy terminations. Regardless of one's personal feelings concerning abortion, investigators working in the many jurisdictions where this procedure is legal can obtain human tissue for research. The use of such tissue must strictly follow all legal and institutional ethical regulations, and individual research protocols must be approved by the relevant institutional review board (typically referred to as the IRB) prior to the initiation of any project. In the United States, only scientists who have previously completed training in the responsible conduct of research (RCR) are permitted to submit protocols for review by IRBs. It is my experience that researchers who work with human and animal tissues are deeply cognizant of the ethical and legal obligations that are an essential part of their research.[11]

The extensive literature on neurogenesis in the mouse and other mammals coupled with access to the sequenced human genome provides the tools needed for efficient studies of the limited samples of human tissue that are available. One of the most important methods is immunohistochemistry on fixed tissue using antibodies that recognize human homologs of well-characterized mouse markers. It is also possible to study mitotic events and cell lineages in cortical slices and in dissociated cortical cell culture. For example, a living cortical slice from a human embryo can be exposed to a GFP-expressing retrovirus cultured and imaged using a confocal microscope.[12] The resulting images, collected at short intervals over several days, can be compiled into time-lapse videos of development. Treatment of living slices with BrdU paired with subsequent fixation and immunohistochemical detection of incorporated BrdU allows identification of proliferating cells. It is also possible to make electrophysiological recordings from slices of developing human cortex.

As in the mouse, ventricular and subventricular zones can be distinguished in the developing human cortex. Studies of fetal monkey brains, however, revealed that the nonhuman primate brain is characterized by an expanded outer region of the subventricular zone referred to as the OSVZ. The OSVZ is even more prominent in the developing human brain than in the monkey brain (fig. 5.13). If it can be shown that the cells of the OSVZ are neural progenitor cells, perhaps the origin of the neuron-rich human cortex are to be found in this band of tissue.

Recall that the cell population of the mouse subventricular zone consists of intermediate progenitor cells produced by asymmetrical division of apical progenitor cells in the ventricular zone and their progeny, either neurons or additional intermediate progenitor cells. Studies of OSVZ cells obtained from gestational Week 14 human fetuses revealed a population of proliferation marker-expressing OSVZ cells with a long, radially oriented basal process that extends to the pia. A long basal process, of course, is one of the defining features of apical progenitor or radial glial cells. But apical progenitor cells also have a short apical process and therefore contact the ventricle as well as the pia. By contrast, the OVSZ cells lack an apical process and therefore form a distinct population of progenitor cells. They are referred to as OSVZ radial glialike cells, or oRG for short. Studies of the daughters of oRG cells dividing in dissociated cell culture revealed that oRG cells can produce either additional oRG cells or cells expressing neuronal markers. A quantitative comparison of the number of proliferating cells in the human ventricular zone or inner subventricular zone versus the OSVZ showed that the OSVZ eventually becomes the dominant site for cortical neurogenesis.

These results highlight the importance and limitations of the mouse model. Without the essential guidance provided by prior studies of the mouse, it would be impossible to design studies sufficiently sophisticated to justify experimental manipulation of human tissue. But it is also the case that, if the goal is to understand human brain development, it is necessary to study the developing human brain.

New insights into the cells of the OSVZ in humans raise many questions. For example, does the OSVZ account for the expansion of the human cortex in particular and for gyrencephalic brain development in general?[13] Do abnormalities in the OSVZ explain known syndromes in human patients? There is also much to interest the cell biologist in the description of the oRG progenitor cells of the OSVZ. Why do some types of progenitor cells retain a connection with the ventricle whereas others do not? At present this question is unanswered.

Despite the differences in progenitor cell populations, the molecular aspects of neurogenesis display stunning conservation across hundreds of millions of years of diverging lineages. The same study that described the unusual

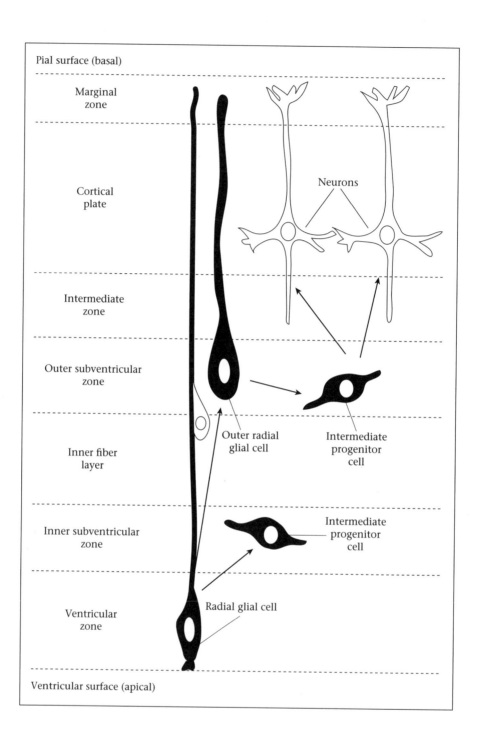

Pial surface (basal)

Marginal
zone

Cortical
plate

Neurons

Intermediate
zone

Outer subventricular
zone

Outer radial
glial cell

Intermediate
progenitor
cell

Inner fiber
layer

Inner subventricular
zone

Intermediate
progenitor
cell

Radial glial cell

Ventricular
zone

Ventricular surface (apical)

morphology of the oRG cells also reported that oRG cells express the human Notch-activated protein HES1. HES1 is encoded by a human member of the gene family that contains the *Drosophila Enhancer of split* gene-complex genes. Like the *Drosophila Enhancer of split* genes, HES1 is also a bHLH gene, and like other bHLH genes, it interferes with signaling by proneural genes. One possible implication of this finding is that Notch signaling may maintain the pool of progenitor cells in the OSVZ.

Microcephaly

Disruption of neurogenesis during development of the human nervous system results in a small brain, a condition known as primary microcephaly or microcephaly vera.[14] All parts of the brain are present, but all are smaller (by two to four standard deviations) than usual. Head circumference can be used as a surrogate measure of brain size, and the initial diagnosis of infants with microcephaly is typically made with a tape measure. Affected individuals display mild to moderate intellectual disability but otherwise have normal patterns of growth in terms of height and weight. Although there are non-genetic causes of congenital microcephaly (including maternal alcohol consumption during pregnancy), primary microcephaly is a recessive genetic condition. Primary microcephaly is rare in most human populations (<1 per 1 million births). The highest incidence, as high as 1 per 10,000 births, is seen in societies in which marriage of cousins is common: two examples are Amish communities in the state of Pennsylvania and communities in the North-West Frontier Province of Pakistan.[15] Studies of affected families from these populations have used pedigree analysis to identify gene loci associated with microcephaly. To date, five genes have been discovered that, when mutated, give rise to primary microcephaly. These genes are referred to as MCPH genes. They likely represent only the tip of a large iceberg, because a substantial fraction of clinical cases are not linked to any of the known loci or identified genes.

Figure 5.13. The outer subventricular zone (OSVZ) of humans. The OSVZ lies beneath the intermediate zone and is separated from the inner subventricular zone by the inner fiber layer. The OSVZ is present in humans and other primates but is not found in mice. The OSVZ contains a mixture of outer radial glial cells (oRG) and their intermediate progenitor cell (IPC) progeny; the IPCs produce neurons, which enter the cortical plate. The multitasking, multitalented radial glial cells, which are the only cells in the cortex that contact both the pial and the ventricular surfaces, divide asymmetrically to produce neurons, oRGs, and IPCs. oRGs and IPCs are present in mice and other mammals, but a distinct OSVZ is observed only in primates. Drawing adapted from Molnár et al. (2011).

The cerebral cortex is particularly affected in primary microcephaly, suggesting that the MCPH genes might be expressed at particularly high levels in the developing cortex. But, surprisingly, this is not the case. Instead, the products of all of the MCPH genes identified to date are expressed in all tissues, not just the developing brain.

The known MCPH proteins share a feature beyond ubiquitous tissue expression—they associate with the centrosome. The centrosome is a protein complex required for assembly of microtubules. Centrosomes associate with the spindle poles during mitosis. It is easy to imagine that abnormal mitosis could lead to defective neurogenesis, but why is the developing nervous system more vulnerable to this mutation than are other tissues? At present there is no clear answer. Clues, however, have come from experimental studies of a specific MCPH protein called ASPM, the product of the *Aspm* (abnormal spindlelike microcephaly-associated) gene.

As one might expect, experimental studies of ASPM function have been performed in mice, not humans. These studies have revealed that ASPM is present in progenitor cells in the ventricular zone of the developing mouse cortex. Because the ASPM protein attaches to the mitotic spindle poles and because knockdown of *Aspm* gene expression alters the orientation of the plane of cleavage, it is possible that this protein is required to maintain the pool of progenitor cells. If this is the case, absence of ASPM could lead to an early end to the build-up of the progenitor cell population, which in turn would reduce the total number of neurons produced. The timing of the shift from symmetric divisions that build the progenitor pool to asymmetric divisions that build neuronal populations is a critical aspect of neurogenesis throughout the nervous system.

Adult Neurogenesis

New neurons are added to the brain after hatching or birth in many animals. In insects, including the fruit fly, neurons are born during both the larval and the pupal stages. The neurons born during the postembryonic period play a vital role in metamorphosis of the insect nervous system (Chapter 9). A striking example is provided by the generation of the optic lobe interneurons during pupal development from two groups of progenitor cells, the outer and inner optic lobe *anlagen* (a German word used by embryologists to indicate the site in the embryo from which an organ forms). These interneurons are not needed by the larva, which does not have compound eyes. But the addition of compound eye–associated interneurons means that more than a third of the neurons in the adult fly brain are born during postembryonic life. Neurogenesis also continues during the larval stages in *C. elegans*. The 222 neurons present at hatching increase to a total of 302 in the adult hermaphrodite and 473 in the adult male.

In both flies and worms, postembryonic neurogenesis transforms a nervous system matched to the needs of one stage of life into a nervous system matched to the needs of the next stage. The changes in the nervous system reflect changes in body structure. This is *not* what neuroscientists typically mean when they use the phrase *adult neurogenesis*.

Continuation of ongoing neurogenesis in immature animals (the equivalent of childhood in humans) is also *not* what neuroscientists mean by this phrase. Neurogenesis in juveniles can be easily understood in terms of putting the finishing touches on the nervous system. In the dentate gyrus of the mouse hippocampus, for example, neurogenesis continues after birth and does not peak until postnatal Day 8.[16]

Adult neurogenesis, as the name implies, occurs later in life. Adult neurogenesis does not build new structures but instead adds neurons to preexisting structures. This can result in the enlargement of the structure that receives the new neurons, but in other circumstances the new neurons may maintain structures by compensating for neuronal death.

There is no adult neurogenesis in *C. elegans*. Cell lineages in *C. elegans* are not open ended: the nematode builds its nervous system, then stops. The short-lived fruit fly also does not provide examples of adult neurogenesis, although both taxon and life span turn out to be poor predictors of whether a given species of insect adds new neurons to the adult brain. For example, crickets add new cells to their brains as adults, whereas long-lived honey bees do not.[17]

Adult Neurogenesis in Zebrafish

Zebrafish and likely most, if not all, other teleost fish have an anatomically widespread capacity to add new neurons to the adult nervous system.[18] Zebrafish produce new neurons along the length of the rostrocaudal brain axis for their entire lives. As in the embryo, a layer of cells associated with the ventricles is the site of proliferation. The ventral ventricular zone produces neurons that migrate into the major telencephalic nuclei and the olfactory bulb. The dorsal ventricular zone contains clusters of proliferating cells along its ventricular surface. Here the majority of the newborn cells express markers associated with glia, but glia marker-negative cells (putative neurons) are also present. These neurons are added to the already well-developed dorsal telencephalon.

Two features of adult neurogenesis in the zebrafish distinguish the phenomenon in this species from that observed in mammals and birds. First, unlike mammals and birds, fish continue to grow throughout their entire adult lives. This growth is likely responsible, at least in part, for the prevalence of adult neurogenesis in teleost fish. Second, zebrafish are notable for their capacity to respond to brain injuries with extensive neurogenesis. For example, zebrafish that received a spinal cord crush injury that resulted in the death of interneurons and motoneurons regained normal swimming function within a month of the injury.[18] This ability to recover from an injury that would be devastating in a human presumably reflects extensive neurogenesis and gliogenesis induced by the injury. Why the central nervous system of fish supports regeneration whereas the central nervous system of mammals does not is an interesting question that, when answered, may lead to new treatments for patients with spinal cord injuries. The complex sequence of events that precedes neurogenesis (disruption of the blood–brain barrier, invasion of macrophages, tissue edema, neuronal death, and proliferation of radial glial cells) means that recovery is unlikely to depend on a single cell type or a single molecule. Even small improvements in regenerative capacity could benefit human patients: that is, a complete cure for all spinal cord injuries may not be attainable, but in some cases it may be possible to use information gained from the study of adult neurogenesis in fish to tip the odds in favor of recovery of some function in humans.

As noted, the progenitor cells that account for adult neurogenesis in the zebrafish are located in the ventricular zone. A careful study of the zebrafish telencephalon that combined markers for glial cells with markers for mitosis and markers for neurogenesis revealed that three cell types are present in the ventricular zone. The majority of the cells were positive for glial markers but not for PCNA, an endogenous marker for mitosis (see Chapter 1). A much smaller group of cells (<10 percent) expressed glial markers and PCNA, whereas an even smaller group of cells expressed PCNA and markers for neurogenesis. A straightforward interpretation of these results is that the first group consists of nondividing radial glial cells, the second group consists of dividing radial glial cells, and the third group consists of direct neural progenitors. By themselves, however, even comprehensive studies based on marker expression cannot reveal if particular nondividing radial glial cells will reenter the cell cycle and under what conditions they are most likely to do so.

If common molecular mechanisms regulate neurogenesis in the embryo and the adult, a strong candidate for regulation of neurogenesis in the ventricular zone of the adult zebrafish is signaling via the Notch pathway. The set of vertebrate Notch receptors (*notch1a*, *notch1b*, *notch2*, and *notch3*) were found to be expressed in the ventricular zone of adult fish but not in cells expressing cell cycle markers. The many Notch ligand-encoding genes (*deltaA*, *deltaB*, *deltaD*, *delta-like4*, *jagged1a*, *jagged1b*, and *jagged2*) displayed the opposite pattern of expression, because they were expressed almost exclusively by dividing cells. If Notch signaling is functioning to suppress neurogenesis, blocking Notch activity will increase the number of dividing cells. Notch activity can be blocked either chemically (using an inhibitor called DAPT, which blocks the generation of the intracellular domain fragment capable of translocating to the nucleus) or genetically. In both cases, the number of dividing cells expressing glial markers increased significantly, in accord with this prediction. These studies provoke the idea that Notch-blocking drugs might be used to turn on neurogenesis to repair damaged brains, even in animals that normally show limited proliferation in adulthood, such as mammals.

Adult Neurogenesis in Mammals

In mammals, adult neurogenesis is restricted to two sites in the telencephalon: the subventricular zone (SVZ) and the subgranular zone of the hippocampus (SGZ) (fig. 5.14). In rodents, neurons born in the adult SVZ migrate into the olfactory bulb via a pathway called the rostral migratory stream (RMS). Neurogenesis in the SVZ is an ongoing process that results in a continuous stream of migrating neurons headed toward the olfactory bulb. The new cells are referred to as neuroblasts as they migrate, a potentially confusing terminology given the widespread use of the same term to refer to non-migrating neural progenitor cells in the *Drosophila* nervous system. Once the new cells arrive at the olfactory bulb, they move into the layers of the bulb, where approximately half (in mice) integrate into existing neural circuitry.

By contrast, neurons born in the SGZ do not migrate long distances. Instead, they mature locally within the dentate gyrus of the hippocampus (fig. 5.15). The percentage of neurons born in the SGZ that survive to become functional granule cells in the hippocampus varies. The list of factors that affect survival includes developmental signals such as Notch, synaptic signals such as glutamate, and experience-dependent signals such as exercise, learning, and stress.[20] The last category of regulatory factors is perhaps the most interesting; experience-based regulation of adult neurogenesis implies

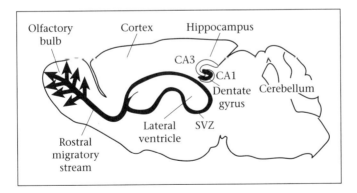

Figure 5.14. Adult neurogenesis in the mouse brain. Neurogenesis in the adult mouse brain is restricted to the subventricular zone (SVZ) of the lateral ventricle, which produces the neurons of the rostral migratory stream, and the subgranular zone (SGZ) of the hippocampus, which contributes new neurons to the dentate gyrus. In this side view of the adult mouse brain, anterior is to the left.

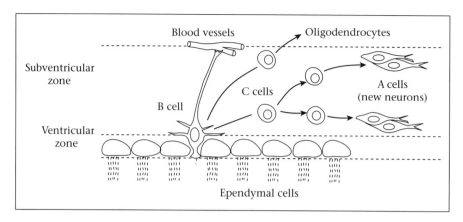

Figure 5.15. The subventricular zone (SVZ) of the adult mouse brain. The B cells found in the lateral ventricles are also called subventricular astrocytes. B cells typically project a long basal process that ends on a blood vessel. B cells divide asymmetrically to produce C cells. Some C cells divide to produce new oligodendrocytes, while others produce A cells (new neurons). Ciliated ependymal cells line ventricles of the brain, as shown. The ependymal cells interact with a network of blood capillaries called the choroid plexus (not shown) to form the cerebrospinal fluid. Drawing adapted from Kriegstein and Alvarez-Buyulla (2009).

that even the brains of identical twins will contain different populations of adult-born neurons, because no two individuals have identical experiences.

Understanding of adult neurogenesis in mammals is based primarily on studies of rodent brains, but there is no doubt that, despite the difficulty of studying the phenomenon, neurogenesis also occurs in the brains of adult humans. In a seminal paper published in 1998, Eriksson and colleagues published a study showing that the hippocampi of human cancer patients treated before death with BrdU contained BrdU-labeled nuclei in both the dentate gyrus and the subventricular zone.[21] Many of the BrdU-positive nuclei in the dentate gyrus were found in cells coexpressing neuronal markers; this double labeling provided essential evidence in support of the claim that the BrdU-positive cells were in fact neurons.

The patients in the Eriksson study received BrdU to help in the staging of their tumors and gave informed consent for the study of their brains after death. Since that time, however, no clear therapeutic benefit has been defined for BrdU, and its use for research in humans is no longer ethically acceptable.[22] Improved understanding of the cellular and molecular biology of adult neurogenesis, however, has led to identification of endogenous markers that can be used to track neurogenesis in post mortem brains. One such marker is DCX, a brain-specific microtubule-associated protein. This protein is the product of the doublecortin gene. The name of this gene refers to an inherited brain disorder of humans (double cortex syndrome) in which an abnormal band of neurons forms between the ventricle and the superficial cortical layers. Patients with severe cortical malformations resulting from mutations in this gene experience seizures and intellectual disability.[23] The extra band of neurons near the ventricle reflects a failure of neuronal migration during the embryonic formation of the cortex and, because of its association with microtubules, indicates the role of the cytoskeleton in migration. Because the DCX protein is expressed by newborn neurons as they begin the process of migration, antibodies to DCX can be used to study neurogenesis in the human brain without the complication of having to introduce a marker such as BrdU prior to death. A study of DCX immunolabeling in human autopsy specimens was performed on patients ranging in age from one day to 100 years old. Even the oldest brains in the study contained some DCX-positive cells in the hippocampus.[24] Although caveats apply to the interpretation of human post mortem tissue, these results are consistent with the view that hippocampal neurogenesis is present throughout adult life in humans. This is a case of a good match between what the mouse model tells us and events well documented to occur in the human brain.

There is, by contrast, little support for the persistence of a human RMS into adulthood. A comprehensive study of the SVZ in a sample of human brains ranging in age from birth to over 80 years showed that neurogenesis is rare in children older than 18 months and, for all intents and purposes, not present in adults.[25] These studies relied on immunolabeling and in situ hybridization to assess the proliferative status and expression of neuronal markers in the SVZ. Expression of Ki-67 (used as a marker for proliferation) and DCX (used as a marker for immature neurons) was detected only in the youngest brains studied. As in rodents, the RMS connected the human SVZ with the olfactory bulb, but this pathway was undetectable after 18 months of age.

An exciting and unexpected finding of this study was the discovery of a previously unknown branch of the RMS: the medial migratory stream or MMS. This branch connects the SVZ with the prefrontal cortex. Although the MMS disappeared after 6 months of age, perhaps this supplement to the histogenesis of the prefrontal cortex accounts in part for the expansion of the human prefrontal cortex relative to other mammals. Studies designed to identify an MMS in mice yielded negative results. It will be of great interest to determine if nonhuman primates have a branched or unbranched RMS. What is the functional significance of the prefrontal cortex? It is the region of the brain responsible for what neuroscientists like to call *executive* functions. Some of the cognitive tasks performed by the prefrontal cortex include planning, appropriate deployment of attention, and impulse control. In short, the prefrontal cortex is a major contributor to our human nature.

Functions and Risks of Adult Neurogenesis

Rodent models have been used to explore the many external factors that regulate adult neurogenesis. The list of such factors is so long that it is difficult to say more at present than that *the rodent brain certainly is plastic*. Studies that assess rate of precursor cell proliferation, survival of newborn neurons, and phenotypic differentiation have revealed that few, if any, experience-based factors affect the rate of neuronal production, which tends to be slow and steady. Many factors, however, have been shown to influence the proportion of neurons produced that survive long enough to become part of functioning neural circuits. Mice (and rats) can also be used to demonstrate the relevance of adult neurogenesis to behavior and cognition. Studies that blocked adult neurogenesis have reported impaired performance on hippocampus-dependent spatial learning and novel object recognition tasks.[26] The impairments are often subtle in that the mice can usually still perform the task but on average do a little less well. It's hard to say what this means to a mouse, but a possible human analogy might be the difference between an A and a B performance on an exam, a difference of keen importance to many students.

The questions raised by the restricted pattern of adult neurogenesis in the mammalian brain may best be addressed by an evolutionary approach. For example, it has been proposed that the reduction of cell proliferation in the nervous system of adult mammals relative to that seen in fish results in a decreased risk of tumor formation. Such scenarios are of more than theoretical interest if therapeutic interventions that boost adult neurogenesis in human patients have the side effect of increased risk for cancer.

Notes

1. Bats are the best-known exception to the rule that neurogenesis occurs in the adult mammalian nervous system. See Amrein et al. (2007).

2. The easiest way to access this information is online via WormBook.org.

3. In insects, imaginal discs are sacs of undifferentiated cells present in embryos and larvae that serve as the source of cells that, during metamorphosis, proliferate to form adult-specific structures such as wings, legs, genitalia, and compound eyes.

4. "Round up the usual suspects" is a key line of dialogue from the Academy Award–winning 1942 film *Casablanca*. It was spoken by Claude Rains in his role as Captain Louis Renault, the French prefect of police in the Vichy-controlled Moroccan city of Casablanca.

5. See Taghert et al. (1984).

4. A neurofilament is an intermediate filament (cytoskeletal protein) found specifically or primarily in neurons. These filaments are important for axonal growth and the regulation of axonal diameter.

5. GABA (gamma-aminobutyric acid) is the major inhibitory neurotransmitter in animal nervous systems.

6. By comparison, the brain of an adult worker honey bee contains approximately 1 million neurons; the central nervous system of an octopus (including the brain, optic lobes, and nervous system of the arms) contains approximately 500 million neurons.

7. See Gohlke et al. (2008) for an example of a combined bioinformatics and microarray study.

8. This method is called RNA-seq despite the fact that it is cDNAs (reverse transcribed from a population of RNAs) that are sequenced. This method uses high-throughput sequencing to generate a list of RNAs present in a sample. This list is called the transcriptome. RNA-seq is also called whole-transcriptome shotgun sequencing.

9. An adherens junction is a complex of proteins that connect adjacent cells in epithelial tissues. Cadherins and catenin-family proteins form the junctional complex.

10. See Götz and Huttner (2005).

11. In the United States, the National Institutes of Health (NIH) Revitalization Act of 1993 regulates many aspects of fetal tissue research. The text of this legislation is available online through the Library of Congress THOMAS page. See the link provided in the online resources for this chapter.

12. For an example of the application of this technique, see Hansen et al. (2010).

13. Gyrencephalic brains have a cortex characterized by folds and convolutions. Humans, other primates, and dolphins have gyrencephalic brains, as do many other large animals. The opposite of a gyrencephalic brain is one with a smooth cortex. Such a brain is called lissencephalic. Small rodents have lissencephalic brains.

14. Patients with microcephaly vera have a small but *architecturally normal* brain. Some degree of intellectual disability is usually present. See Woods et al. (2005) for additional information on clinical features of microcephaly vera.

15. See Thornton and Woods (2008) for additional references.

16. See Kempermann (2011).

17. For many comparative references, see Cayre et al. (2002).

18. For a comprehensive review of adult neurogenesis in zebrafish, see Kizil et al. (2012).

19. See Hui et al. (2010).

20. See the *cloud of regulation* in figure 1 of Kempermann (2011) for a summary of the many factors demonstrated to regulate adult neurogenesis.

21. A seminal paper is one that becomes a classic in its field, often because it changes the way a particular scientific community thinks about a topic. A seminal paper is typically highly cited, even if it becomes less and less read as the years pass. The article by Eriksson et al. (1998) is an excellent example of a seminal paper, because it provided the first demonstration using a modern cell-labeling technique that new neurons are born in the adult human brain.

22. See Curtis et al. (2011).

23. See the link to rare diseases provided in the online resources to learn more about double cortex syndrome in humans.

24. See Knoth et al. (2010).

25. This comprehensive study analyzed tissue removed from human brains during neurosurgery in combination with autopsied brains. Numerous antibodies were used to characterize the SVZ and the RMS, including antibodies to Ki-67, Doublecortin, and the glial markers glial fibrillary acidic protein (GFAP), vimentin, and polysialylated neural cell adhesion molecule (PSA-NCAM). See Sanai et al. (2011). Glial markers are discussed in Chapter 8.

26. For example, see Lafenêtre et al. (2010). In this study, adult neurogenesis in the hippocampus was reduced by a genetic manipulation (mice expressed activated Ras under the direction of the neuronal Synapsin I promoter); neurogenesis was as-

sessed by incorporation of BrdU. The test of memory was the ability to recall an object examined ten minutes before the test. The exercise of running in a wheel rescued both hippocampal neurogenesis and novel object recognition.

Investigative Reading

1. The doublecortin gene in humans encodes a microtubule-associated protein expressed during neurogenesis. The protein is called double cortex (DCX). Antibodies that recognize this protein to can be used as markers for the recent occurrence of neurogenesis because they label young neurons. It is also known that mutations in this gene, located on the X chromosome, cause lissencephaly in males and double cortex syndrome in heterozygous females. You generate a doublecortin knockout mouse to serve as a model for study of the human disorder. What research method do you use to show that your knockout is a true knockout? How do you study cortical layering in your mutant mice? What do you predict is the phenotype of your knockout?

Corbo, Joseph C., Thomas A. Deuel, Jeffrey M. Long, Patricia LaPorte, Elena Tsai, Anthony Wynshaw-Boris, and Christopher A. Walsh. 2002. *Journal of Neuroscience* 22: 7548–57.

2. Notch and its ligands, including Delta, are broadly expressed in the developing fly nervous system. The effects of Notch activation are often dependent on the molecular context (i.e., what other molecules are present). In *Drosophila,* the glycosyltransferase enzyme Fringe glycosylates sites on the Notch extracellular domain, thereby altering the ability of Notch to bind to its ligands. The effect of Fringe activity is predictable but variable: Delta binds with higher affinity to Fringe-modified Notch, but binding of other ligands to Notch is decreased. Fringe modification of Notch plays a role in establishing compartment boundaries and gene expression stripes in the developing fly nervous system. The vertebrate homolog of Fringe is called lunatic fringe (Lnfg). Injected morpholino oligonucleotides were used to knock down expression of Lnfg in zebrafish embryos (30 and 36 hpf, hours post fertilization). Lnfg knockdown resulted in an increased number of cells initiating neurogenesis in the developing neural tube. You want to know if Lnfg is coexpressed with *deltaA,* the zebrafish Notch ligand that is expressed at low levels in progenitors and upregulated in differentiating neurons, but you do not have access to antibodies specific for the Lnfg and deltaA proteins. What method do you use to determine expression of these genes at a cellular level of resolution?

Nikolaou, Nikolas, Tomomi Watanabe-Asaka, Sebastian Gerety, Martin Distel, Reinhard W. Köster, and David G. Wilkinson. 2009. *Development* 136: 2523–33.

3. Despite the many advantages of using mice as models to study development and disease, some common human afflictions do not occur spontaneously in rodents. This is true of certain types of brain tumors. In this regard, dogs and cats are more similar to humans in terms of tumor incidence, tumor-induced changes in the histology of the brain, and response to treatments such as radiation or chemotherapy. One potentially important side effect of cancer treatment on the human brain is disruption of adult neurogenesis. You propose to develop a dog model to study the consequences of loss of adult neurogenesis, and you decide to focus on the RMS. There is, however, no published documentation that the RMS even exists in dogs. Outline a research strategy that could be used to establish the existence of the RMS in dogs.

Malik, Saafan Z., Melissa Lewis, Alison Isaacs, Mark Haskins, Thomas Van Winkle, Charles H. Vite, and Deborah J. Watson. 2012. *PLoS ONE* 7: e36016.

4. *Challenge Question:* In the developing *Drosophila* nervous system, the presence of Numb protein inside a cell inhibits Notch signaling and therefore blocks the cell that contains Numb protein from differentiating a Notch-dependent phenotype. This makes it possible for one mitotic division of a neural progenitor cell to produce two different types of neurons: if one daughter receives Numb and the other does not, only the latter adopts the Notch-dependent fate. A third gene, called *sanpodo,* is expressed in both of the daughter cells produced in an asymmetrical division. Develop a model of the interactions among the *Notch, Numb,* and *sanpodo* genes and their gene products that takes into account the following experimental observations: (1) in flies that do not express the *sanpodo* gene, none of the daughter cells resulting from asymmetric divisions adopt the Notch-dependent fate; (2) expression of *sanpodo* in epithelial cells undergoing Notch-mediated lateral inhibition increase the number of bristle sensory organs relative to the wild type; (3) flies engineered to overexpress Notch no longer require *sanpodo* to produce daughter cells with a Notch-dependent fate; and (4) the effectiveness of Notch overexpression in converting cells to the Notch-dependent fate is significantly enhanced in flies that do not express the *sanpodo* gene.

Babaoglan, A. Burcu, Kate M. O'Connor-Giles, Hemlata Mistry, Adam Schickedanz, Beth A. Wilson, and James B. Skeath. 2009. *Development* 136: 4089–96.

SIX

Later Events

Not All Animals Are Segmented

Now it is time to say a temporary goodbye—really more of a "see you later"—to *C. elegans*. This is not unfair, because nematodes are unsegmented, and the first key event discussed in this chapter is regionalization of the developing nervous system by acquisition of segmental identity. The second topic of this chapter, the formation of laminar structures in the nervous system, uses the cerebral cortex as an example and therefore focuses on the mammalian brain.

Regionalization in the *Drosophila* Nervous System

To recap, in the course of Chapter 4 the *Drosophila* embryo defined its anterior–posterior axis through the actions of *bicoid* and a group of terminal genes. Now, with neurogenesis under way (Chapter 5), it is time to fill in the spaces between the head and the tail with appropriate thoracic and abdominal segments. (The body of the adult fly has three thoracic, eight abdominal, and one terminal abdominal segment in addition to its head, which is formed by the fusion of seven segments.) The process of segmentation precedes the acquisition of segment identity (segment differentiation). Initiation of segmentation is one of the final tasks undertaken by the maternal genes. The process begins in the syncytial blastoderm and continues beyond the transition to the cellular blastoderm. Three broad categories of genes regulate segmentation: gap genes, pair-rule genes, and segment polarity genes. If you have ever seen a striped *Drosophila* embryo, you were almost certainly looking at the expression profile for one or more segmentation genes.[1]

The experimental data confirming the basic mechanisms of segmentation were published in the 1990s, but the story begins a decade earlier with a 1980 article in the journal *Nature* by Christiane Nüsslein-Volhard and Eric Wieschaus titled "Mutations affecting segment number and polarity in Drosophila."[2] In this article the authors reported the results of a project designed to identify mutations (and gene loci) that impact segmental patterning. They attributed mutations that resulted in the loss of a region of the body to a category of genes called gap genes, mutations that caused problems in development of alternating segments to pair-rule genes, and mutations that caused defects in every segment to segment polarity genes (fig. 6.1). All of the genes in these categories are zygotic genes, and they are expressed in a temporal hierarchy: gap gene expression precedes (and is required for) pair-rule gene expression, which in turn precedes (and is required for) segment polarity gene expression.

Gap Genes

Gap gene expression begins before cellularization of the blastoderm (Chapter 3). Gap genes are expressed in separate domains (each larger than the segments that will eventually form) along the anterior–posterior axis. Gap gene expression is either activated or repressed by maternal genes: the earlier gradients of gene expression that characterize the young embryo (e.g., *bicoid* in the head, *caudal* in the tail) organize the pattern of gap gene expression. There is experimental evidence for 14 *Drosophila* gap genes.[3]

Three of the best-studied gap genes are *hunchback, krüppel,* and *giant. Hunchback* and *krüppel* were identified in the screen performed by Nüsslein-Volhard and Wieschaus; *giant* was identified in a follow-up screen focused on the X chromosome.[4] In *hunchback* mutants, the head and thorax are missing and abdominal segments 7 and 8 fuse; mutations of *krüppel* result in deletion of all three thoracic and five anterior abdominal segments, which are replaced by a mirror-image duplication of the posterior abdomen; mutations of *giant* cause the loss of abdominal segments 5–7 plus defects in the head. All three of these genes encode transcription factors.

Regulation of *giant* serves as our example of how gap genes work. At the head end of the embryo, expression of high levels of *bicoid* and *hunchback* gene products induces anterior expression of *giant.* By contrast, near the tail, *giant* gene expression is activated by caudal protein. The resulting two stripes of *giant* expression predict the phenotype of loss-of-function mutants, which show defects in the head and the absence of several posterior abdominal segments.

Gap gene stripes have sharp boundaries, although the patterns of gap gene expression can overlap. The boundaries are the result of interactions between pairs of gap genes involving mutual repression of gene expression. Typically, the interacting pairs of gap genes are not those expressed in adjacent stripes. For example, *giant* gene expression inhibits the expression of *krüppel,* which is not expressed by any adjacent nuclei.

Pair-Rule Genes

Expression of pair-rule genes is first evident during cellularization of the embryo. The expression of the pair-rule genes results in formation of the zebra stripe scaffold that patterns the segments as they develop.[5] Nine genes with

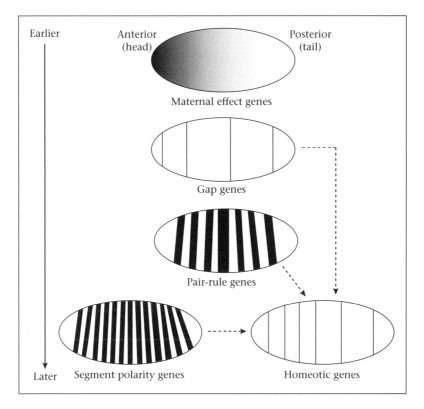

Figure 6.1. Expression of segmentation genes in *Drosophila* embryos. These schematic depictions are based on experimental analyses of gene expression using in situ hybridization and immuno-labeling. The broadly expressed maternal effect genes activate gap genes, which in turn activate the expression of pair-rule genes in periodic stripes. This results in the subdivision of the embryo into alternating stripes of segment polarity gene expression. As the dashed arrows indicate, the patterned expression of the homeotic genes is regulated by the combined actions of the gap genes, the pair-rule genes, and the segment polarity genes. Note that the bands are sometimes defined by overlapping combinations of gene expression and sometimes by discrete patterns.

pair-rule properties have been described. Note that there is an imperfect alignment between the segments visible on the outside of the fly body and the gene expression–defined segmentation of internal structures such as the ventral nerve cord. The patterns defined by segmentation gene expression have been named parasegments to distinguish them from the visible body segments. Parasegments form earlier than visible segments, and they join the posterior half of one segmental compartment with the anterior half of the adjacent posterior segment.[6] To keep things simple, we will refer to parasegments as segments in our subsequent discussion of how pair-rule and segment polarity genes regionalize the developing *Drosophila* nervous system (fig. 6.2).

Even-skipped serves as our example of a pair-rule gene. It is considered a primary pair-rule gene because of its early expression and its pattern of seven

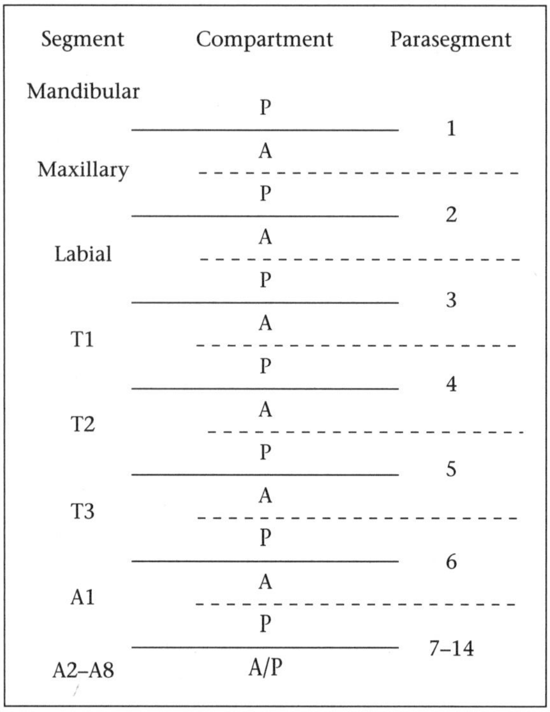

Figure 6.2. Parasegments in *Drosophila* embryos. The familiar segmented insect body plan of head segments, thoracic (T) segments, and abdominal (A) segments is clearly evident in *Drosophila* embryos, but the visible segmental divisions of the epidermis are not aligned with underlying patterns of gene expression, which reveal that the fundamental unit of organization is the parasegment: the posterior compartment of one segment and the anterior compartment of the next posterior segment. Compartment abbreviations: A, anterior; P, posterior.

Segment	Compartment	Parasegment
Mandibular	P	
		1
	A	
Maxillary	– – – – – – –	
	P	
		2
	A	
Labial	– – – – – – –	
	P	
		3
	A	
T1	– – – – – – –	
	P	
		4
	A	
T2	– – – – – – –	
	P	
		5
	A	
T3	– – – – – – –	
	P	
		6
	A	
A1	– – – – – – –	
	P	
		7–14
A2–A8	A/P	

stripes and also because its pattern does not depend on the prior expression of any of the other pair-rule genes. Even-skipped protein functions as a transcriptional repressor. The promoter region of the *even-skipped* gene contains binding sites for bicoid protein as well as for gap gene proteins. The combined actions of these genes regulate the expression of *even-skipped* and, because their expression varies along the anterior–posterior axis of the embryo, the position of the resulting stripes of even-skipped protein. This type of regulation of gene expression is often referred to as combinatorial, a term used by mathematicians to refer to the arrangement of elements in sets. In biology, this term or the related term *combinatorial code* refers to regulation of gene expression by a particular combination of proteins. Careful analysis of numerous segmentation gene mutants revealed that the combinatorial code for each of the even-skipped stripes is unique. Once pair-rule genes are expressed, one of the mechanisms that maintains pair-rule gene expression is positive feedback. For example, even-skipped proteins have been shown to bind to sites in the even-skipped promoter.[7]

Segment Polarity Genes

The expression of pair-rule genes results in formation of 14 stripes defining repeated units. Now we are poised for expression of the third category of segmentation genes: the segment polarity genes. Expression of the segment polarity genes is characteristic of the cellular blastoderm stage. The result of segment polarity gene expression is definition of anterior and posterior compartments (segment polarity) in each of the segments (see fig. 6.2). *Engrailed* is the canonical segment polarity gene of *Drosophila*. It functions in tandem with *invected,* another segment polarity gene, but for simplicity the present account focuses on *engrailed.*

The *engrailed* gene encodes a transcription factor. Expression of the long list of genes known to be downstream from *engrailed* defines the posterior compartment in each repeated unit. Mutations that do not express *engrailed* have a striking defect: the entire segment adopts anterior cell fates.[8] The within-segment expression of *engrailed* is regulated by the pair-rule gene *even-skipped:* expression of the *engrailed* gene is high only in *even-skipped*-expressing cells. The expression of *engrailed* by posterior compartment cells activates transcription of the *hedgehog* (*hh*) gene in those cells. As we shall soon see, hedgehog protein is a secreted morphogen required for many subsequent events in development. We therefore need to give hedgehog signal transduction pathways our close attention.

Hedgehog was discovered in the 1980 Nüsslein-Volhard and Wieschaus screen for body plan mutants in *Drosophila*. The name reflects the phenotype of defects in embryonic cuticle that results from disrupted expression of *hedgehog*. Embryonic cuticle is secreted by a monolayer of epidermal cells. There are two types of epidermal cells: one type secretes smooth cuticle, while the other secretes short, hairlike structures called denticles. Different categories of denticles can be identified on the basis of size and orientation, and the position of particular denticles on the cuticle in each body segment is stereotyped. Mutations that severely reduce or eliminate the activity of *hedgehog* are embryo-lethal, but before the embryo dies it secretes an abnormal, unsegmented cuticle characterized by numerous incorrectly oriented, pointy denticles. Such embryos slightly resemble European hedgehogs.[9]

The protein encoded by *hedgehog* is now known to be a member of a family of proteins also referred to as hedgehog (hh). Multiple homologs of *Drosophila* hedgehog are present in vertebrates. Secreted hedgehog protein binds to a plasma membrane receptor called Smoothened. When Smoothened is activated by hedgehog, a protein called cubitus interruptus enters the nucleus and activates hedgehog-regulated genes. One gene activated by hedgehog via cubitus interruptus signaling is Wingless, the *Drosophila* Wnt protein. Like hedgehog, Wingless is a secreted morphogen. As previously described (Chapter 4), Wnts signal through Frizzled receptors to activate β-catenin. In Wingless target cells, β-catenin activates expression of Engrailed and hedgehog. The result is a signaling loop that stabilizes the transcriptional activities of cells with either anterior or posterior fates. Gradients of hedgehog and Wingless in each segmental compartment permit differentiation within the larger longitudinal body plan.

We could stop right here if all it took to make a fruit fly was a chain of equivalent body segments. But the process already described is insufficient to produce anything resembling an actual fly. As was noted earlier (and as we have all known since we studied butterflies in primary school), the insect body is divided into three parts: head, thorax, and abdomen. Each part has characteristic structures and functions, including segment-specific populations of neurons. How do we go from a chain of stripe-based, more or less equivalent segments to segments appropriate for the different body parts?

The answer is that the cells in the developing segments, including the cells that will form the nervous system, possess molecular addresses. Such an answer might seem to simply kick the can on down the road—instead of definitively answering questions regarding the origin of differentiated body parts, we instead say that the best we can do is to identify a set of regulatory genes (the molecular addresses) for which we will, in turn, need to identify downstream targets, many of which are also likely to be regulatory genes. But when we consider the complexity of the final product of a fertilized egg—an animal with a nervous system!—in the context of the tools available to the organism (genes and proteins), it quickly becomes apparent that the answer could hardly be otherwise. The good news for the student is that the common origin of the animals means that the same tools are used over and over again in all different types of animal embryos.

Homeodomains and Hox Genes

Many of the genes regulated by gap genes and pair-rule genes are members of the homeodomain (HD) gene family. Homeodomain proteins are transcription factors that share a conserved DNA-binding domain, the homeodomain. The homeodomain is encoded by a 180–base pair sequence called the homeobox. The term homeobox is often contracted to Hox. Another commonly used term for homeodomain genes is Hox genes.

Hox genes and their roles in development were discovered in *Drosophila*. In fruit flies, mutation of Hox genes results in dramatic phenotypes involving transformation of one body part to another. For example, fruit flies, like all other dipteran insects, have a single pair of wings: famously, loss of function mutations in a Hox gene called *Ultrabithorax* result in development of a second pair of wings.[10] Another well-known example of body part transformation involves the Hox gene called *Antennapedia*. In *Antennapedia* mutants, the antennae are transformed into a pair of legs. The general term for a gene that transforms one body part into another when mutated is homeotic gene.

The terminology can be confusing because homeotic genes are Hox genes, but not all Hox genes are homeotic genes. This means that some Hox genes have functions other than regulation of segment identify. For now we focus on the Hox genes that are also homeotic genes. In *Drosophila*, two gene complexes contain most of the known homeotic genes. These are the *Anten-*

napedia complex and the *bithorax* complex. These complexes are both found on Chromosome 3 and are sometimes referred to collectively as *the* homeotic complex. The homeotic Hox genes are expressed in the embryo in a linear array of domains along the anterior–posterior body axis. Remarkably, the order of the Hox gene sequences along Chromosome 3 matches the order of Hox gene expression in the body! For example, the three genes of the *bithorax* complex are *Ultrabithorax (Ubx)*, *abdominal A (abdA)*, and *Abdominal B (AbdB)*. On Chromosome 3, *Ubx* precedes *abdA,* which in turn precedes *AbdB*. The same sequence—*Ubx* gene expression followed by *abdA* gene expression followed by *AbdB* gene expression—is seen in the developing abdomen, with *Ubx* expressed at the boundary between the thoracic and abdominal compartments and *AbdB* expressed in the posterior compartment. The same correspondence between position on the chromosome and position in the body is also characteristic of the other homeotic Hox genes (fig. 6.3).

Because homeotic Hox genes are transcription factors, understanding acquisition of segmental identity requires identifying the genes that have their expression regulated by Hox genes. The identification of genes downstream of Hox genes has proved challenging. Consider the following: the homeodomain of Hox proteins always binds to a specific DNA sequence called the Hox response element, or HRE. But the highly conserved homeobox sequences, regardless of the particular gene under consideration, in turn encode highly conserved homeodomain sequences, which in turn bind to the same HREs. This means that the homeotic Hox proteins likely acquire specificity of transcriptional regulation via interactions with other DNA-binding factors. Such complex transcriptional regulatory mechanisms are not unusual. In such a situation, however, the biologist cannot rely solely on

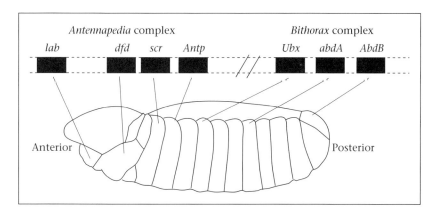

Figure 6.3. The *Antennapedia* and *bithorax* complexes in *Drosophila.* The linear order of the genes on the chromosome matches the order (from anterior to posterior) in which their transcripts are expressed in the body. Gene symbols: *lab, labial; dfd, deformed; scr, sex combs reduced; Antp, Antennapedia; Ubx, Ultrabithorax; abdA, abdominal A; AbdB, Abdominal B.* Drawing adapted from Gilbert (2010), 234.

straightforward bioinformatics-based strategies of identifying genes downstream of HREs. Instead, the effects of Hox gene expression on transcription must start with samples from living tissue. Fortunately, tools that can be used for such analyses are now available. For example, a recent search for direct transcriptional targets of the Ubx protein combined a yellow-fluorescent protein-tagged protein trap line of *Drosophila* with chromatin immunoprecipitation and microarray analysis of gene expression.[11] The result was the identification of over a thousand genes associated with Ubx binding.

Homeotic mutations are most easily identified in epidermal structures. This reflects not only the obvious nature of appendage transformations (e.g., a leg where we expect an antenna) but also the visible patterning of the cuticle with denticles and sense organs. The central nervous system of fruit flies is also segmented in terms of its neuronal populations. The embryonic segments of the nervous system are referred to as neuromeres to differentiate them from the visible epidermal aspects of the body segments. One might reasonably ask if the homeotic genes expressed in epidermal structures are also expressed in the neuromeres. It is possible that the acquisition of segmental identity in the nervous system follows (and depends on) the earlier acquisition of segmental identity by the epidermis. Many studies, however, have shown that acquisition of segmental identity by the neuromeres is an autonomous event that requires expression of the homeotic homeodomain genes in the nervous system as well as in the adjacent epidermis.[12]

Regionalization in the Vertebrate Nervous System

Can we apply our knowledge of fruit fly segmentation mechanisms to understand the origins of segmented structures in other animals, including vertebrates? To a surprising extent, the answer is yes. In this case, however, our starting point is the undifferentiated neural tube, not a syncytial blastoderm. The forming central nervous system already knows its head from its tail (Chapter 4): as cell proliferation commences (Chapter 5), it must sort out dorsal from ventral and make the neurons and glia appropriate for the different levels of the nervous system, from forebrain to the caudal spinal cord. The signaling systems responsible for both types of developmental decisions are well characterized in vertebrates. Both were first described in *Drosophila*, and both have already been introduced in this chapter. The signal for ventralization is a member of the Hedgehog gene family called Sonic hedgehog (*Shh*); the signals for segmental identity are homeodomain proteins.

The division of the neural tube into dorsal and ventral compartments is easily appreciated in the context of spinal cord development (fig. 6.4). When viewed in cross section, the gray matter at the center of the spinal cord resembles the outline of a butterfly. The upper (dorsal) pair of wings contains sensory interneurons; the lower (ventral) wings contain the motoneurons.

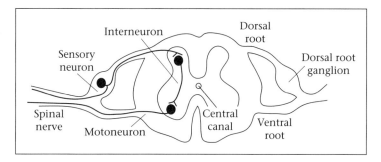

Figure 6.4. Organization of the vertebrate spinal cord. A simple neural circuit is depicted. Incoming (sensory) and outgoing (motor) axons fuse to form the spinal nerves. The dorsal root ganglia contain the somata of sensory neurons. Motoneuron somata are found in the ventral horn (*hindwing*) of the butterfly-shaped central gray region; the somata of interneurons are located in the dorsal horn. The surrounding white matter contains the myelinated axons of spinal neurons. The cerebrospinal fluid–filled central canal runs the length of the spinal cord; it is continuous with the ventricular system of the brain. All levels of the spinal cord have the same basic dorsal-ventral organization.

The first step in the formation of the dorsal and ventral compartments of the spinal cord is the flattening of the ventral cells of the neural tube into a structure called the floor plate. As neurogenesis proceeds (Chapter 5), the floor plate and adjacent regions produce motoneurons. The question that must now be asked is *what signal induces the floor plate?*

Studies on development of the spinal cord in amphibian embryos performed during the early twentieth century identified the notochord as the source of the floor plate–inducing signal.[13] The notochord is well known to biology students as one of the defining features of the phylum Chordata.[14] It is not a part of the nervous system—in fact, it is not even ectoderm. The notochord is a rod of mesoderm that forms during gastrulation. It defines the long axis of the body and is located dorsal to most structures but, importantly, lies just ventral to the neural tube. If the notochord is removed, motoneurons are not produced; conversely, if a supernumerary notochord is added dorsal to the neural tube, a second floor plate is formed and motoneurons are produced instead of sensory neurons. In the early 1990s, investigators inspired by studies in *Drosophila* searched for vertebrate homologs of *Hedgehog* by screening zebrafish and rat embryonic cDNA libraries for similar sequences. These screens were successful. The sequences retrieved in these screens were used to localize *Hedgehog*-like mRNA in embryos by in situ hybridization. Studies of rat embryos showed abundant expression of *Hedgehog*-like mRNA in the notochord at the time of floor plate induction.[15] Expression of the rat gene encoding a vertebrate *Hedgehog* in frog embryos (achieved by injecting rat cDNA directly into frog embryos) resulted in ectopic expression of floor plate markers.

Hedgehogs as Morphogens

We now know that, as in *Drosophila,* Hedgehog proteins act as morphogens in the developing vertebrate nervous system. As in *Drosophila,* vertebrate Hedgehogs bind to Smoothened-type receptors (Patched is a specific vertebrate receptor for Sonic hedgehog), in turn activating transcription factors that control transcription of Hedgehog-responsive genes. But, as is typical when we make vertebrate–*Drosophila* comparisons, vertebrate genomes contain multiple homologs of the Hedgehog signaling pathway components. I should also note that the actions of Hedgehogs in vertebrate development are not restricted to the nervous system. Although these topics won't be covered further here, Hedgehog signaling is also essential for development of the heart, gut, bladder, prostate, pancreas, gonads, taste buds, ears, teeth, and bones.[16]

Three subgroups of vertebrate Hedgehogs have been defined: Sonic hedgehogs, Desert hedgehogs, and Indian hedgehogs. Birds and mammals, including mice and humans, have one gene in each of the subgroups, but zebrafish have additional Hedgehog homologs (one in the Sonic hedgehog group and two in the Indian hedgehog group) as a result of genome duplication during evolution of their lineage. Despite the charm (at least to some) of the *original vertebrate hedgehog* gene names, usage in the literature is shifting to the simpler designation of *Shh* genes. As appropriate, both the original names and their streamlined versions will be given in this account.

Sonic hedgehog is the best studied of the vertebrate Hedgehog homologs because of its role in dorsal–ventral patterning of the nervous system. In mice, Sonic hedgehog is expressed in the notochord between 8 and 11 dpc (Theiler Stages 12–18). Knockout mice that lack all Sonic hedgehog signaling die at birth. As would be expected from the widespread requirement of *Sonic* hedgehog in developing tissues, knockout embryos display many defects, and their growth is retarded. Detailed analysis of neural tube development in the knockouts provided strong evidence that Sonic hedgehog secreted by the notochord induces differentiation of ventral cell types, starting with the floor plate, as no spinal motoneurons were produced in the knockouts.[16] Subsequent studies have shown that abnormal Sonic hedgehog signaling during human development is associated with numerous birth defects, including holoprosencephaly (described in Chapter 4) and polydactyly (extra digits on the hand or foot). Note that abnormal signaling may result not only from mutations in the Sonic hedgehog gene itself but also from mutations in other pathway components, such as the Patched receptor or Sonic hedgehog–activated transcriptional regulators. It is unlikely that a human embryo with a null mutation in the Sonic hedgehog gene or its receptor could survive.

What about zebrafish, whose genome encodes not one but two genes in the Sonic hedgehog family: Sonic hedgehog itself and Tiggy-Winkle hedgehog? A detailed analysis of the expression of these genes (now referred to as *Shh a* and *Shh b*) by in situ hybridization revealed that both are expressed early in embryonic life in association with the neural tube.[18] Tiggy-Winkle hedgehog is expressed slightly earlier (at 50 percent epiboly, starting in the embryonic shield) than Sonic hedgehog (at 60 percent epiboly). Both are eventually expressed along the anterior–posterior axis, but the two patterns of expression are subtly different: Sonic hedgehog mRNA is present in the notochord and the ventral region of the neural tube, whereas Tiggy-Winkle hedgehog expression is restricted to the ventral neural tube. By the end of somite formation, the floor plate is evident, and both genes are expressed in that tissue. Separate injections of mRNA of Sonic hedgehog and Tiggy-Winkle hedgehog yielded similar outcomes: in both cases, ectopic expression of ventral markers was induced. Both of the zebrafish Sonic hedgehog genes likely function to establish the dorsal–ventral axis of the developing nervous system; as in mammals, these genes also regulate many other aspects of development, including the development of eyes and scales.

In all cases, Hedgehog gene products function as secreted proteins that act as morphogens (fig. 6.5). The evidence for this mode of action is strong. For example, a knock-in mouse line was created in which the wild-type Sonic hedgehog gene was replaced with a Sonic hedgehog::GFP fusion gene designed so that the protein product would fluoresce after posttranslational processing and secretion. The pattern of fluorescing GFP was compared with the distribution of mRNA for the fusion gene: in this way, the site of Sonic hedgehog production and its subsequent location in embryonic tissue could be independently assessed. Antibody markers were used to identify populations of neural (Chapter 5) and glial (Chapter 8) progenitor cells. As expected, Sonic hedgehog mRNA was first detected in the notochord (E8.5) and then in the notochord and ventral midline of the neural tube (E9.5), indicating that the floor plate had been induced. At E8.5, fluorescent GFP was confined to the notochord, but at E9.5 and E10.5 a clear gradient of fluorescent GFP was evident, coincident with the emergence of the ventral pattern of cell populations, including the compartment called pMN that is the source of the ventral motoneurons.

Hox genes also regulate the developing vertebrate brain. As described in Chapters 2 and 3, the hindbrain or rhombencephalon develops as a set of repeated segments called rhombomeres (fig. 6.6). Rhombomeres are similar to the neuromeres of the embryonic *Drosophila* nervous system in that they are transient structures that are not visible in the mature hindbrain. Like the *Drosophila* neuromeres, they leave behind rhombomere-specific sets of neurons. How the initially uniform epithelium of the neural tube is transformed into rhombomeres has been intensively studied as a model of how parts of the vertebrate brain acquire regional identity.

The structure of the mature hindbrain reflects its origin in seven consecutive rhombomeric compartments. These are designated r1, r2, r3, and so on, working backward from the posterior border of the midbrain. The distinctive identity of each rhombomere is particularly evident in its complement

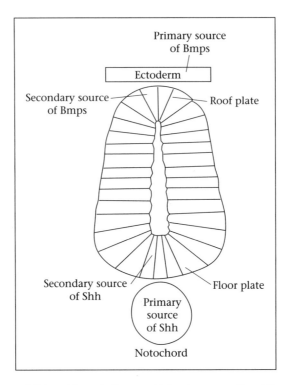

Figure 6.5. Sources of Shh and Bmps in the neural tube. A cross section of the neural tube and adjacent structures is shown. The secretion of Shh by the adjacent notochord, a mesodermal derivative, creates a ventral-high/dorsal-low gradient of Shh in the neural tube *and* induces the synthesis and secretion of Shh by the cells of the floor plate. The secretion of Bmps by the adjacent epidermis, which is an ectodermal derivative, creates a dorsal-high/ventral-low gradient of Bmps in the neural tube *and* induces the synthesis and secretion of Bmps by the cells of the roof plate. The divisions of the neural tube along the dorsal-ventral axis indicate the location of different progenitor cell domains. The different domains express different combinations of transcription factors and consequently produce different populations of neurons.

of motoneurons. For example, the axons that control the eye muscles form the abducens nerve (the VIth cranial nerve); the motoneurons that give rise to these axons are found in r5. Other rhombomeres contain other populations of motoneurons.

As we might expect from the description of segmentation in *Drosophila*, patterns of gene expression mark the rhombomeric boundaries. The most important of these genes are the Hox genes, expressed in predictable, partially overlapping patterns in the developing hindbrain. The aforementioned abducens motoneurons arise in reflection of the simultaneous expression of Hox genes from groups 2 and 3. r5 is the only rhombomere with this pattern of Hox gene expression. Deletion of specific Hox genes results in fusion of the developing rhombomeres and a loss of rhombomeric identity that causes

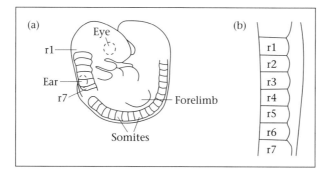

Figure 6.6. Rhombomeres in the hindbrain of a 9.5 dpc mouse embryo (a). The vertebrate hindbrain is a segmented structure formed of units called rhombomeres (b). The different cranial nerves (not shown) are associated with different specific rhombomeres: for example, the Vth nerve projects from the first, second, and third rhombomeres (r1, r2, and r3). The borders of Hox gene expression mark the boundaries of the rhombomeres. The Hox genes of mice are on separate chromosomes; their linear order on their respective chromosomes is co-linear with their pattern of expression in the developing rhombomeres. For a view of rhombomeres in the chicken embryo, see figure 2.5. Drawings adapted from Alexander et al. (2009).

defects in cranial nerve development. Remarkably, as is the case in *Drosophila*, the Hox genes of vertebrates are organized as linear arrays (3′–5′) on chromosomes that match the anterior-to-posterior pattern of gene expression in the embryo. The 3′ Hox genes are expressed in the developing hindbrain, whereas the 5′ Hox genes are expressed only in posterior structures (for example, in the tissue that becomes the tail in the mouse).

The number of Hox genes is larger and their organization is more complicated in vertebrates than in *Drosophila*. As a result of chromosomal duplications, multiple Hox gene clusters are found on multiple chromosomes (four in the mouse and the human, seven in the zebrafish). In mammals, the four clusters are designated A, B, C, and D. The corresponding genes on each chromosome are called paralogs (for example, A1 and B1 designate paralogs on different chromosomes that have the same position in the cluster);[20] the set of corresponding paralogs across chromosomes is referred to as a paralogous group (A1, B1, C1, and D1). Multiple paralogs and multiple members of paralogous groups must be knocked out (or in) to decipher the Hox gene code for each rhombomere. This redundancy is likely the reason that frank homeotic mutations such as *Antennapedia* or *Ultrabithorax* are rare to nonexistent in vertebrates. Note that many of the paralogous groups are incomplete because of gene losses that occurred later in the evolution of the animals than the duplication of the clusters. Mammalian genomes contain 39 Hox genes spread over 13 paralogous groups (there is no C1 in the paralogous group described earlier). The seven clusters found in the zebrafish are thought to represent an additional genome duplication event followed by the loss of one of the clusters (4 + 4 = 8, 8 − 1 = 7).[21] Despite our focus on their roles in the hindbrain, it is important to reiterate that Hox genes pattern many regions of the developing vertebrate embryo.

Given the importance of Hox genes in formation of the rhombomeres, the logical follow-up question is what factors regulate expression of Hox genes in this tissue? Here the fruit fly and the vertebrates apparently diverge, although much remains to be learned about Hox gene expression in vertebrates. Instead of using pair-rule genes and segment polarity genes to slice and dice ever smaller compartments in the nervous system, as described in Chapter 4 the vertebrate embryo appears to rely instead on retinoic acid to regulate critical aspects of Hox gene expression in the rhombomeres.

The source of retinoic acid is the paraxial mesoderm, found adjacent to the neural tube. A gradient is present, for retinoic acid concentrations are higher in posterior regions than in anterior regions (see Chapter 4). Many of the Hox genes have upstream promoters that contain retinoic acid response elements, implying that the expression of these genes can be directly regulated by binding to retinoic acid. A critical role for retinoic acid in the initiation of rhombomeric Hox gene expression has been demonstrated in both zebrafish and mice. For example, the zebrafish gene *hoxd4a* is normally expressed in the most posterior rhombomeres (r6 and r7) a little after 10 hpf. Treatment of embryos with drugs that inhibited the expression of the synthesizing enzyme retinaldehyde dehydrogenase or that blocked retinoic acid receptors reduced the expression of *hoxd4a* and resulted in abnormal development of the posterior rhombomeres.[22] Conversely, implantation of tiny beads coated with retinoic acid close to the yet-to-be-formed hindbrain at 6 hpf resulted in precocious expression of *hoxd4a* near the bead. These bead implant experiments support a model in which rhombomere fates are successively induced, in a concentration-dependent manner, in response to production of retinoic acid by the adjacent mesoderm.

Hox genes delight biologists for many reasons. These gene complexes have been conserved across vast reaches of animal evolution, retaining both their homeodomain and their functional role in the regulation of the body plan. This suggests that understanding the evolution of Hox genes will produce insights into the evolution of animal morphology. The conserved spatial colinearity of the Hox genes on their chromosomes (3′–5′) and their expression in the body (anterior to posterior) inspire studies of the cis-regulatory modules (noncoding elements of the genome, such as enhancers) responsible for the regulation of these complex genetic loci. This story, much of which remains to be told, is too long to compress into an introductory textbook. But one more aspect of the Hox tale needs to be recounted before we shift our focus to formation of laminar structures such as cortex.

Comparative genomics studies of Hox genes revealed that, compared with most other bilaterian animals, *C. elegans* has surprisingly few Hox genes. This is true even when the comparison groups are other nematode species!

In addition to having a smaller number of Hox genes (six), *C. elegans* also suffers from a *broken complex*. This means that the gene expression profile is not colinear with the array of genes on the chromosome. There is evidence that the ancestors of *C. elegans* had a well-organized and complete set of Hox genes. In addition, studies of Hox gene expression in *C. elegans* embryos have revealed that expression of Hox genes depends on cell lineage rather than position. This conclusion is based on experiments in which migration of specific cells was blocked by drugs that disrupt the cytoskeleton. Cells that ended up in the wrong place relative to the body axis still displayed their customary pattern of Hox gene expression.[23]

Students sometimes wonder if invertebrates provide us with examples of simpler or lower patterns of development as opposed to the more complex and higher patterns displayed by vertebrates. Comparative genomics allows us to appreciate the false nature of such distinctions. Is regional specification in the nervous system of *C. elegans* by cell lineage simpler than in Hox gene-dependent animals? Does your answer to this question change when you know that the mechanisms in *C. elegans* are derived from (by which I mean appeared later in evolution than) Hox-dependent mechanisms? Does your answer change when you learn that some chordates (members of the phylum that contains the vertebrates) have also experienced a loss of *Hox* genes and a breakdown of the *Hox* cluster? More pertinent categories than higher and lower might be *ancestral* (earlier in phylogeny) versus *derived* (later in phylogeny) and *conserved* (reflecting the ancestral state of a trait) versus *divergent* (reflecting a specialized state of a trait). It is also useful to keep in mind that some of biology's favorite models have diverged from other groups in their lineage. For example, *Drosophila* is a member of the insect order Diptera, characterized by having only two wings instead of the typical four. In another well-known example, primates, particularly humans, have significantly larger brains than one would predict based on knowledge of other mammals. All things considered, it is probably better to say that *C. elegans* appears to have unusual (even odd) mechanisms for specifying cell fate at different positions in its body rather than that it has a primitive mechanism.

Histogenesis of the Mammalian Cortex

Development of the neocortex of mammals is the final topic of this chapter. Instead of asking how different parts of the developing nervous system acquire a regional identity, we now examine the construction of a single structure.

The cerebral cortex is the outermost part of the cerebral hemispheres in vertebrates. It faithfully follows all the twists and turns of the gyri and sulci of the surface of the brain. The older (speaking in evolutionary terms) parts of the cerebral cortex are referred to as allocortex. The major structures of the allocortex in mammals are the medial temporal lobes, the piriform lobe, and the entorhinal cortex. The hippocampus is also considered part of the allocortex, although its laminar structure contains fewer layers than the other regions of the allocortex.

The neocortex is found in the dorsal region of the cerebral cortex. As its name indicates (the prefix *neo-* comes from the Greek word *neos,* meaning new), it is the most recent part of the cortex to evolve. The neocortex is a six-layered structure present in all mammals but not in other vertebrates. It is disproportionately enlarged in humans, even relative to other primates (fig. 6.7). The layers of the neocortex are numbered using Roman numerals. The outermost (most superficial) layer just under the pial surface is designated Layer I; in the mature nervous system, this layer contains few neurons. Layers II and III contain small and medium-sized pyramidal neurons; layer II also contains a substantial population of stellate neurons. These layers are the source and target of many corticocortical connections, including interhemispheric connections. Layer IV contains a mix of stellate neurons and pyramidal cells and is the primary target of subcortical afferents (for example, from the thalamus). Layer V contains the largest pyramidal cells of the cortex and is the primary source of subcortical efferents (e.g., to the brain stem and spinal cord). The innermost (deepest) layer is Layer VI, which is composed of smaller neurons that send efferents primarily to the thalamus. Layer VI sits atop the subventricular zone, which in turn sits atop the ventricular zone. In addition to numbers, the layers also have names: the molecular layer (Layer I), the external granular layer (Layer II), the external pyramidal layer (Layer III), the internal granular layer (Layer IV), the internal pyramidal layer (Layer V), and the multiform or fusiform layer (Layer VI). All regions of the mammalian neocortex have six layers, but different regions differ in the details of cytoarchitecture, permitting boundaries between adjacent cortical regions to be discerned by the histologist. When the species and brain region being investigated are clear from context, the term *cortex* is often used as a synonym for *neocortex.*

The source of the neurons and glial cells that form the cortex is, as for the other regions of the brain, the neurogenic zone of the neural tube epithelium. The specific regions of the neural tube that give rise to neocortex are the telencephalic vesicles. The general definition of a vesicle is a small sac that contains a liquid. The sacs of the telencephalic vesicle form after the closure of the neural tube. The earliest vesicles to form in the brain are referred to as the primary vesicles, and they mark the three great subdivisions of the developing brain: prosencephalon, mesencephalon, and rhombencephalon. The telencephalic vesicles form as a result of the subdivision of the prosencephalic vesicle into the telencephalic and diencephalic regions. The next stage, neurogenesis in the ventricular and subventricular zones, was described in Chapter 5.

Reflection on this process leads naturally to the conclusion that the cortex is built from the inside out by adding layers on top of the initial layer. The concept that lamination results from an orderly process of radial migra-

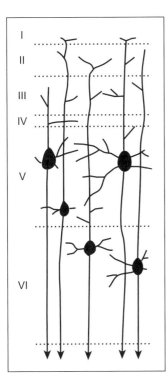

Figure 6.7. Structure of the mammalian neocortex. It is a six-layered structure, with the layers defined by the size and density of the neurons they contain. The cellular composition of the layers varies from cortical region to cortical region. The slice of schematic cortex depicted here is an example of motor cortex, which is sometimes referred to as agranular cortex because, unlike sensory cortex, it lacks a well-developed layer of relatively small granule cells (Layer IV). Instead of abundant granule cells in Layer IV, the motor cortex is characterized by large pyramidal neurons found in Layer V. These pyramidal neurons are often referred to as Betz cells. The Betz cells are corticospinal neurons: they project their axons to the anterior horn of the spinal cord. The motor cortex controls voluntary movements.

tion has been confirmed by birthdating studies showing that the oldest neurons migrate the shortest distance from their site of origin in the ventricular zone. This view, however, is incomplete. At least two other populations of cortical neurons—GABAergic interneurons and the Cajal-Retzius cells of the marginal zone—migrate into the neocortex from other sites of origin. The interesting life histories of these neurons are described at the end of the account of how the basic layers develop.

The mature six-layered neocortex is built on a scaffold of ephemeral structures: the preplate, the subplate, the cortical plate, and the intermediate zone. The preplate is formed of the first neurons born in the ventricular zone. These neurons migrate only a short distance, basically out of the ventricular zone (away from the ventricle). As development continues, the preplate divides into a superficial marginal zone and a deeper layer called the subplate. The marginal zone is home to the Cajal-Retzius cells, which can be distinguished from other cortical neurons and glial cells because of their horizontal orientation. The subplate contains some of the firstborn neurons of the underlying ventricular zone. As neurogenesis and radial migration of neurons proceed, the subplate becomes separated from the marginal zone by insertion of a new population of neurons called the cortical plate. Simultaneously, axons growing into the cortex separate the subplate from the proliferative, intermediate progenitor cell–filled subventricular zone. The cell-sparse (but process-rich) region that divides the subplate from the subventricular zone is called the intermediate zone. The cortical plate thickens by addition of new neurons born in the ventricular and subventricular zones. Cortical plate neurons eventually sort into layers and become the heart of the cortex. The subplate will eventually disappear as a discrete structure, but many subplate neurons will persist to become part of the mature cortex.

GABAergic interneurons are the primary local circuit interneurons of the neocortex. GABAergic neurons can be identified as they migrate within the growing neocortex. They are simultaneously immunopositive for the inhibitory neurotransmitter GABA and two transcription factors: Dlx1/2, a homeodomain protein, and Mash1, the bHLH protein described in Chapter 5. The expression of Mash1 links these cells to Mash1-expressing progenitors in the ventricular and subventricular zones of the neocortex. Studies of cell migration in living slices of embryonic human forebrain tissue have indicated that about two-thirds of the GABAergic neurons in the neocortex are born in the neocortex and migrate radially away from their site of origin. But studies of living slices of developing neocortex revealed the existence of a population of GABAergic neurons migrating tangentially (as opposed to radially) through the intermediate zone. These GABAergic neurons express Dlx1/2 but not Mash1.[24] Slices of neocortex cultured in isolation without other brain tissues have significantly fewer GABAergic cells than larger explants. Careful studies using explants (pieces of living tissue placed into an artificial medium) of fetal human neocortex revealed that the Dlx1/2-positive, Mash1-negative GABAergic neurons are produced in a structure called the ganglionic eminence. The ganglionic eminences are temporary proliferative subcortical regions of the ventral telencephalon. In addition to producing a substantial fraction of the cortical GABAergic neurons, the ganglionic eminences also contribute neurons to other brain regions. If we never studied the developing human brain, we would be unlikely to suspect the existence and function of the ganglionic eminences.

Another population of horizontally oriented neurons found in young neocortex is found just under the pial surface in the marginal zone. As was the case of the GABAergic neurons, the horizontal orientation is a clue. It signals that these neurons are migrating tangentially within the cortex rather than radially. These are the Cajal-Retzius cells.[25] Cajal-Retzius cells are present before the formation of the cortical plate, in the early preplate, and they cover the developing cortex. They are produced in a transient embryonic structure (notice a pattern here?) called the hem of the cortex. The hem of the cortex is located at the intersection of the prospective hippocampus and the choroid plexus, the specialized region of the pia mater that produces cerebrospinal fluid. As Cajal-Retzius cells mature, they extend axons that form a tangled network at the boundary between the marginal zone and the cortical plate. They are such a prominent part of the young cortex that it is astonishing to learn that they soon will all be gone. In humans, almost all of the Cajal-Retzius cells are gone by Week 35 of gestation; in mice, they will be gone by the end of the first postnatal week. Before they go, however, they will have contributed in an important but somewhat mysterious way to the formation of the cortex.

The Cajal-Retzius cells are the major producers of reelin, an extracellular matrix glycoprotein. During the earliest stages of cortical development, Cajal-Retzius cells are the only sources of reelin. The phenotype of mice lacking reelin (as in the case of the spontaneously arising reeler mutation) is highlighted by disorganization of laminar brain structures (defects in the cerebellum are the source of the characteristic reeling gait of these mutants).[26] The cortex of reeler mice is said to have an inverted structure because later-born neurons appear to get stuck in the cortical plate behind the earlier-born neurons. Mutations in the human reelin gene result in lissencephaly, cerebellar hypoplasia (too few neurons in the cerebellum), and cognitive deficits.[27]

Reelin exerts effects on migrating cortical neurons and radial glial cells by activating an unusual signaling pathway. The receptors for reelin are receptors for lipoproteins of the low-density lipoprotein or LDLR type. LDLRs are found in the plasma membrane of many cells. Their typical function is to bind to low-density lipoproteins (LDLs) and then bring them into the cell. Because LDLs are carriers for cholesterol, LDLRs are an important part of the mechanism by which blood levels of cholesterol are regulated. That the receptor for reelin is an LDLR was discovered when it was observed that knockout mice lacking two LDLR receptors, VLDLR and apolipoprotein E receptor 2, have the reeler phenotype. As noted, these lipoprotein receptors are often thought of primarily in the context of lipid transport and metabolism, but at the cell surface they bind and internalize a surprisingly long list of ligands, including reelin. Cultured cells transfected with the gene for VLDLR internalized reelin protein, while non-transfected control cells could not.[28] Other studies have shown that internalized reelin increases the tyrosine phosphorylation of the intracellular protein Dab1. All components of this signaling pathway, from reelin to Dab1, are present in the developing cortex, and it has been shown that tyrosine phosphorylation of Dab1 is required for transduction of the reelin signal. What is the cellular function of Dab1? It appears to lack enzymatic or DNA-binding activity. Instead, Dab1 is an example of an adaptor protein. Adaptor proteins contain multiple protein-binding motifs capable of linking protein-binding partners together for the purpose of forming larger signaling complexes.

Given this wealth of information, it may be surprising that the function of reelin and the role of Cajal-Retzius cells in cortical development remain mysterious. The titles of several recently published articles in this field make the challenge clear: "Puzzling out the reeler brainteaser"; "Go or stop? Divergent roles of Reelin in radial neuronal migration"; and my personal favorite,

"What does Reelin actually do in the developing brain?"[29] Understanding the molecular mechanisms that support the development of the neocortex remains a major goal of neuroscience.

It is sobering how much remains to be learned about the development of the neocortex. Research on the molecular basis of cortical development, however, has the potential to provide tools to answer one of the most interesting questions in biology: how did the human brain evolve?

Notes

1. You were looking at mRNA expression patterns based on in situ hybridization, a protein expression pattern based on immunohistochemistry, or flies genetically engineered so that a marker such as GFP was under the control of a segmentation gene promoter.

2. See Nüsslein-Volhard and Wieschaus (1980).

3. See the page of The Interactive Fly on zygotically transcribed genes for further discussion (http://www.sdbonline.org/fly/aimain/5zygotic.htm).

4. See Wieschaus et al. (1984). Note that the journal in which that was published (*Roux's Archives of Developmental Biology*) was given new life as *Development, Genes, and Evolution* in 1996; for an interesting discussion of this transition from the editorial perspective, see p. 321 of *Roux's Archives of Developmental Biology* 205 (1996).

5. Sean Carroll (1990) provides a thoughtful discussion of zebra stripes in *Drosophila* embryos.

6. The parasegment concept was defined by Martinez-Arias and Lawrence (1985). A brief, thoughtful discussion of segments and parasegments can be found on p. 227 of Gilbert (2010).

7. See Frasch et al. (1988).

8. For a full conversion from posterior to anterior fate, both *engrailed* and *invected* loss of function are required.

9. A hedgehog is a spiny mammal in the subfamily Erinaceinae. Hedgehogs are found in Europe, Africa, and Asia but not in North and South America.

10. A good source of information about the four-winged fly is the online account of E. B. Lewis's Nobel Prize–winning research (Nobelprize.org). Lewis shared the 1995 Nobel Prize in Physiology or Medicine with Christiane Nüsslein-Volhard and Eric Wieschaus, whose work has already been mentioned several times in this textbook.

11. See Choo et al. (2011).

12. For example, Ghysen et al. (1985) used a monoclonal antibody that labeled developing wing and leg neuromeres in thoracic ganglia to show the independence of nervous system and epidermal homeotic transformations.

13. Johannes Holtfreter's studies between 1930 and the 1960s established the significance of the notochord in development of the nervous system. A full account of Holtfreter's life and work is available as part of *Biographical Memoirs*, published by the U.S. National Academy of Science. See Gerhart (1998).

14. The defining features of chordates are the notochord, a hollow dorsal nerve cord, pharyngeal gill slits, and a postanal tail.

15. See figure 2 in Roelink et al. (1994).

16. See figure 2 in Varjosalo and Taipale (2008).

17. See Chiang et al. (1996).

18. See Ekker et al. (1995) . This article provided the first report of *Tiggy-Winkle hedgehog,* a gene found only in fish.

19. See Chamberlain et al. (2008).

20. Homologs are genes that are related by descent from the same ancestral DNA sequence. When homologous genes are found in the genomes of different species, they are referred to as orthologs and are expected to perform the same function across the different species. When multiple homologous genes are found in the genome of a species, they are assumed to have arisen as a result of a gene duplication event and may have evolved new functions. Both individual genes (a piece of a chromosome) and whole chromosomes can be duplicated.

21. A cluster is defined as all of the Hox genes found on a single chromosome. See Woltering and Durston (2006).

22. See Maves and Kimmel (2005).

23. The drugs used to disrupt migration are widely used in cell biology research: cytocholasin D (which inhibits actin polymerization) and colchicine (which depolymerizes microtubules). See Cowing and Kenyon (1996).

24. See Letinic et al. (2002).

25. It appears to be a tradition for neuroscientists to refer to the Cajal-Retzius cells as enigmatic. See Soriano and del Río (2005).

26. The name of the mouse strain is *reeler* is; *reelin* is the name of the mutated gene in this strain; reelin is the name of the protein absent in reeler mutants. The *reeler* mouse first appeared in 1948 in a colony of inbred mice at the University of Edinburgh, Scotland, as a spontaneous mutant. Mutations in every component of the *reelin* signaling pathway reproduce the *reeler* phenotype.

27. See Hong et al. (2000).

28. The transfected cells were immortalized COS cells, which are widely used for in vitro signal transduction assays. This cell line was developed by immortalizing a cell line developed from kidney cells of the African green monkey. Expression constructs with the appropriate promoter replicate extremely well in COS cells. See D'Arcangelo et al. (1999) for experiments in which transfection of COS cells with VLDLR allowed the COS cells to internalize reelin.

29. See Luque (2007), Zhao and Frotscher (2010), and Hattori (2011).

Investigative Reading

1. You speculate that Notch- and reelin-signaling pathways interact to control cell fate in the developing mouse cortex. What experimental approaches do you use to test your hypothesis?

Keilani, Serene, and Kiminobu Sugaya. 2008. *BMC Developmental Biology* 8: 69.

2. Millipedes (myriapods) are characterized by long, segmented bodies, but little is known about the molecular basis of development in these noninsect arthropods. Based on studies of embryogenesis in the fly *Drosophila,* an insect arthropod, what genes would you choose to study in

millipede embryos? What patterns of gene expression do you expect to find?

Janssen, Ralf, Graham E. Budd, Nikola-Michael Prpic, and Wim G.M. Damen. 2011. *EvoDevo* 2: 5.

3. *Challenge Question:* Exposure to ultraviolet (UV) radiation crosslinks DNA and proteins. Purified nuclei from *Drosophila* of different genotypes were exposed to UV. Chromatin was extracted from the irradiated nuclei and immunoprecipitated using an antibody specific to the Engrailed protein. The immunoprecipitated chromatin was used to construct a library enriched in genomic sequences that bind to the Engrailed transcription factor in *Drosophila* embryos. The immunoprecipitated DNA was then used to construct a genomic library enriched for the DNA fragments that were bound to Engrailed in the nuclei of the fly embryos that donated their chromatin. What can you learn by using this genomic library in your research? (A genomic library is a population of bacteria that each express a DNA fragment packaged into a cloning vector that was used to transfect the bacteria.)

Solano, Pascal Jean, Bruno Magat, David Martin, Franck Girard, Jean-Marc Huibant, Conchita Ferraz, Bernard Jacq, Jacques Demaille, and Florence Machat. 2003. *Development* 130: 1243–54.

SEVEN

Becoming a Neuron

Axons, Dendrites, and the Formation of Synapses

A newborn neuron is roughly spherical. A neuroanatomist would say that it is all soma. Its lack of an axon and dendrites prevents it from fulfilling its destiny of joining a neural circuit. An important stage of neural development is therefore the extension of threadlike processes outward from the soma. This stage is followed by formation of specialized junctions between the axons of one neuron and the dendrites of another. These junctions, called synapses, constitute the polarized communication network that is the basis of behavior and thought.

The development of axons and dendrites—referred to collectively as processes—distinguishes neurons from all other cell types in the body. At first glance, a process appears to consist entirely of plasma membrane wrapped tightly around a cylinder of cytoskeleton. Because processes are so thin—on average no more than a couple of micrometers in diameter—there is little room for cytoplasm and organelles. The spare appearance of neuronal processes, however, is deceptive. Within each axon and dendrite there is a steady flux of molecules, organelles, and components of plasma membrane. This traffic is both anterograde (outward from the soma) and retrograde (toward the soma). By this means the soma supplies the needs of the tips of its processes and allows the nucleus to sample signals from a distance.

A typical neuron develops a single axon and multiple dendrites. Mature axons and dendrites can be distinguished in terms of their structure, function, and molecular components. Axons have a relatively uniform diameter. In large animals, axons can be quite long.[1] Axons often branch near their distal tip before forming the structures known as axon terminals or boutons. The axon terminal forms the presynaptic half of a synapse. It is specialized for release of neurotransmitter but can also take up molecules from the synaptic cleft. By contrast, dendrites are usually significantly shorter than axons. In vertebrate neurons, dendrites can arise at many points on the soma (see Chapter 3 for a brief discussion of neuronal morphology). In contrast to axons, dendrites taper. The defining structure of the dendrite is the dendritic spine, the postsynaptic component of many synapses. The plasma membrane of dendrites contains receptors for the neurotransmitters released by axon terminals. A surprising feature of dendrites is that they often contain ribosomes, which allows them to synthesize proteins.

Many variations on this basic theme of neuronal morphology can be found, particularly if one looks beyond vertebrates (see Chapter 3; fig. 7.1). Most invertebrate neurons produce a single process that branches a short distance from the soma, with one branch becoming the axon and the other forming a dendritic arbor. Some sensory neurons lack dendrites, whereas others lack axons. Yet even these apparently monopolar neurons are fundamentally polarized in terms of structure and function.

One of the interesting questions related to process outgrowth is how a neuron that forms many, many processes manages to produce only a single axon. It is obviously extremely difficult to study the earliest steps in the development of axons and dendrites in the intact vertebrate central nervous system. New neurons are produced in specialized neurogenic regions, and most migrate to reach their final locations. During migration, neurons develop a leading and a trailing process, one of which likely becomes the axon, but until recently following migrating neurons in the developing brain was nearly impossible. As a result, much of what is known about process outgrowth comes from a model system. In this case, however, the model system is not a more tractable species but instead a method for culturing neurons obtained from the brains of late-stage rodent embryos. Young neurons from many different regions of the central and peripheral nervous systems grow processes in culture, but the most popular in vitro system for studying neuronal polarity is that used intensively by Banker and colleagues.[2] These investigators have used rat hippocampal neurons to examine in detail the morphological and molecular changes that occur after hippocampal neu-

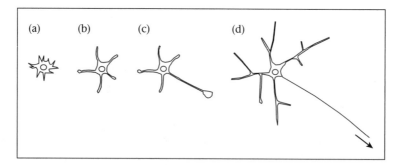

Figure 7.1. The polarized structure of neurons. This structure develops in a stereotyped sequence that is exaggerated in embryonic neurons grown in culture. First, short, stubby processes form (a). Then immature processes called neurites extend (b). Next, one of the neurites becomes clearly identifiable as a thin, fast-growing axon (c). The sequence continues with further axon outgrowth and dendritic branching (d). Drawing adapted from Barnes and Polleux (2009).

rons are dissociated and allowed to settle onto an adhesive substrate. It is no exaggeration to state that much of what we *know* about the growth of axons and dendrites is based on studies of cultured hippocampal neurons.

This chapter begins with an account of the mechanisms by which neurons establish their axons and dendrites that draws heavily on these in vitro studies of mammalian neurons. We then shift our attention to how a growing axon navigates to its target. Both in vivo and in vitro methods have contributed to our understanding of targeting. The chapter concludes with an overview of synaptogenesis. Here we again rely heavily on a single model. This model is again not a species but a specific version of the synapse: the neuromuscular junction.

The Decision to Grow a Process

We take it for granted that neurons have processes, but what signal triggers initial process outgrowth? Does it originate at the cell surface or come from inside the new neuron? Why does a neurite extend from certain locations on the cell surface and not from others? These questions take us back to the earliest postmitotic hours in the lives of all cells. Neurons are extremely, spectacularly polarized, but polarity is not unique to neurons. For example, the apical and basolateral domains of epithelial cells are as critical to their function in tissues as axons and dendrites are to the function of neurons. What factors allow an individual cell to play out, on its own tiny stage, the processes that result in the polarized body plan of the bilaterians? Are the points on the soma of neurite outgrowth random or determined? Is the decision to grow a particular neurite local or global?

Most research on the development of neuronal polarity takes as its starting point the stage at which multiple neurites are already present. For example, after several hours in culture, rat hippocampal neurons extend and retract multiple minor processes. Some hours later, one of these minor processes extends more rapidly than the others. It is this aggressively extending process that becomes the axon. Studies of the young axon can teach us how axons grow but do not help us understand the origin of neurites.

The question of neurite origin is by no means answered, but studies on the development of polarity by epithelial cells and on initiation of neuronal differentiation in a rat tumor cell line converge on the protein encoded by the adenomatous polyposis coli (*APC*) gene.[3] The protein encoded by this gene, APC, is a large protein with multiple domains. It is expressed by many cell types. Because one of the functions of APC is regulation of the stability of β-catenin, we infer that Wnt signaling is likely involved in the earliest neuronal polarization decisions (fig. 7.2).

In this context, APC is best thought of as forming the backbone of a multiprotein complex that regulates the phosphorylation of β-catenin, which in turn regulates whether β-catenin is ubiquitinated, which in turn regulates

Wnt ABSENT	Wnt PRESENT
Disheveled INACTIVE	Disheveled ACTIVE
GSK3 ABSENT	GSK3 INACTIVE
β-catenin STAYS IN APC COMPLEX	β-catenin LEAVES APC COMPLEX
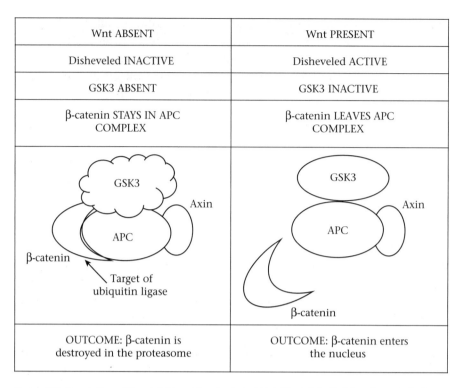	
OUTCOME: β-catenin is destroyed in the proteasome	OUTCOME: β-catenin enters the nucleus

Figure 7.2. Regulation of the stability of β-catenin by Wnt. The left half of the diagram depicts cellular events in the absence of Wnt; the right half shows the changes that occur when Wnt binds to its membrane receptor, Frizzled (not shown in this diagram; see figure 4.3a for additional details).When Wnt is not present, the kinase GSK3 is active and as a result β-catenin is maintained in a complex with other cytoplasmic proteins, including APC and Axin. When β-catenin is part of this APC-based complex, it is a target of ubiquitin ligase. The result is that β-catenin does not accumulate in the cytoplasm because once it is ubiqutinated it is degraded in the cytoplasm by the proteasome. When Wnt is present, the GSK3 inactivated and β-catenin can dissociate from the APC-based complex. Because it is no longer a target of uqibuitin ligase, it accumulates in the cytoplasm. Some of the accumulated β-catenin enters the nucleus, where it complexes with nuclear proteins to regulate gene transcription.

whether β-catenin is destroyed in the proteasome.[4] APC has multiple binding sites for β-catenin. In the absence of Wnt signaling, β-catenin bound to APC is phosphorylated by GSK3 and becomes a target for ubiquitin ligase. Wnt signals (such as Disheveled) inactivate GSK3, allowing unphosphory-lated β-catenin to enter the nucleus and activate TCF or LEF transcription factors. The other relevant property of APC is its association with cytoskeletal elements, particularly microtubules.

Development of polarity by epithelial cells depends on extracellular contacts, because epithelial cells growing in isolation lack their normal polarity. Contacts with other cells mediated via E-cadherin signaling provide a spatial cue that leads, on the other side of the plasma membrane, to assembly of a local signaling network based on receptor-mediated reorganization of the cytoskeleton. The localized enhancement of signaling facilitated by the reorganized cytoskeleton leads, in turn, to further changes in the cytoskeleton. Eventually, apical and basolateral proteins segregate and the epithelial cell establishes its polarized identity. APC is a candidate regulator of the process of polarization because of its ability to link the Wnt signaling network to the cytoskeleton. Evidence for this role includes visualization of APC in complexes with β-catenin and microtubules in protrusions formed by cultured epithelial cells.[5] Why consider a model for initiation of neuronal polarization based on cell adhesion–based spatial cues? One important reason is that neurons in vivo undergo polarization in the context of interactions with other cells.

Another model for the initiation of neuronal polarization comes from the study of nerve growth factor–induced differentiation of PC12 cells. Nerve growth factor is a neurotrophic protein secreted by neuronal targets that exerts effects on neuronal survival and differentiation. The PC12 cell line was initiated from a neuroendocrine tumor of the rat adrenal medulla. PC12 cells grow and divide in culture without added nerve growth factor but do not extend processes in the absence of nerve growth factor. The changes that occur when nerve growth factor is added to the medium are dramatic: the cells stop dividing and extend processes similar to those produced by cultured neurons.[6] Suppression of APC protein expression by transfection of PC12 cells with antisense mRNA targeted to APC blocked the expected growth of neurites after nerve growth factor application. Treatment of PC12 cells with antisense mRNA *after* neurites had been induced to grow had no effect on growth, suggesting that APC regulation might be specifically required for the initial decision to grow. One hypothesis is that the sites of neurite outgrowth correlate with the position of the APC-based multiprotein complexes.

Microtubules, Actin, and Growth Cones

The tips of growing axons end in paddle-shaped structures called growth cones. Growth cones enable neurons to form neural circuits. Growing dendrites also form growth cones, but dendritic growth cones are smaller and less well studied than axonal growth cones. Videomicroscopy of growth cones in living tissue shows that growth cones make a series of decisions that allow them to move toward their target while simultaneously lengthening their trailing axon. This amazing biological process is analogous to a railroad track's building a new rail line to a specific but distant destination without a map or GPS. The major components of the track and the leading growth cone are the two key families of proteins that form the eukaryotic cytoskeleton—the tubulins and the actins.

α-tubulin and β-tubulin are globular proteins present in all eukaryotes.[7] Together an α-tubulin and a β-tubulin molecule form a heterodimer called a

Figure 7.3. Structure of microtubules. Microtubules are biopoly-
mers formed by the assembly of
α-tubulin and β-tubulin protein
monomers. Note the opposing plus
and minus ends; polymerization
occurs more quickly at the plus end.

tubulin subunit (fig. 7.3). Tubulin subunits assemble into strands called protofilaments. Protofilaments (typically 13 in animals) come together to form
long, straight cylinders called microtubules. The assembly of tubulin subunits into microtubules is dependent on guanosine triphosphate (GTP).
Microtubules are the largest component of the eukaryotic cytoskeleton, with
an external diameter of 25 nanometers and a hollow core of 14 nanometers.
Microtubules formed of tubulin serve many functions in cells, including
maintenance of cell shape, organelle positioning, and transport of proteins
and organelles within the cell. They form the centrioles of centrosomes and
the axonemes (central cylinders) of cilia and flagella. In cells undergoing
mitosis they form the mitotic spindle; in postmitotic neurons they provide
support for axons and dendrites. The ability of microtubules to multitask
reflects their susceptibility to posttranslational modifications (such as acetylation and detyrosination) and their interactions with diverse microtubule-
associated proteins (MAPs). Kinesins and dyneins are two important
microtubule-associated motor proteins that bind to tubulin and generate
forces that can result in molecular movement. Nonmotor MAPS often stabilize or destabilize assembled microtubules. Endogenous stabilizers include
tau proteins, which are abundant in neurons, and two proteins called simply MAP1 and MAP2. There are also drugs such as taxol that stabilize microtubules.[8] Endogenous MAPs that act as destabilizers include stathmin and
katanin. The presence of stabilizing and destabilizing MAPs allows microtubules to respond to the changing needs of cells.

Microtubules are present in axons and dendrites, but the neuron (and
hence the neuroscientist) can distinguish between processes on the basis of
microtubule organization. Microtubules grow at only one end, referred to as
the plus end. In axons, the plus ends of the microtubules all point in the
same direction—away from the soma. In dendrites, the plus ends point both
toward *and* away from the soma. Another difference between axons and

dendrites is the relative abundance of specific MAPs. Axons contain tau pro-teins but little or no MAP2, whereas dendrites contain MAP2 but lack tau. Specific antibodies to tubulins and the various MAPs can be used to map the distribution of these proteins in cells.

Actin is possibly the most abundant protein found in eukaryotic cells. It has been estimated that actin constitutes between 5 and 15 percent of total cellular protein. Most animal genomes contain multiple actin genes with similar structures. Actin is a globular protein. It is referred to as G-actin when in its monomeric form and F-actin when polymerized into filaments (fig. 7.4). Because monomeric actin is an ATPase, the hydrolysis of ATP is an im-portant aspect of actin polymerization. Actin filaments organize into a two-stranded helix 7 nanometers in diameter. F-actin is a polarized filament, with distinct fast-growing (plus) and slow-growing (minus) ends. Elongation of the filament occurs at the fast-growing plus end.

Actin filaments can form linear or branched bundles but can also be found in patches associated just under the plasma membrane in a region of the cell sometimes referred to as the cell cortex. Endogenous molecules regulate the polymerization and depolymerization of F-actin, some by interacting with the filaments and some by regulating the availability of G-actin for polymer-ization. Actin is as necessary for cell motility as it is for support of cell struc-

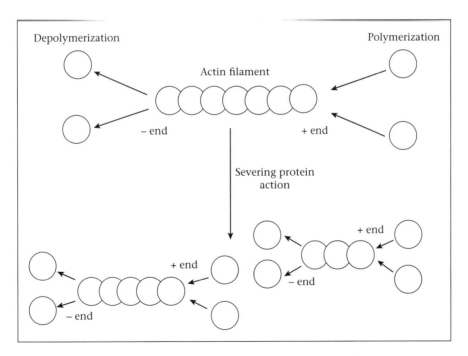

Figure 7.4. Actin filaments. These filaments are formed by the assembly of actin monomers. Note the opposing plus and minus ends. Actin-severing proteins can reorganize actin filaments. Drawing based on http://vcad-hpsv.riken.jp/en/research/outline/member/g009.php.

ture. Both functions of actin are represented in neuronal growth cones (described in the next section). In mature neurons, actin is distributed throughout the cell and its processes but is especially enriched in dendritic spines. Antibodies can be used to map the location of actin in cells, but many anti-actin antibodies do not discriminate G-actin and F-actin. Another label for actin that is widely used is phalloidin, one of the toxins found in the death cap mushroom, *Amanita phalloides*.[9] Phalloidin binds selectively to F-actin from all species and is commercially available conjugated with many different fluorophores.

Studies of tubulin and actin in growth cones have revealed that they share a common structure in all species and all types of neurons.[10] The ends of growing axons flatten into paddles tipped with fingerlike extensions called filopodia. Neat bundles of F-actin support each filopodium. The webs (think of a duck's foot) between filopodia contain networks of branched F-actin. The weblike regions of growth cones are called lamellopodia or veils. The filopodia and lamellopodia are collectively referred to as the peripheral domain or the P-zone of a growth cone. The central region (central domain or C-zone) of a growth cone is dominated by the ends of the microtubules that support the axon along its length. Most axonal microtubules end short of the filopodia, but a small number follow the F-actin bundles all the way to their ends. Filopodia are the major points at which growth cones attach to their substrates, which implies a role for actin in growth cone motility. But studies have shown that growth cones do not move if microtubules are not present. Coordination of the microtubule and actin elements of the growth cone cytoskeleton is required for an axon to grow toward its target.

For didactic purposes only, it is helpful to think about axon elongation and growth cone motility as separate phenomena. These processes have been intensively studied in vitro and have been found to be shared by growth cones of many species. Elongation of a growing axon occurs in three stages, each sufficiently distinct to have its own name: protrusion, engorgement, and consolidation. A simplified account follows. During protrusion, the actin-rich peripheral zone extends as a consequence of increased actin polymerization. During engorgement, microtubules move deeper into the growth cone; vesicles and other organelles enter the veils of the lamellopodia. During consolidation, the proximal part of the growth cone becomes cylindrical and wraps closely around the microtubules as a result of actin depolymerization. Time-lapse studies of living growth cones revealed that the protrusion–engorgement–consolidation cycle depends on the active polymerization and depolymerization of actin in the filopodia and on forces generated by myosin. The force generated by the interaction of myosin and actin is a backward pull. How can a backward pull advance a growth cone? The answer is that it can do so only when the filopodia are attached so firmly to the substrate that this attachment overcomes the force of the pull.

Direct observation of fluorescent actin and tubulin in elongating axons has shown that microtubules are added at the distal end and that actin filaments form in the filopodia. Drugs that interfere with the polymerization and/or depolymerization of actin and tubulin have striking effects on axon outgrowth. Cytochalasins are compounds produced by fungi that bind to the growing ends of F-actin and prevent subsequent polymerization or depolymerization. Application of cytochalasins slows axon elongation. By contrast, stabilization of microtubules with taxol does not block axon elongation. The opposite effect (slowing of the rate of elongation) is observed when a drug such as nocodazole that depolymerizes microtubules is applied. The effects of these drugs are interpreted as mimicking, in a crude fashion, the actions of endogenous regulators of F-actin and microtubule stability. A major take-home message from these studies is that axon elongation requires a balance between cytoskeletal stability and instability. This message is reinforced by the finding that applications of these drugs to one side or the other of growth cones can cause growth cones to turn, whereas a uniform application of the same drugs to both sides results in straight growth.[11]

Importance of Primary Cultures in Developmental Neuroscience

Many, if not most, cell biologists use continuous (immortal) cell lines as their primary research tool. Neuroscientists are unusual in their preference for primary culture, which involves removing bits of nervous tissue from living animals, dissociating the cells present in that tissue, and observing them in culture until they die after a few days or a few weeks. The technically challenging approach of primary culture is popular for a single reason—very few continuous cell lines produce axons, dendrites, and synapses. I previously noted that hippocampal cultures, originally prepared from rat embryos, have contributed much to our understanding of process outgrowth. Extending this model to mice ensures its continued relevance. For example, cultures can be prepared from transgenic mice with targeted mutations or prepared from the brains of GFP expressing mice, which allows the efficient visualization of the complete set of processes grown by cultured neurons.

Multipolar Neurons in Drosophila

Drosophila neurons were once regarded as poor subjects for studies of neuronal polarity because they are almost all unipolar. Most insect brain neurons clearly have dendritic arborizations distinct from a single axon, but the dendrites branch off from a primary neurite that is also the origin of the axon (fig. 7.5). There are, however, genuinely multipolar neurons in the insect peripheral nervous system. Dendrites project directly from the somata of these neurons. These multipolar sensory neurons are found under the cuti-

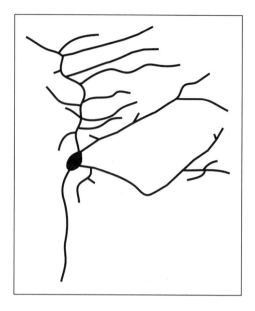

Figure 7.5. Example of a multipolar sensory neuron associated with the body wall of larval *Drosophila*. Polarized neurons with dendrites extending directly from the soma are unusual in insects; compare the neuron shown here with the unipolar neuron depicted in Fig. 3.8c. This neuron is an adaptation of the dendritic arbor of the neuron ddaE, as imaged during the third larval stage by Williams and Truman (2005).

cle and function as touch-sensitive proprioceptors. They are said to *tile* the body wall because the dendritic branches of individual cells do not overlap.[12] These neurons can be used to explore the differences between unipolar and multipolar neurons in the fruit fly and to explore how neurons develop distinct axons and dendrites in different animal taxa (keeping in mind that the vast majority of animals on the planet have the unipolar neurons found in the fly brain, not the multipolar vertebrate neurons found in the textbooks). One example of an intriguing difference between *Drosophila* and vertebrates comes from studies of microtubule polarity. As noted, in vertebrate axons the plus ends of microtubules point outward, whereas the orientation of microtubules in dendrites is a mix of plus and minus ends pointing outward. A study of microtubule polarity in *Drosophila* revealed a surprising difference in the orientation of microtubules in dendrites only: in fly dendrites, all of the microtubules are oriented with the minus end pointing outward.[13]

Axon Path Finding

Netrins

How do growth cones know where to go? How do they know when they are there? These are two of the fundamental questions of developmental neuroscience. Substantial progress in the identification of the relevant molecules awaited the advent of modern molecular biology paired with modern cell labeling and neuroimaging techniques. Today the list of chemical signals known to direct growth cone movement is long. The list of receptors for these signals is equally long. Signals can be either attractive ("grow here") or repulsive ("keep off the grass"). Insoluble and/or immobilized signals are transmitted via contact with the substrate, which can be the surface of another cell or the extracellular matrix. Alternatively, water-soluble signals diffuse from their sources to provide long-range directional cues. It is possible for a single signal, depending on context, to attract some growth cones and repel others. Growth cones function as little sensing devices that choose where to lead their axon based on the interaction of multiple ligands (the cues) with multiple receptors.[14] The signals received by the growth cone receptors must ultimately couple to regulators of the cytoskeleton.

The family of netrin proteins provides good examples of chemical cues that can act over long and short distances and, in combination with other cues, can both attract and repel growth cones. Netrins were discovered in *C. elegans* by application of a forward genetics approach to mutant hermaphrodites with abnormal locomotion on a solid medium. The category of mutations responsible for abnormal locomotion is named *unc,* for *uncoordinated.*[15] Some of these mutants have muscle abnormalities, but others have subtle neuroanatomical defects. It was analysis of a subset of these neuroanatomical defects that led to the first description of a netrin (UNC-6, encoded by the *Unc-6* gene) and a netrin receptor (UNC-40, encoded by the *Unc-40* gene).

In *Unc-6* mutants, the affected neurons grow axons, but the axons do not follow their usual pathways. Among the neurons affected are the mechanosensory PDE neurons. The somata of the two PDE neurons are located on the sides of the worm, midway between the vulva and the rectum. Each PDE neuron contributes a ciliated dendrite to a cuticular sensillum and sends a bifurcating axon to run anteriorly and posteriorly in the ventral nerve cord. The axon of a PDE neuron typically grows straight down the lateral epidermis to enter the ventral nerve cord directly (fig. 7.6).

In the 1980s it was discovered that some *C. elegans* sensory neurons can be filled with the fluorescent tracer fluorescein merely by immersing living worms in fluorescein solutions.[16] The dye presumably enters these neurons via exposed receptor cilia. Because only a subset of neurons is labeled by this method, the path taken by individual PDE axons can be examined in living animals (if all neurons were labeled, the entire nervous system would glow and it would be impossible to track individual axons). In approximately one-third of *Unc-6* mutants (adult hermaphrodites) examined, the PDE axons failed to reach the ventral nerve cord because they did not grow in a ventral direction. Even in worms in which the wandering axons eventually reached the ventral nerve cord, the axons took variable and unusual paths to their goal. Comparable defects were found in another *unc* mutant, *Unc-40.*

Follow-up studies revealed that the UNC-6 protein shares functional domains with laminin, a component of the basal lamina, the layer of extracellular matrix secreted by epithelial cells. The story of UNC-6 then took a surprising turn. To understand how the homology of UNC-6 to the vertebrate netrins was recognized, we need to consider the formation of structures called commissures.

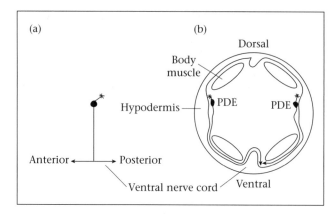

Figure 7.6. Outgrowth of PDE neurons in *C. elegans*. The axons of the PDE neurons grow along the hypodermis, cross lateral nerve processes, and then travel between the ventral body muscles and the hypodermis to join the ventral nerve cord located at ventral midline, where they split to send an axon in both the anterior and posterior directions. A schematic diagram of a single PDE neuron is shown in (a); a cross section through the middle region of a second-stage larva is shown in (b). In the ventral nerve cord, the PDE processes form synapses with other ventral nerve cord processes. Adapted from Hedgecock et al. (1985).

Animals with bilaterally symmetrical bodies often have a need to transfer information from one side of the body to the other. This is achieved through commissures. A commissure is any bundle of nerve fibers that passes across the midline from the left side to the right side of the nervous system and vice versa. Neurons with axons that do not cross the midline are described as having ipsilateral (same-side-of-the-body) projections. Commissural neurons have contralateral (opposite-side) projections.

Commissural neurons are found in the vertebrate brain (think of the corpus callosum!) and the spinal cord. Many studies of the initial outgrowth of the axons of commissural neurons have used the spinal cord as a model (fig. 7.7). The dorsal spinal cord contains commissural interneurons that, during embryogenesis, extend axons to the floor plate at the ventral midline of the forming spinal cord. (The neuroepithelial cells that constitute the floor plate were introduced in Chapter 6 as the source of Sonic hedgehog (Shh) signals.) Studies in which short lengths of the developing rat spinal cord (collected from E11 embryos) were cultured with and without a piece of the floor plate (an explant) revealed that something secreted by the floor plate cells both promoted and guided the outgrowth of the axons of the dorsal commissural interneurons. A more robust version of this in vitro system based on slightly older (E13) rat spinal cords was used as an assay to identify the attractant substances. Extracts of homogenates of floor plate cells had the same attractant power as explants, as did homogenates of the embryonic brain. The proteins that are now called netrin-1 and netrin-2 were eventually purified from fractions prepared from thousands of homogenized embryonic chick

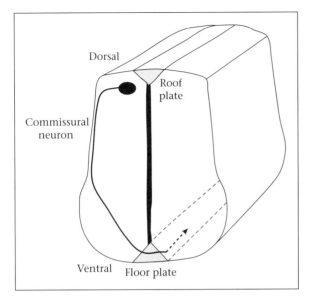

Figure 7.7. Outgrowth of commissural axons in the developing vertebrate spinal cord. The general path taken by the axons of commissural neurons is indicated. These axons extend to the ventral compartment of the developing spinal cord, then cross the midline via the ventral commissure before traveling along the long axis of the spinal cord. The roof plate is an important source of Bmps; the floor plate is a source of Shh and netrin-1.

brains, which could be obtained at lower cost than a comparable supply of rat brains.[17] Amino acid sequencing of the netrin proteins followed by cloning of their genes revealed the homology of the chick netrins to the UNC-6 protein of *C. elegans.* As is the case for the effects of UNC-6 in *C. elegans,* the effects of the netrins are restricted to specific populations of axons growing in the developing spinal cord. For example, neither motor axons originating from neurons in the ventral compartment of the spinal cord nor sensory axons originating from neurons in the dorsal root ganglia were attracted by netrins.

Subsequent studies revealed the homology of *C. elegans* UNC-40 and a vertebrate protein called DCC. The full name of the gene that encodes DCC is deleted in colorectal carcinoma; this name reflects the initial discovery of DCC not in the nervous system, but rather in cells that line the colon, and it is a reminder that many genes and gene products that neuroscientists like to claim as their own also have important functions in other organ systems. As would be predicted, the netrin receptors UNC-40 and DCC are expressed by the growing axons rather than their targets. Netrins are now known to have many functions (and other receptors) in the developing nervous system, but their signature action, conserved across a vast swathe of animal evolution, is attraction of growing axons to the midline.[18]

The spinal cords of transgenic mice deficient in *netrin-1* expression developed almost normally, but close examination revealed a phenotype consistent with the proposed role for netrin-1 as a cue for growing commissural axons: homozygotes had a significant reduction in the thickness of the ventral commissure of the spinal cord compared with both wild-type and heterozygote animals (fig. 7.8).

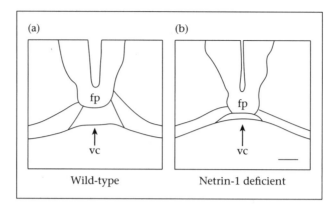

Figure 7.8. Reduction of the ventral commissure in netrin-1-deficient animals. Drawings based on histological cross sections through the ventral commissure and floor plate of the spinal cord in wild-type (a) and homozygous mutant (b) netrin-1-deficient mice. Scale bar indicates 30 micrometers. fp, floor plate; vc, ventral commissure. Adapted from Serafini et al. (1996).

Pioneers and Fascicles

The first axon to grow to a particular location in the nervous system is often referred to as the pioneer axon. This designation indicates that, unlike all subsequent axons that traverse the same route, the pioneer must rely entirely on cues that originate in other tissues or other types of neurons. Follower axons are envisioned as having a somewhat easier time of it, because they can simply grow alongside the pioneer axon. It is a characteristic of nervous systems that axons form bundles, which, depending on their location, are referred to either as tracts (within the central nervous system) or as nerves (projecting to or from the central nervous system). Because another term for tracts is *fascicles,* axons that grow together are described as fasciculated.[19]

Fasciculation results from interactions between molecules expressed on axon surfaces. Such molecules can be thought of as providing guidance information to follower axons (although this does not preclude the possibility that follower axons also respond to the same cues that attracted the pioneer axon). The general term for molecules that produce fasciculation is *cell adhesion molecules,* or CAMs. Many different types of CAMs have been discovered. Some CAMs, particularly those in the immunoglobulin G superfamily, bind to each other. This type of binding is referred to as homophilic, which means that the same molecule serves as both ligand and receptor. (In this lingo, binding mediated by different ligands and receptors is referred to as heterophilic.) The immunoglobulin G–type CAMs were first discovered in the vertebrate nervous system, but many subsequent studies have focused on the role of CAMs in the developing insect nervous system. *Drosophila* has two immunoglobulin G–type CAMs, fasciclinI (FasI) and fasciclinII (FasII). Studies using antibodies specific to FasI or FasII have revealed that FasI is expressed primarily by commissural axons, whereas FasII is expressed primarily by longitudinal axons (fig. 7.9). Misexpression of fasciclins in the *Drosophila* embryo does not block axon outgrowth, but some defasciculation is evident in mutants.

Congenital Mirror Movements

In this chapter the netrins and the fasciclins of *Drosophila* were used as examples to suggest the richness of the signaling environment that growth cones sample as axons grow to their targets. If you are reading carefully you will have noted that reduced expression of these signals perturbs development but does not bring it to a halt. These findings imply that myriad signals guide the growth of axons to their targets. Many of these other signals are known, but many remain to be discovered. The field is sufficiently mature, however, to allow neuroscientists to ask if defects in the development of the human nervous system can be attributed to failures of axon guidance or cell adhesion.

The functional consequences of mutations in genes involved in axon guidance could be significant if the axons never reach their normal targets. An example of such a mutation involves DCC, the netrin receptor. The resulting human syndrome is congenital mirror movements. The term *mirror movements* refers to unwanted, involuntary movements performed on one side of the body mirroring intentional movements performed by the other side of the body. The corticospinal pathways controlling voluntary move-

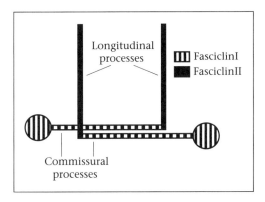

Figure 7.9. Expression of fasciclinI and fasciclinII by developing interneurons in the segmental ganglia of insect embryos. This schematic diagram of growing axons is based on the pattern of immunolabeling obtained with antibodies specific to the membrane-associated glycoproteins fasciclinI and fasciclinII. These interneurons express fasciclinI on the commissural segments of their axons and fasciclinII on the longitudinal segments. The fasciclins and other membrane-associated proteins mediate the formation of tracts within the central nervous system. Drawing adapted from Harrelson and Goodman (1988).

ment in humans are commissural—that is, stimulation of the primary motor cortex of the right side of the brain results in movements on the left side of the body and vice versa. In patients with congenital mirror movements, stimulation of one side of the brain results in synchronized movements on both sides of the body.[20] One reason that this might occur is a partial failure of corticospinal neurons to be attracted to the midline, which is a necessary precondition for axons to grow across the midline. Studies of three unrelated families in which multiple individuals have congenital mirror movements identified mutations in the *DCC* gene as the most likely cause of the symptoms.[21] This is evidence consistent with the view that netrin signaling is a component of commissural axon growth in humans, as is the case in all other animals that have been studied.

Roundabout and Slit

So many attractive and repulsive cues guide axon growth cones that it is difficult to stop at netrin. But how can one select just one more cue from such a lengthy list? In this case I opt for one of my personal favorites: the now classic story of the roundabout receptor and its partner, a ligand called Slit.

The *roundabout* gene (the gene symbol is *robo*) was discovered in a large-scale, multiyear forward genetics screen for mutations affecting axon growth in *Drosophila* embryos.[22] Over 13,000 mutant embryos were screened for abnormal growth patterns using a monoclonal antibody specific to central nervous system axons. The mutations were generated by treating male flies with a chemical mutagen and mating them with virgin females from carefully selected genetic lines. Once stable lines had been obtained, embryos were collected, immunolabeled, and examined en masse as wholemounts under a stereomicroscope. The screeners were searching for deviations from the stereotyped, ladderlike scaffold of longitudinal and commissural axons typical of wild-type embryos.

One of the genes discovered in this screen turned out to be *robo*. The *robo* mutants were fairly normal, and some even survived to adulthood, but immunolabeling with the axon-specific monoclonal antibody revealed an interesting phenotype: *robo* mutants had fuzzier and thicker commissures (and thinner longitudinal tracts) than normal. This phenotype reflected a propensity for longitudinal axons to enter commissures and cross the midline, a place to which no self-respecting growth cone should ever lead a longitudinal axon.

Robo, the protein encoded by the *robo* gene (now called Robo1 because additional Robos have since been discovered) is now known to be another member of the immunoglobulin superfamily of proteins. It is expressed on the growth cones of longitudinally growing axons but not on the growth cones of axons in the process of growing across the midline. It is also expressed on growth cones of commissural axons as they leave the midline and grow toward their targets (that is, commissural axons first need to be attracted to and then repelled from the midline). Its ligands are the Slit proteins, which are secreted by midline cells.[23] Slit ligands, characterized by leucine-rich repeat domains, repel axon growth cones.

Robo receptors and Slit ligands are also present in vertebrates. The mouse genome encodes three Robos and three Slits. In the embryonic nervous system, Robos are expressed by neurons, whereas Slits are expressed by the cells that form the floor plate at the ventral midline of the developing spinal cord. Slit triple mutants (knockouts lacking any expression of Slit1, Slit2, and Slit3) displayed a phenotype of disorganized commissures.[24] Studies using a tracer (the lipophilic DiI) to track the movements of growth cones revealed that many growth cones appeared to *stall* at the midline. A similar phenotype was seen in mouse embryos lacking expression of the Robo1 and Robo2 receptors. In addition to providing another striking example of conservation of function across animal phyla, the story of Robo and Slit also provides a reminder of the molecular redundancy characteristic of vertebrates, for it proved necessary to delete all six alleles encoding the three Slit proteins from the mouse genome before defects in axon growth cones were evident.

The third mammalian Robo (Robo3, also called Rig1) is less closely related to the *Drosophila* Robos than are Robo1 and Robo2. Although Robo3 is, like the other two Robos, a receptor for Slit, it has an additional mechanism of action. Mice lacking Robo3 have a different phenotype than mice lacking Robo1 and Robo2: they do not form ventral commissures in either the hindbrain or the spinal cord.[25] Subsequent studies led to the hypothesis that growth cones lacking Robo3 are prematurely responsive to the repellant effects of Slit ligands—that is, that Robo3 regulates the responses of the other Robos to Slits

Analysis of the causes of a rare autosomal recessive genetic disorder of humans has demonstrated the importance of human Robo3 in the development of the nervous system. The disorder is horizontal gaze palsy with progressive scoliosis (HGPPS). Among other traits, individuals with this condition lack the ability to make coordinated horizontal eye movements. This ability, which we typically take for granted, depends on axons that cross the midline to connect the abducens and oculomotor nuclei. These axons travel in a tract called the medial longitudinal fasciculus. Magnetic resonance imaging of HGPPS patients revealed anatomical abnormalities in the ventral brain stem consistent with a failure of axons to cross the midline.[26] Follow-up electrophysiological studies confirmed the failure of axon tracts to make the crossing.

The mutation responsible for HGPPS had been previously localized to Chromosome 11. DNA analysis of HGPPS patients revealed that all had mutations located in the portion of the *Robo3* gene that codes for the extracellular domain of the Robo3 protein. Unaffected relatives of these patients did not have mutations in the *Robo3* gene. It's hard to think of a tool of modern developmental neuroscience that did not contribute to this exciting discovery of the causes of this previously puzzling human disorder.

Synaptogenesis

The Neuromuscular Junction

When a growing axon reaches its target, the growth cone at its tip transforms into an axon terminal and a synapse forms. Synapses are complex, asymmetrical cell junctions found only in the nervous system. The plasma membranes of the pre- and postsynaptic cells (either a neuron and a muscle fiber or a neuron and another neuron) are apposed but separated by a tiny space called the synaptic cleft.[27] The typical synapse is specialized for chemical communication. Small-molecule neurotransmitters—for example, acetylcholine, glutamate, and GABA—are released from the presynaptic axon terminal, diffuse across the synaptic cleft, bind to receptors in the postsynaptic membrane, and modify the membrane potential of the postsynaptic cell. The information transfer that occurs at the synapse is the raison d'etre of the nervous system![28] The formation of synapses is naturally highly characteristic of the early stages of nervous system development, but synapses are formed, modified, and eliminated over the lifetime of an individual.

The neuromuscular junction is the synapse formed between the axon terminals of motoneurons and the surface of skeletal muscles. Compared with synapses in the central nervous system, the neuromuscular junction is larger and more accessible. Although the neuromuscular junction is specialized, it shares the basic elements of a synapse with other synapses. At a minimum, the presynaptic compartment contains the structures required for neurotransmitter synthesis and release; the postsynaptic compartment contains neurotransmitter receptors and all of the molecular machinery required to tether the receptors to the synapse.

The basic structure of the neuromuscular junction is similar in all animals, but the details are sufficiently different that I need to specify that here I am describing a vertebrate neuromuscular junction. In vertebrates, each muscle fiber (the name given to the individual muscle cells that make up a muscle organ) is innervated by the branched terminals of a single motor axon. The neurotransmitter released from the axon terminals is acetylcholine, which binds to acetylcholine receptors in the plasma membrane of the muscle fiber.

In the 1950s the venom of a poisonous snake, the Taiwanese banded krait (*Bungarus multicinctus*), was discovered to contain a peptide neurotoxin that binds selectively and irreversibly to nicotinic acetylcholine receptors. This toxin (named α-bungarotoxin) causes muscle paralysis in krait bite victims but has proved a valuable tool for study of the neuromuscular junction. This is because clusters of acetylcholine receptors define the location of neuromuscular junctions. Application of labeled α-bungarotoxin therefore reveals the distribution of acetylcholine receptors on the muscle fiber.

Studies using α-bungarotoxin revealed that acetylcholine receptors are present and cluster prior to innervation of the muscle but that clustering is enhanced when the axon terminal of a motoneuron arrives. This implies that a signal from the axon terminal promotes receptor clustering and starts the formation of a synapse. The glycoprotein agrin has been identified as one of the signals provided by the axon terminal. Agrin is synthesized by motoneurons, transported from somata to the axon terminals, and released onto the surface of the muscle fibers. This release of agrin results in the appropriate postsynaptic clustering of the acetylcholine receptors. Biochemical analyses revealed that agrin's actions reflect its ability to induce the tyrosine phosphorylation of MuSK, a muscle-specific kinase. Agrin regulates MuSK but does not bind MuSK directly. Instead, MuSK is part of a larger protein complex that includes a lipoprotein receptor-related protein called Lrp4 (fig. 7.10). Current evidence is consistent with a model that defines Lrp4 as the agrin-binding site. Transgenic mice with no agrin never moved and were stillborn or died shortly before birth on E18. The primary defect appeared to

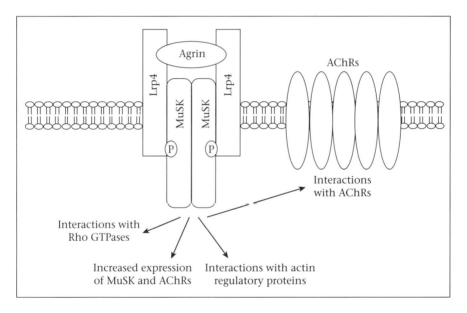

Figure 7.10. Agrin, MuSK, and Lrp4 at the vertebrate neuromuscular junction. Agrin, produced by motoneurons, activates MuSK, a receptor tyrosine kinase produced by muscle by binding Lrp4, a member of the low-density lipoprotein receptor gene family. This signaling is required for the formation of the vertebrate neuromuscular junction, especially the clustering of the acetylcholine receptors (AChRs). P indicates the location of a regulatory phosphate group; Rho GTPases are a family of proteins that regulate the actin cytoskeleton of cells, including nerve and muscle cells. Drawing adapted from figure 3 of Wu et al. (2010).

be a reduction in the postsynaptic clustering of acetylcholine receptors, which led to failure to form functional neuromuscular junctions.[29] Defective clustering of acetylcholine receptors has also been reported for MuSK and Lrp4 knockout mice.

Understanding the importance of the agrin–MuSK–Lrp4 interaction for the development and maintenance of neuromuscular junctions has led to an understanding of the etiology of myasthenia gravis in a subset of affected individuals. Myasthenia gravis is an autoimmune disorder that affects voluntary muscles. The symptoms vary from individual to individual, but all are an indication of muscle weakness: drooping eyelids, blurred or double vision, slurred speech, difficulty chewing and swallowing, fatigue of arm and leg muscles, and difficulty breathing. The majority (85 percent) of patients with myasthenia gravis produce antibodies against the acetylcholine receptor. The antibody-mediated destruction of their receptors is responsible for their weakness. A substantial proportion of those who test negative for acetylcholine receptor antibodies, however, test positive for antibodies against MuSK. This indicates that inappropriately clustered acetylcholine receptors can result in insufficient muscle innervation to the extent that normal function is impaired.

Central Synaptogenesis

Some central synapses are cholinergic, but the major excitatory neurotransmitter in the central nervous system is glutamate. For most glutamatergic synapses, the postsynaptic partner is a dendritic spine. Is clustering of the receptors for glutamate—AMPA receptors and NMDA receptors—relevant to the formation of glutamatergic synapses?

Experimental evidence supports the hypothesis that receptor clustering is a universally important feature of synaptogenesis. One example comes from the study of synapse formation in cultured hippocampal neurons. As discussed earlier in this chapter, cultures of hippocampal neurons prepared from embryonic or neonatal rats or mice have been used to study the development of neuronal polarity. The dramatic events of initial process outgrowth occur during the first 3–4 days in culture. If the cultures are carefully maintained, however, the neurons will continue to grow and can survive for 3–4 weeks. During the second and third weeks of culture, the neurons growing on a coverslip interact and dendritic spine and synapse formation begin. What was initially an in vitro system for the study of neuronal polarity becomes an in vitro system for the study of synaptogenesis. In these neurons, contact between an axon and a dendrite triggers dendritic spine formation. The spine then matures as a functional postsynaptic signaling complex, acquiring clustered NMDA and AMPA receptors and changing morphologically from a thin filament to a stout, mushroom-shaped spine (fig. 7.11).

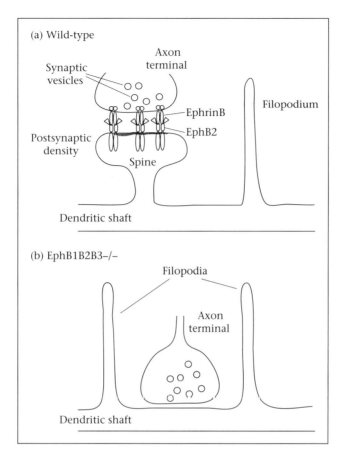

Figure 7.11. Ephrin–EphB signaling in hippocampal neurons. Cultured wild-type hippocampal neurons form mushroom-shaped spines that receive synapses from other cultured hippocampal neurons (a). The spine heads contain polymerized actin (not shown) and an obvious postsynaptic density. Some filopodia are present. By contrast, hippocampal neurons obtained from triple EphB1B2B3–/– knockout mice do not form mushroom-shaped spines, although occasional synapses on the dendritic shaft are observed and many filopodia are present. Drawing adapted from Henkemeyer et al. (2003).

The maturation of the dendritic spines of cultured hippocampal neurons is dependent on the signaling of proteins called ephrins via receptor tyrosine kinases called Ephs. The ephrins are membrane-bound ligands expressed primarily by axons. Conversely, Eph receptors have been localized to dendrites. Many different Eph receptors are expressed in the hippocampus. These include three Type B Ephs (EphB1, EphB2, and EphB3), which are activated by binding to Type B ephrins. To assess the role of EphB receptors in receptor clustering and synaptogenesis, investigators used triple mutant mice lacking all three of the EphB receptors.[30] Cultures of hippocampal neurons were established from wild-type mice and the triple mutants and allowed to grow for up to 4 weeks. The mutants failed to form spines, and immunolabeling using antibodies directed to the AMPA and NMDA glutamatergic receptors revealed a lack of receptor clusters. Follow-up in vivo analyses revealed a similar but less severe phenotype in the intact hippocampus of the mutant mice, supporting the importance of ephrin–EphB signaling for synaptogenesis but also indicating that, in vivo, other signals contribute to the process.

Consideration of the formation of glutamatergic synapses in the central nervous system brings this chapter nearly full circle. The close coupling of dendritic spine morphogenesis and synaptogenesis again reveals the critical role played by regulators of the neuronal cytoskeleton in the development of the nervous system.

Notes

1. How long is the longest axon? In tall humans, the axons of the spinal motoneurons that innervate muscles in the foot can be a meter or more in length. It is easy to imagine that very tall animals (think of giraffes) and very long animals (think of whales) have even longer axons. Very large extinct animals (think of dinosaurs) may have had the longest axons that ever existed on our planet.

2. An excellent and entertaining account of early studies of neuronal polarity is provided by Craig and Banker (1994). Detailed accounts of methods for culturing neurons are provided by Banker and Goslin (1998).

3. Mutations in the human APC gene cause familial adenomatous polyposis, a condition in which large numbers of polyps (abnormal growths of tissue in a mucous membrane) form in the colon. These polyps inevitably progress to colon cancer. The APC gene product was therefore originally described in the literature as a tumor suppressor. Sporadic mutations in the APC gene are now also linked to cancer in other organs, including the liver, ovary, and testis. Note that only a small proportion of patients with colon cancer have familial adenomatous polyposis, indicating that acquired gene mutations likely account for a majority of these cancers.

4. When a protein becomes conjugated to ubiquitin, it is typically said to be ubiquitinated, but it can also be said to be ubiquitinylated—the two terms mean the same thing. Ubiquitination or ubiquitinylation regulates protein turnover in cells. Ubiquitin does not degrade proteins directly but rather tags them so that they can be recognized and subsequently destroyed by the proteasome. The proteasome is a complex of enzymes that destroys proteins by breaking peptide bonds. The 2004 Nobel Prize in Chemistry was awarded to Aaron Ciechanover, Avram Hershko, and Irwin Rose for their discovery of ubiquitin-mediated protein degradation.

5. Studies showing clusters of APC and β-catenin at microtubule tips in cell protrusions were performed in Madin-Darby canine kidney (MDCK) cells. This immortalized cell line was developed in 1958 from dog kidney tissue and is widely used to study the cell and molecular biology of epithelial cells. See Faux et al. (2010).

6. The authors of the paper reporting the establishment of the PC12 cell line noted that one of the "numerous problems in neurobiology and neurochemistry"

these cells could be used to address was "initiation and regulation of neurite out-growth." See Greene and Tischler (1976), 2428.

7. Tubulin seems so perfect that it is hard to think of these proteins by any other name, but until the designation *tubulin* was established in 1968 they were called by various names: spactin, flactin, and tektin.

8. Taxol (also called paclitaxel) is a drug originally extracted from the bark of the uncommon Pacific yew tree (*Taxus brevifolia*) but now produced by plant cell fermentation from needles of the more common English yew (*Taxus baccata*). It is a powerful microtubule stabilizer. When it stabilizes the microtubules of the spindle fiber, it prevents the completion of mitosis. As a result, it is used as an antiproliferative drug for the treatment of cancer. Taxol is also used by neuroscientists to study the effects of microtubule stabilization on the growth of axons and dendrites.

9. Phalloidin is toxic and must be handled carefully in the laboratory, but it is not the cause of death in the unwary who ingest death cap mushrooms. These mushrooms contain small cyclic peptides called amatoxins. These toxins inhibit eukaryotic RNA polymerase II and thereby block the synthesis of RNAs and, ultimately proteins. The liver and kidneys are the most severely affected organs, and death typically occurs as a result of liver failure. Amatoxins extracted from *Amanita phalloides* are used in research as selective inhibitors of RNA polymerase II and to block protein synthesis.

10. The largest known growth cones are found in cultures prepared from large neurons of the marine mollusk *Aplysia californica*. Under some conditions, growth cones wider than half a millimeter can be observed. See Lovell and Moroz (2006).

11. See Buck and Zheng (2002).

12. See Grueber et al. (2005).

13. See Rolls (2011).

14. See Tessier-Lavigne and Goodman (1996).

15. The *Unc-6* mutation was described by Sidney Brenner in one of the original publications describing the use of *C. elegans* to investigate the genetic basis of behavior. See Brenner (1974).

16. The method of filling sensory neurons in *C. elegans* by immersion in solutions containing fluorescein dyes is described by Hedgecock et al. (1985). Note that the method fills only amphid and phasmid chemosensory neurons in wild-type *C. elegans* but will also fill PDE mechanosensory neurons in mutants with abnormal mechanosensilla.

17. See Serafini et al. (1994). The name of what is now known to be a small family of proteins characterized by their shared homology to the N-terminus of laminin and their shared ability to guide axons to the ventral midline is derived from the Sanskrit root of *netr,* translated into English as "one who guides."

18. The *Drosophila* netrin homologs, Netrin-A and Netrin-B, are expressed by glial cells and neurons located at the ventral midline of the developing nervous system. The *frazzled* gene encodes the *Drosophila* netrin receptor. There are many commissural neurons in the insect nervous system; mutations in either the netrin-encoding or the *frazzled* gene perturb the formation of commissures.

19. A fascicle is a little fasces. In ancient Rome, the fasces was a bundle of birch sticks used as a symbol of a magistrate's authority. The tradition of using the fasces as a symbol of governmental authority has continued into modern times, and the aware observer can find depictions of fasces in symbols associated with many levels of gov-

ernment in the United States. The fasces was also famously adopted as a symbol of power by twentieth-century European fascists, particularly in Italy.

20. See Farmer et al. (1990).

21. See Srour et al. (2010).

22. See Seeger et al. (1993) for a description of the large-scale screen that led to the discovery of *robo*. Another famous gene described in this paper is *commissureless*. The authors provide an exceptionally clear account of how the screen was designed and performed.

23. Thirteen alleles of Slit were recovered in the original axon guidance screen reported by Seeger et al. (1993). The initial publication resulting from this screen, however, focused on *robo* and *comm,* in part because of the lack of good midline markers. See Brose et al. (1999) and Kidd et al. (1999).

24. See Long et al. (2004).

25. See Sabatier et al. (2004).

26. See Jen et al. (2004). This amazing paper is authored by an international team of 35 investigators who used the following research methods to link the human *Robo3* gene to human horizontal gaze palsy with progressive scoliosis: pedigree analysis, magnetic resonance imaging, evoked potential recordings, linkage analysis using microsatellite markers, bioinformatics, RT-PCR, DNA sequencing, and in situ hybridization analyses of *Robo3* gene expression in fetal human brains.

27. Because synapses are typically studied using electron microscopy, which is performed using tissue exposed to various chemical fixatives, it is difficult to estimate the width of the synaptic cleft in living tissue. A typical estimate based on fixed specimens is 20–25 nanometers.

28. A French phrase that means, roughly, "reason for being."

29. See Gautam et al. (1996).

30. See Henkemeyer et al. (2003).

Investigative Reading

1. The protein Semaphorin 3A acts as a repulsive guidance cue by binding to receptors on growth cones. The response of growth cones to Semaphorin 3A is referred to as collapse: the growth cone loses its filopodia, retracts, and temporarily stops advancing. You are studying the mechanisms of growth-cone collapse in cultured dorsal-root ganglion neurons obtained from E13.5–E14 mouse embryos. Your goal is to understand how the signal transduction events initiated by Semaphorin 3A induce the reorganization of the growth cone's cytoskeleton. After growing your cells in culture overnight, you apply Semaphorin 3A to induce a controlled collapse of your growth cones. What tools do you use to study the resulting changes in the actin cytoskeleton of the growth cones?

Brown, Jacquelyn A., and Paul C. Bridgman. 2009. *Developmental Neurobiology* 69: 633–46.

2. There is evidence that several phases of development of the vertebrate neuromuscular junction are negatively regulated by the neurotransmitter

acetylcholine, which is released by developing motoneurons even before they reach their muscle fiber target. Transgenic mice that do not express the enzyme responsible for acetylcholine synthesis (choline acetyltransferase) have a phenotype of increased motor endplate width, indicating a failure to restrict the distribution of acetylcholine receptors to appropriate regions of the muscle fiber surface. They also display increased branching of their motor axon terminals relative to wild-type mice. What methods do you use to ask the following question: are the negative effects of acetylcholine on development of the neuromuscular junction mediated via acetylcholine receptors? If the answer to this question is yes, how do you determine if the relevant receptors are presynaptic or postsynaptic?

An, Mahru C., Wiechun Lin, Jiefei Yang, Bertha Dominguez, Daniel Padgett, Yoshie Sugiura, Prafulla Aryal, Thomas W. Gould, Ronald W. Oppenheim, Mark E. Hester, Brian K. Kaspar, Chien-Ping Ko, and Kuo-Fen Lee. 2010. *Proceedings of the National Academy of Sciences, U.S.A.* 107: 10702–7.

3. The phosphoprotein synapsin I is enriched in synaptic vesicles. You wish to test the hypothesis that, in addition to regulating activity-dependent neurotransmitter release at mature synapses, synapsin I also plays a role in the development of new synapses. Devise an experiment to test your hypothesis in a mammalian model.

Chin, Lih-Shen, Lian Li, Adriana Ferreira, Kenneth S. Kosik, and Paul Greengard. 1995. *Proceedings of the National Academy of Sciences, U.S.A.* 92: 9230–34.

4. You are interested in using the zebrafish model to understand the role of netrin proteins in axonal guidance. At the time you begin your project, you know that the genomes of the human and the mouse each contain three secreted members of the netrin family but that the only netrins known to be expressed in zebrafish are orthologs of netrin-1. How do you find out if any additional netrin family members are present in zebrafish? Describe at least one method that does not involve sacrificing any fish.

Park, Kye Won, Lisa D. Urness, Megan M. Senchuk, Carrie J. Colvin, Joshua D. Wythe, Chi-Bin Chien, and Dean Y. Li. 2005. *Developmental Dynamics* 234: 726–31.

EIGHT

Glia

Glia and Neurons

If brains could compete for Oscars, every year the award for best cell in a supporting role in nervous tissue would be won by glia.[1] In fact, in terms of supporting roles in nervous systems, glia (also called glial cells—the two terms are equivalent) have no competition worthy of the name. In addition to providing structural support and trophic factors, glial cells form the blood–brain barrier, regulate calcium flux, assist in the maintenance of water balance, and modulate synaptic transmission. In both vertebrate and invertebrate nervous systems, neurons and their processes are tightly intertwined with glia and their processes. In vertebrate nervous systems, glial cells wrap axons to form myelin, a layer of insulation that increases the conduction rate of action potentials, arguably one of the major evolutionary innovations that *allow vertebrates to be vertebrates*. Like neurons, glia produce highly branched processes, and like neurons, the embryonic origin of glia is the neural tube neuroepithelium. Despite these shared features, however, glial cells are not just another type of neuron. They lack fast ion currents, neurotransmitters packaged in vesicles, axons, and dendrites. Glial cells are essential components of nuclei in the central nervous system and of ganglia in the periphery, but they do not themselves form clusters. In contrast to neurons, differentiated glial cells are not inevitably postmitotic.

The dispersed distribution of glia in neural tissue slowed biochemical and molecular analysis of glial cell contents. Further, because glia are not organized into polarized input–output circuits and do not typically produce action potentials, glia do not easily reveal their secrets to the electrophysiologist. Ablation is uninformative because most populations of neurons do not survive the loss of their glia.[2] Research on glial differentiation and function was also hampered by a relative lack of glial-specific markers.

Recent advances in our understanding of glial function reflect application to the study of neural tissue of modern tools including transgenic animals, confocal microscopy, calcium imaging, and RNAi.[3] In addition, old prejudices that invertebrate glia have little to tell us about vertebrate glia are fading, inspiring new studies of glial cells in *C. elegans* and *Drosophila*. The clinical importance of glial cells is well recognized. Many injuries to the brain trigger reactive gliosis, a phenomenon also known as glial scar formation. Changes in glial cell protein expression and glial proliferation associated with reactive gliosis impede recovery from brain injury. It has also been estimated that as many as 95 percent of human brain cancers are glial in origin.

Glia in *C. elegans*

Until quite recently, the cells that are now described as glia in *C. elegans* were customarily referred to as glial-like. The grudging granting of glial status to these cells reflects how different they are from vertebrate glia and our current poor characterization of their function. If they don't look like glial cells, can they act like glial cells? A case can be made that the answer to this question is yes.

Of the 959 somatic cells of the adult hermaphrodite, 302 are neurons and only 50 are glia. As discussed later in this chapter, by contrast glia are a dominant cell type in the vertebrate nervous system. Of the 50 *C. elegans* glia, 24 are categorized as sheath glia and 26 as socket glia. With the exception of four sheath glia cells that wrap the circumpharyngeal nerve ring, all of the glial cells of *C. elegans* are associated with sensory synapses. The sheath glia wrap (ensheathe) sensory neuron dendrites; the socket glia do the same, but they also form channels through the cuticle that allow chemosensory dendrites to come into direct contact with odorants, salts, amino acids, and other small molecules.

The amphids are the main chemosensory organs of *C. elegans* (Chapter 3). An amphid is formed of 12 neurons. Each amphid also contains one glial cell called a sheath cell and one glial cell called a socket cell. Processes of the sensory neurons and the glial cells join at an opening in the cuticle at the anterior tip of the worm to form what might be called a nose (fig. 8.1). The extensions of the amphid cells are arranged so that glial cell processes line a channel that connects the sensory surface to the body surface. The channel contains the ciliated dendrites of the sensory neurons, and it is in this environment that the sensory neurons taste the world beyond the cuticle.

In first-stage larvae, the sheath glial cells of the amphid can be ablated by using a focused laser microbeam or by expressing a gene encoding a toxin under the control of a sheath glia–specific promoter.[4] The sensory neurons of the amphid survived this manipulation, indicating that the sensory neurons of *C. elegans* are not dependent on their glia for survival. The structure and function of the sensory neurons, however, were disrupted.[5] Notably, the ciliated dendrites of several sensory neurons had reduced branching. This is significant because the odorant receptors have a ciliary location; in effect, reduced branching means that there is a reduction in the area of the sensory

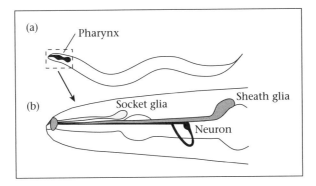

Figure 8.1. Amphid sensory organ of *C. elegans.* As shown in (a), which depicts an adult hermaphrodite, the amphid sensory organs are located at the head end of the worm. Each of the two bilaterally symmetric amphids comprises 12 neurons (one of which is depicted here) and two glial cells, the sheath cell and the socket cell (b). Neuronal and glial processes converge at the anterior tip. The glial processes form a channel through which amphid neuron dendrites extend sensory cilia. Drawings adapted from Oikonomou and Shaham (2011).

surface. Glia-ablated larvae were unable to swim toward two odors that intact larvae find highly attractive.[6] Other defects resulting from glia ablation included failure to detect a concentrated salt solution and failure to respond to an osmotic stimulus with the expected increase in intracellular calcium in the sensory neuron that mediates this behavior (fig. 8.2).

The study of glial function in *C. elegans* is at its dawn. At present, relatively little is known about development of glial cells in this species. This model has potential to help define molecular signaling pathways that regulate the close association of growing glial and neuronal membranes, particularly in the context of sensory neuron development and function.

Glia in *Drosophila*

The glia:neuron ratio in the nervous system of *C. elegans* is low compared with that of our other models, less than 0.2 (calculated on the basis of 50 glial cells and 302 neurons in the adult hermaphrodite). In mammals this ratio is estimated to be closer to 1.0.[7] In *Drosophila* this ratio likely falls somewhere in between. Until recently, studies of glia in *Drosophila* focused primarily on the embryonic origins of glial cells, but new findings suggest that fly glia might provide useful models at the cellular and molecular levels for reactive gliosis and gliomas, two phenomena of enormous clinical significance.

Three major categories of *Drosophila* glia have been described, primarily on the basis of their locations in the central nervous system: surface glia, cortex glia, and neuropil glia (fig. 8.3). Each of these categories can be subdivided. For example, the category of surface glia comprises two distinct cell populations called the outer perineurial glia and the inner subperineurial

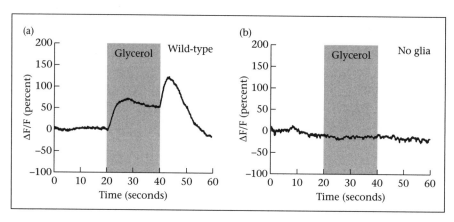

Figure 8.2. Glial cell requirement for wild-type intracellular calcium responses in *C. elegans* amphid sensory neurons. The neuron ASH is a multimodal sensory neuron that responds to various stimuli, including nose touch and changes in the osmolarity of the environment. As shown in (a), the ASH neuron of wild-type *C. elegans* responded to the application of glycerol (a hyperosmotic stimulus) with an increase in intracellular calcium. No change in intracellular calcium was produced by the same stimulus in a worm in which the sheath glia was previously ablated (b). Intracellular calcium was detected using a genetically encoded sensor protein called G-CaMP. G-CaMP fluoresces when it binds calcium. The measure ΔF/F is the ratio of the change in fluorescence intensity divided by the baseline fluorescence intensity. Graphs adapted from Bacaj et al. (2008).

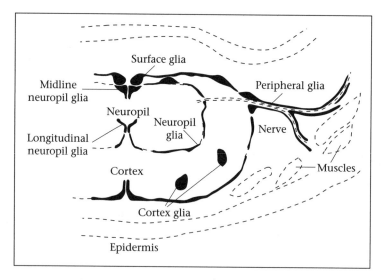

Figure 8.3. Distribution of the major categories of *Drosophila* glia in the embryonic ventral nerve cord. The schematic diagram shows a cross section through the ventral nerve cord at the level of a segmental ganglion in an embryonic Stage 16 embryo. The somata of the segmental neurons are found in the cortex (not shown); synaptic connections are made in the central neuropil. Drawing adapted from Hartenstein (2011).

glia. The processes of glial cells, which can be extensive, wrap all peripheral nerves and sensory organs. The glia that form this outer glial cell layer in the periphery are sometimes referred to as wrapping glia.

The origins and differentiation of specific examples of neuropil glia have been extensively studied (think of this approach as finding models within a model—even the relatively small number of glia in the *Drosophila* nervous system is too complicated to study all at once). We will turn our attention to the neuropil glia in a few paragraphs. Less is known about the origins of the other categories of glia, but their functions are too important to omit from even a brief account.

A layer of subperineurial surface glial cells ensheathes the *Drosophila* central nervous system. These cells are so flattened that they appear two dimensional, and they are so large that it takes fewer than 50 of them to surround one hemisphere of the adult brain. The subperineurial glial cells form the functional blood–brain barrier of the *Drosophila* nervous system. The blood–brain barrier restricts movement of large molecules into the central nervous system, thereby maintaining the stability of the extracellular environment of nervous tissue. The barrier is formed by the cell junctions that connect adjacent subperineurial glial cells. These junctions are called pleated septate junctions (fig. 8.4). Septate junctions significantly reduce the permeability of the layer of cells they connect. This type of junction is characterized by expression of proteins called neurexins. The central nervous systems of Stage 17 *Drosophila* embryos carrying a null mutation for one of the neurexins (neurexinIV) were found to be significantly more permeable to fluorescent dextrans than were wild-type embryos of the same stage, indicating that the blood–brain barrier had not formed properly.[8] In fact, the permeability of the blood–brain barrier of the *neurexinIV* mutants was similar to that in embryos lacking the entire complement of subperineurial glial cells.[9]

The neuropil glia of *Drosophila* surround the central neuropil of the brain and all of the segmental ganglia. The processes of some of these neuropil glial cells penetrate deeply into the neuropil. Although the processes of these glia are often thin, they define distinct compartments in the neuropil. The neuropil glia are diverse in shape and location. The longitudinal glia are the best-studied example of neuropil glia. The longitudinal glia are found in every thoracic and abdominal neuromere of the developing central nervous system. Their function is to wrap the connectives, the fascicles of neuronal processes that connect the segmental ganglia of the ventral nerve cord. At present, we have limited knowledge of the function of longitudinal glia in the adult fruit fly beyond the standard functions of supporting, sustaining, and insulating the connectives they ensheathe from the extracellular environment. Neuroscientists, however, have been able to study the development of this subcategory of neuropil glia by exploiting knowledge of their stereotyped location within the thoracic and abdominal neuromeres of the embryonic nervous system.

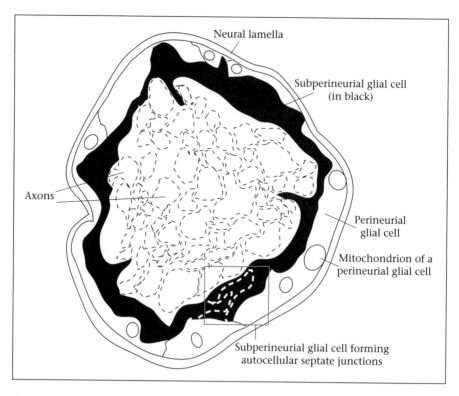

Figure. 8.4. Septate junction in a *Drosophila* nerve. Subperineurial glial cells form autocellular septate junctions in *Drosophila* nerves, creating a functional blood–brain barrier. In a drawing based on an electron micrograph of a cross section through a larval nerve, one subperineurial glial cell (in black) completely encircles a bundle of axons. Studies at higher magnification revealed that septate junctions form where the ends of the subperineurial glial cell meet. A small number of perineurial glia are found immediately beneath the acellular neural lamella. The extensive processes of another, repo-immunopositive *wrapping* glial cell, entwines the axons (not shown). The entire nerve is less than 10 micrometers in diameter. Drawing adapted from Stork et al. (2008).

The longitudinal glia of *Drosophila* originate in the neurogenic region of the ectoderm. The longitudinal glia are produced by divisions of blast cells that have separated from epidermis-forming regions of embryonic ectoderm. Some of the longitudinal glia are produced by dedicated glioblasts, but others are the products of neuroblast lineages. (Neuroblasts that produce both cell types are sometimes referred to as neuroglioblasts.) Each thoracic and abdominal neuromere in the embryonic fly contains 10 pairs of glia progenitor cells: 4 glioblasts and 6 neuroglioblasts. The signal that results in adoption of a glial cell fate is expression of a gene called *glial cells missing* (*gcm*). *gcm* is a transcription factor transiently expressed by almost all glial cells (a small subset of midline glial cells are the exception). Several proglial genes downstream of *gcm* have been identified. One of them, the persistently expressed *reversed polarity* gene, has provided *Drosophila* researchers with an invaluable marker for glial cells.

The *reversed polarity* gene is almost always referred to as *repo*. Its protein product is a DNA-binding homeodomain protein (Chapter 6) that promotes a glial fate for cells derived from ectoderm. Initial expression of the *repo* gene in differentiating longitudinal glia cells is evident at embryonic Stage 10 in cells close to the nascent connective in each thoracic and abdominal hemisegment. Expression of *repo* immediately follows expression of *gcm* in these same cells at Stage 9: if *gcm* expression is blocked, the neuroectoderm produces only neurons, and the glialess embryos die.

The *repo* gene is named for the phenotype produced by a hypomorphic mutation of the *repo* gene. The identification of the genetic locus that contains the *repo* gene provides an excellent example of the power of the forward genetics approach described in Chapter 3. A team of investigators was searching for recessive autosomal P-element-induced mutations that affect the adult visual system. Flies were screened by recording the electrical responses of their compound eyes to light using a simple method called an electroretinogram. Flies with the mutant allele responded to light with an electrical signal opposite in polarity to that of wild-type flies.[10] Further investigation revealed that the defect was not in the photoreceptor cells themselves, which produced a normal receptor potential. Instead, the defect was in the number and organization of glial cells in the brain optic lobes.

Null mutations in the *repo* gene are embryonic lethal, and all glia in the central and peripheral nervous systems (with the exception of midline glia) can be identified by their expression of *repo*. Lines of *Drosophila* have been developed that express markers such as Gal4 or GFP under the control of the *repo* promoter. As described in Chapter 1, intracellular injection of tracers (such as fluorescent dyes) into putative progenitor cells in these flies can be used to trace the lineage of the longitudinal glial cells.[11] These studies have revealed that even the longitudinal glia progeny of a single glioblast differentiate distinct gene expression phenotypes. The functional significance of this diversity is currently unknown.

The position of the longitudinal glia in each hemisegment and their close association with axons growing out from the ganglia led to the hypothesis that longitudinal glia play a role in axon guidance. Surprisingly, experimental ablation of the longitudinal glia revealed that these cells play a minor role in axonal pathfinding, for the axons that form the connectives continued to grow in their absence. But careful analysis of a slightly later stage in development revealed defects in fasciculation and defasciculation, implying that interactions with longitudinal glia are critical for the development of axon bundles. Analysis of a later stage revealed another, possibly even more important, role for longitudinal glia in development. Recall (from Chapter 7) that the axons of pioneer neurons form a scaffold that can be used by follower neurons as they extend their axons. In contrast to pioneer neurons, many follower neurons did not survive ablation of longitudinal glia. This result suggests that some populations of *Drosophila* neurons depend on interactions with glia for trophic factors. A trophic role for axon-wrapping glia in the fly central nervous system has parallels in the function of oligodendrocytes in the vertebrate nervous system.

To recap, there are three major categories of glia in the *Drosophila* nervous system: surface, cortical, and neuropil. At present, the best-studied glial populations in this insect in terms of development and function are the segmentally repeated neuropil glia of the thoracic and abdominal neuromeres. These cells are similar to neurons in that they have their origin in embryonic neuroectoderm and are produced by dedicated glioblast or glioneuroblast progenitor cells. They are also similar to neurons in that members of the same lineage can differentiate diverse complicated phenotypes. Expression of a unique transcription factor, encoded by the gene *gcm,* is sufficient to trigger differentiation as a glial cell. One of the targets of the gcm transcription factor, *repo,* is a useful marker for glia at all stages of *Drosophila* development.

We now shift our focus to the development of the glial cell populations of vertebrates, although we return to the fruit fly at the end of this chapter to show that studies of *Drosophila* have the potential to improve understanding of glial abnormalities in our own brains.

Glia in Zebrafish

The different vertebrate glia have been given many names. These rich cellular taxonomies reflect both the creativity of anatomists and the gap between studies of mammals and those of nonmammals prior to the development of molecular tools for study of the nervous system. The maturation of zebrafish as a model for neuroscience has fostered a greater appreciation of both the unity and the diversity of vertebrate glia.

A useful first approach to organizing information about vertebrate glia is to think in terms of broad morphological and functional categories. Four categories emerge from this perspective: surface-lining, axon-wrapping, dispersed stellate, and phagocyte-like. Note that these terms are not the name of any specific type of differentiated glial cell. But the rich diversity of cell types described in the literature is encompassed by these categories, in much the same way that the general terms *motoneuron, interneuron, sensory neuron,* and *neurosecretory cell* encompass the rich diversity of neuronal populations.

Two of the glial categories receive no further coverage here, despite their functional significance: the surface-lining glia (ependymocytes and tanycytes) and the phagocytelike glia (microglia). The surface-lining glia function to maintain homeostasis within the nervous system. Microglia are the products of a nonectodermal immune cell lineage beyond the scope of this text. Populations of both types of cells are established early in the development of vertebrate nervous systems. It might seem plausible that the surface-lining glia form the blood–brain barrier of the nervous system, as was the case for the surface-lining glial cells (subperineurial glia) of *Drosophila*, but this is not the case. Instead, the endothelial lining of the blood vessels of the brain constitutes the blood–brain barrier of vertebrates.[12]

The remaining categories of glia—axon-wrapping and dispersed stellate—are sometimes described using the term *neuronal companions*. This term captures well the ubiquitous nature of axon-wrapping and dispersed stellate glial cells in vertebrate nervous systems. It is true that neurons meet other neurons directly at synapses without any glial cells in between, but it is also true that the neuronal membrane–glial membrane interface is vastly larger than the neuronal membrane–neuronal membrane interface.

The white matter of the vertebrate brain is composed of axons and the glia that wrap them to form myelin. In the white matter of the central nervous system of vertebrates, the axon-wrapping glia are called oligodendrocytes; in the peripheral nervous system, axons are myelinated by glia called Schwann cells. The dispersed stellate cells are called astrocytes. Not all astrocytes are star shaped, as the term *stellate* implies. In fact, we have already discussed an example of a non-star-shaped astrocyte, the radial glial cell, in the context of neurogenesis (Chapter 5). We will soon have more to say about oligodendrocytes and astrocytes. But because the literature on glial cells in zebrafish is small relative to the literature on glial cells in mice, we will defer this discussion until the mouse section of this chapter. There is, however, a fish tale to tell before moving on. It is the story of how the axons of the zebrafish lateral line system develop their association with axon-wrapping glia.

The lateral line system of fish (and other aquatic animals, including amphibian larvae) consists of sensory organs distributed in an orderly row—the *line*—extending along both sides of the body. The sensory organs are called neuromasts and, depending on the species, are located either on the body surface or in a subepidermal channel just below the surface of the skin (fig. 8.5). A neuromast contains sense cells (15–20 per neuromast) and support cells. Each sense cell extends a structure called a kinocilium that depolarizes or hyperpolarizes the sense cell depending on the direction of kinocilium deflection.[13] The result is that the fish is able to sense the direction of water movement. Fish have many uses for this information: for example, lateral line input enables fish to maintain their orientation in a water current, school with other fish, detect prey, and evade approaching predators. The posterior neuromasts on either side of the body are innervated by bipolar sensory neurons that have their somata located in a ganglion found just behind the ear. The ganglion where these somata reside is called the posterior lateral line ganglion.

The posterior lateral line ganglion, like many other sensory ganglia in vertebrates, develops from a region of the embryonic epidermis called a placode (in this case, the posterior lateral line placode).[14] Studies using marker genes

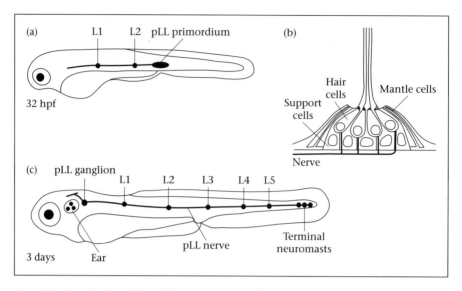

Figure 8.5. Posterior lateral line (pLL) sensory system of zebrafish. At 32 hours post fertilization (hpf), the posterior lateral line primordium has migrated down the trunk and two neuromasts (L1 and L2) have formed (a). Each neuromast organ contains sensory neurons and various supporting cells, including glia (b). Three days post fertilization, the pLL system consists of neuromasts L1–L5 and several terminal neuromasts (c). The pLL ganglion is located posterior to the developing ear. An afferent nerve (not shown) that runs parallel to the efferent nerve innervates the hair cells. Drawings adapted from Ma and Raible (2009).

expressed in neural crest cells suggest that, as is the case for most elements of the peripheral nervous system, the lateral line organs derive from the neural crest and not the neural tube.[15] The posterior lateral line placode develops around 18 hpf (hours post fertilization). It consists of a rostral cluster of sensory neurons and glial precursors and a caudal cluster called the lateral line primordium. The neurons in the rostral cluster stay put, but the caudal cluster begins to migrate away from the placode in the direction of the tail at approximately 22 hpf. The migration of the primordium across the somites is completed by the end of embryonic life (48 hpf). The cells of the primordium divide as they migrate; some do not complete the migration but instead are left at specific locations along the sides of the body to form an initial set of neuromasts: five along each flank, and two to three in the tail. More neuromasts are added as the fish grows, not only during larval life but also as growth continues in adulthood. A large adult zebrafish (roughly 30 millimeters in length) can have more than 500 neuromasts on each side of its body.

Here we focus on the early hours of initial lateral line development. As the primordium migrates, the growth cones of the axons of the sensory neurons and the precursors of their ensheathing glial cells migrate along with them. If appropriate markers are introduced, this movement of growth cones and glial cell precursors can be tracked in the living embryo. Careful observation paired with selective ablation of glia or the sensory neurons revealed the following.[16] First, the glial cells always trail the growing axons—they never jump out ahead. Second, the lateral line nerve grows normally even in embryos in which the glia have been ablated. Third, the glial cells stall or get stuck when the nerve is missing. A final experiment confirmed the dependence of glial migration on signals from the axons. Genetic manipulations produced embryos with abnormally patterned somites. In these embryos, the lateral line axon is misrouted, and the glial cell precursors tamely followed the wandering axons. The conclusion is similar to that reported for pioneering axons entering the connectives in the embryonic fruit fly: the glial cells (in this case, the glial cell precursors) are not required for axonal pathfinding. In fact, the reverse appears to be true.

But this is not the end of the story. Despite normal initial outgrowth of the lateral nerve in the absence of the usual glial complement, things did not go so well at later times. After 72 hpf, the individual axons of the lateral line nerve no longer formed a cohesive bundle recognizable as a nerve. They began to defasciculate. This finding is also similar to the result obtained in fruit flies lacking longitudinal glia.

It is so customary to encounter axons in bundles—tracts within the central nervous system, nerves in the peripheral nervous system—that we rarely stop to wonder how the integrity of these bundles is maintained. As indicated, glial cells are likely to be an important element in the maintenance of axon bundles. The study of lateral line organ formation in the zebrafish provides a reminder that the interaction between neurons and glia is a well-regulated two-way street.

Glia in Mice

The axon-wrapping and dispersed stellate cells of the mammalian central nervous system are, respectively, oligodendrocytes and astrocytes. They are sometimes collectively referred to as macroglia. Recent studies of the transcriptome—the complete set of genes being expressed—of differentiated oligodendrocytes, astrocytes, and neurons have demonstrated that, in terms of gene expression, these are three distinct cell types. This means that, in the adult nervous system, oligodendrocytes and astrocytes are no more alike than oligodendrocytes and neurons or astrocytes and neurons.[17] All three cell types share a common origin in the neural tube. Networks of secreted signals and transcription factors must operate during development to produce the right mix of neurons, astrocytes, and oligodendrocytes. Many of the same signals important in neurogenesis and in the regionalization of neuronal populations are also important in gliogenesis and in the differentiation of mature glial phenotypes. As you read the sections that describe the origins of oligodendrocytes and astrocytes, think in terms of overlapping yet distinct transcriptional networks. At present we are just beginning to appreciate the extraordinary complexity of nervous system transcriptomes.

Oligodendrocytes

Oligodendrocytes are the myelinating cells of the vertebrate central nervous system. Myelin-wrapped axons display saltatory conduction, with action potentials generated at breaks in the myelin called nodes of Ranvier. Myelin is a mixture of lipids and proteins, all of which are produced by oligodendrocytes. Some of the abundant components of myelin are myelin basic protein, myelin oligodendrocyte glycoprotein, and proteolipid protein. Oligodendrocytes are present in all locations in the central nervous system. All mature oligodendrocytes myelinate axons. Some oligodendrocytes may wrap many internodes (the distance from one node of Ranvier to another) on multiple fine axons, whereas oligodendrocytes that wrap large-caliber axons may myelinate only a single internode of a single axon.

In the mammalian brain and spinal cord, the onset of gliogenesis begins as the ferocious early phase of neurogenesis wanes. The origin of oligodendrocytes in the same ventricular and subventricular zones of the brain and spinal cord that produce neurons led to development of a model in which the same stem cells that initially produce neurons later switch to making glia.

Oligodendrocyte precursor cells (OPCs) are produced by divisions of stem cells in defined regions of the ventricular zone. They then migrate to other brain regions. Some migrating OPCs proliferate, giving rise to additional OPCs. OPCs become postmitotic upon arrival at their final destination. At this point they are no longer OPCs but premyelinating oligodendrocytes. If appropriate cues are present, premyelinating oligodendrocytes begin the synthesis of the lipids and proteins that allow them to myelinate axons. Myelination is regulated by back-and-forth signaling between axons and oligodendrocytes. This signaling involves a blend of secreted molecules, cell-surface interactions, and an ability on the part of oligodendrocytes to detect the level of electrical activity in an axon.

Here we focus on the generation of OPCs in the mouse central nervous system, using the ventral progenitor domain of the mouse spinal cord as our example (fig. 8.6). As described in Chapter 6, the ventricular zone is divided into progenitor domains that, at different stages of development, produce specific cell types. Recall that these progenitor domains are patterned by dorsal–ventral gradients of morphogens; two of the cues are Shh from the floor plate and Bmp from the roof plate. The pMN progenitor domain generates motoneurons from E9 through E10.5, then switches to producing oligodendrocytes from E12.5 through E14.5. At later stages of development, including during postnatal life, other progenitor domains continue the production of oligodendrocytes. Currently, the switch from neurogenesis to gliogenesis is best understood for the pMN domain, but similar regulatory

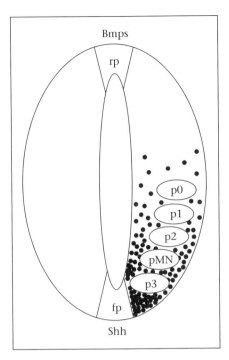

Figure 8.6. Oligodendrocyte origins in the spinal cord of a mouse. A highly schematized cross section of the embryonic spinal cord shows the roof plate (rp) at dorsal midline and the floor plate (fp) at ventral midline. The roof plate is a source of bone morphogenetic proteins (Bmps); the floor plate is a source of Sonic hedgehog (Shh), which regulates transcription in a concentration-dependent fashion in the ventral spinal cord of the mouse. The gradient of Shh is fancifully indicated by the black dots. The pMN progenitor domain is a source of motoneurons from E9 through E10.5, then switches to producing oligodendrocytes. The p0, p1, p2, and p3 progenitor domains produce interneurons. Drawing adapted from Rowitch (2004).

pathways also play out in other progenitor domains in the spinal cord and forebrain.

It is a general rule that neurogenesis precedes gliogenesis. Another general rule is that the same proliferating cells in the progenitor domains produce both neuronal and glial progenitor cells.

The generation of OPCs and hence production of oligodendrocytes is dependent on Shh. OPC generation in the spinal cord at E12.5 does not begin if Shh has not been present. This reflects the dependence of the pMN progenitor domain on the ventral-to-dorsal Shh gradient. But Shh is not a specific pro-oligodendrocyte signal. We need to look further downstream.

In the pMN, Shh triggers expression of a set of homeodomain proteins: Pax6, Nkx6.1, and Nkx6.2. These, in turn, drive expression of Olig proteins, including Olig proteins encoded by the genes Olig1 and Olig2. Oligs are members of the bHLH family of transcription factors. Mutant mice lacking both Olig1 and Olig2 proteins do not produce the expected complement of pMN oligodendrocytes. But because they also produce no motoneurons, it must be the case that the switch from neurogenesis to gliogenesis requires regulatory factors in addition to the Oligs.

A case can be made that among the most important of these additional factors are the neurogenins, Ngn1 and Ngn2. Like the Oligs, Ngn1 and Ngn2 are bHLH transcription factors. Experimental analyses of the pMN progenitor domain have revealed that Olig2 promotes the expression of Ngn2 and that spinal cord motoneurons are produced when the Oligs and the Ngns are both expressed. By contrast, downregulation of Ngn expression combined with continued expression of the Oligs results in a shift to the production of OPCs and hence oligodendrocytes.

Identification of the interaction between the Oligs and the Ngns brings us closer to the neurogenesis-to-gliogenesis switch. Yet the nature of the switch remains unknown. The problem is not that researchers have failed to produce candidate switches. Instead, the issue is that too many good candidates have applied for the job. The following brief discussion moves from the tried-and-true candidates (i.e., Notch signaling) to the up-and-coming (epigenetic mechanisms). And, as will be seen shortly, a group of proteins called Sox proteins also help to flip the gliogenesis switch.

An analysis of genes expressed in the developing chick spinal cord after treatment with activators of Shh signaling revealed that the Notch ligand Jagged2 was upregulated.[18] The investigators performing this analysis then performed parallel studies in chick and mouse embryos to show that Jagged2 was coexpressed with Olig2 in a salt-and-pepper pattern in the pMN domain, that Jagged2 laterally inhibits Olig2-expressing progenitors from becoming motoneurons, and that depletion of Jagged2 from the pMN domain resulted in premature production of OPCs (that is, OPCs were produced at a time when motoneurons were still being generated). A final, compelling piece of evidence in favor of a role for Notch signaling in the regulation of oligodendrogenesis exploited the finding that knockout mice lacking Hes5, a bHLH protein family member expressed in the pMN domain and known to be a Notch target, produce excess oligodendrocytes (fig. 8.7). Could knock-in expression of Hes5 reverse the effects of Jagged2 depletion? The answer was a definite yes. These results were interpreted as indicating that Notch signaling, activated by the ligand Jagged2, blocks OPC production in the pMN domain during the phase of motoneuron neurogenesis. In this analysis, the switch from producing neurons to producing glia resides in the molecular mechanism that downregulates Jagged2 expression.

The term *epigenetic* is applied to factors that regulate gene expression without alteration of the underlying primary DNA sequence. Examples of epigenetic regulation of gene expression include DNA methylation and modification of nucleosomal histones by acetylation and deacetylation. Other epigenetic mechanisms such as microRNAs operate at the posttranscriptional level. microRNAs (miRNAs) are a class of small, noncoding RNA molecules. miRNAs are about 22 base pairs in length. As explained in Chapter 1, they are derived from larger RNA molecules through cleavage by a ribonuclease (RNase) called Dicer. Their function in cells is to bind to the 3′-untranslated region of specific RNA targets. The results of this binding are either degradation of the target RNA or repression of its translation—in other words, miRNAs block the final step in gene expression. A study in which Dicer was knocked out only in Olig1-expressing cells of the embryonic mouse spinal cord resulted in a dramatic blockade of oligodendrogenesis. Strikingly, the production of moto-

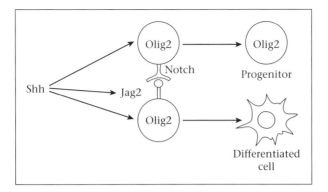

Figure 8.7. Notch signaling in the pMN domain. The expression of Notch ligand Jagged2 (Jag2) in the pMN domain prevents adjacent progenitors from differentiating, initially as motoneurons and subsequently as oligodendrocytes (Olig2). Drawing adapted from Rabadán et al. (2012).

neurons was not affected in these knockouts (fig. 8.8).[19] This study, which indicates that miRNAs are essential for the switch from neurogenesis to gliogenesis, is consistent with reports that miRNAs control the timing of developmental events. Other epigenetic factors regulating oligodendrocyte specification and maturation are likely to be identified in the future. The same could be said about most aspects of development.

Sox proteins are another class of potential regulators for the switch from neurogenesis to gliogenesis. Sox proteins are members of a large superfamily called high-mobility group (HMG) proteins. The most famous member of this superfamily is Sry, the male sex-determining gene of mammals. In fact, the name of the Sox group of HMG proteins derives from <u>S</u>ry-related box genes. Sox proteins are transcription factors. They depend on non-Sox protein partners for target specificity and, in a partner-dependent fashion, activate or repress target gene expression. Phylogenetic analysis of the HMG domains of the Sox proteins has defined subgroups of Sox proteins (SoxA–SoxJ). Sox proteins are expressed in many, many tissues during development and, as a consequence, are thought to regulate many, many developmental events. The SoxE group, which in mice consists of Sox8, Sox9, and Sox10, appears to have a close relationship with oligodendrogenesis. All three of these genes are expressed in pMN during the production of OPCs, and Sox10 continues to be expressed as the premyelinating oligodendrocytes differentiate their capacity to myelinate.

Phenotypic analysis of SoxE knockout mice illustrates the challenges that result from the ability of closely related genes to compensate for a knocked-out gene. Both Sox9 and Sox10 knockout mice display temporary reductions in the number of OPCs and differentiating oligodendrocytes, whereas Sox8 knockouts have no apparent deficits in either OPCs or oligodendrocytes. But Sox8/Sox9 double knockouts have no OPCs at all! This suggests that Sox8

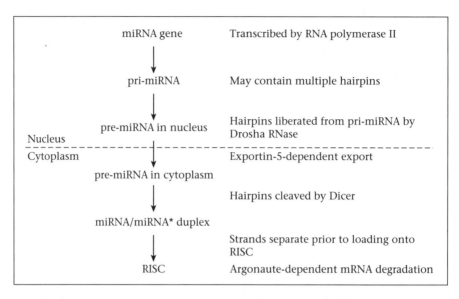

	miRNA gene	Transcribed by RNA polymerase II
	↓	
	pri-miRNA	May contain multiple hairpins
	↓	
	pre-miRNA in nucleus	Hairpins liberated from pri-miRNA by Drosha RNase
Nucleus		
Cytoplasm	↓	Exportin-5-dependent export
	pre-miRNA in cytoplasm	
	↓	Hairpins cleaved by Dicer
	miRNA/miRNA* duplex	
	↓	Strands separate prior to loading onto RISC
	RISC	Argonaute-dependent mRNA degradation

Figure 8.8. The role of the Dicer enzyme in production of microRNAs (miRNAs). miRNAs are short RNA molecules (typically 22 nucleotides) found in eukaryotic cells; they are required for the developmental switch from neurogenesis to gliogenesis in the mouse spinal cord. The function of the Dicer RNase enzyme is to remove the hairpin loop to form the duplex, which then separates into two miRNA strands that mediate the destruction of complementary mRNAs and hence control translation. Dicer knockout mice (see text) cannot produce the normal complement of spinal cord oligodendrocytes and astrocytes, implying that miRNAs regulate the developmental transition from neurogenesis to gliogenesis. miRNA/miRNA* are the two strands of the double-stranded RNA product of Dicer processing. miRNA (without an asterisk) is the strand that eventually enters the RNA-induced silencing complex. miRNA* (with an asterisk) is the opposite strand that does not enter the silencing complex; it is degraded. mRNA, messenger RNA; pre-miRNA, precursor microRNA; pri-miRNA, primary microRNA; RISC, RNA-induced silencing complex.

can compensate for a lack of Sox9 and that the SoxE subgroup proteins are part of the neurogenesis-to-gliogenesis switch in the pMN. A Sox8-Sox10 double mutant resulted in a different phenotype—OPCs were generated, but the differentiation of OPCs as myelinating oligodendrocytes was blocked. It appears that one pair of Sox (Sox8 and Sox9) is required for OPC generation, whereas a different pair (Sox8 and Sox10) is required for production of mature oligodendrocytes.

The Sox story does not end here. There is evidence that SoxD subgroup proteins (Sox5 and Sox6) downregulate the expression of Sox10. Knockout mice lacking Sox5 or Sox6 expression and especially double knockouts lacking both display a phenotype of precocious oligodendrocyte development. Sox5 and Sox6 are expressed in OPCs but downregulated in terminally differentiating oligodendrocytes. Incorporation of these data into our switch metaphor requires us to imagine a two-stage switch. Such a sequence is entirely consistent with a view of cell differentiation as a series of decisions, with each decision having its own controls.

Astrocytes

Astrocytes are the stellate glial cells of the vertebrate central nervous system, but description of their morphology as star-shaped does not in any way do them justice. Some mature astrocytes have so many processes radiating outward from a central cell body that a more apt descriptive term would be *supernova glial cells* (fig. 8.9). The extent of an individual astrocyte's processes is often huge: an astrocyte in a mouse brain may encompass as much as 80,000 cubic micrometers of brain volume. Astrocytes are found everywhere in the mature nervous system and are sometimes said to "tile" the nervous system (in the same way that the multipolar sensory neurons of larval fruit flies were said to "tile" the body wall in Chapter 7) because the domains defined by the processes of adjacent astrocytes barely overlap in healthy brains. A variety of cell junction types connect astrocytes with blood vessels, neurons, and oligodendrocytes. Astrocytes contact other astrocytes via gap junctions, implying coordination of function across large areas of the brain. Some of the known contributions of astrocytes to nervous system function are the maintenance of structural integrity, maintenance of water balance (via connections to blood vessels), support of the blood–brain barrier, regulation of extracellular calcium and potassium levels, and modulation of synaptic transmission. In the gray matter of the brain, their fine terminal processes are so closely associated with axon terminals and dendrites that the term *tripartite synapse* (for axon terminal, dendrite, and astrocytic glia) is

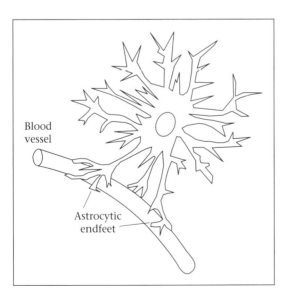

Blood vessel

Astrocytic endfeet

Figure 8.9. Astrocytes. Astrocytes are the highly branched stellate glial cells of the vertebrate nervous system. As shown, astrocytes often contact blood vessels at junctions called endfeet. Drawing modified from http://www.histology.leeds.ac.uk/tissue_types/nerves/Nerve_support_cell.php.

sometimes used. In white matter, astrocytic processes are associated with the nodes of Ranvier. Astrocytes are possibly the most abundant type of cell in the mammalian brain.

Several useful but imperfect markers have aided neuroscientists in studies of astrocytes. They are imperfect in the sense that some do not mark all astrocytes at all stages of astrogliogenesis and some may also mark nonglial populations. Foremost among these markers is glial fibrillary acidic protein (GFAP). GFAP is an intermediate-filament protein that is part of the cytoskeleton.[20] In the nervous system GFAP is primarily expressed by astrocytes, but it is also expressed in many nonneural tissues including kidney and bone. GFAP-expressing astrocytes can be identified by immunocytochemistry and in situ hybridization. Transgenic mouse lines have been developed that express GFP under the control of a GFAP promoter. GFAP is essential for the normal function of astrocytes, and humans born with mutations in this gene display symptoms of Alexander disease, a progressive and lethal disorder characterized by abnormally large brains, spasticity, and intellectual disability. Another intermediate-filament protein, vimentin, is also expressed by astrocytes and is sometimes used as a marker for astrocytes. A soluble calcium-binding protein, S100B, is expressed almost exclusively by astrocytes, but it does not mark all astrocytes. Despite this limitation, antibodies to S100B are also used as markers for astrocytes.

Astrocytes originate in the ventricular zones of the embryonic brain and spinal cord. They are generated by radial glial cells, and, as in the case of oligodendrogenesis, generation of astrocytes follows the initial period of neurogenesis that builds the nervous system. In mice, the maturation of astrocytes continues after birth. Most astrocytes do not acquire their mature morphologies until postnatal weeks 3–4. Newly generated astrocytes maintain a simple, radial glial cell–like morphology as they migrate away from the ventricular zone. Migration typically ends when the leading process makes a connection with a blood vessel. How astrocytes achieve their even distribution throughout the brain is unknown.

The early stages of astrocytogenesis parallel early events in neurogenesis and oligodendrogenesis. Progenitor cells acquire their positional identities in response to signals arising from different sources. These spatial signals (Shh, Bmp) regulate the expression of transcription factors, which in turn regulate the expression of more transcription factors. Members of the HD and bHLH protein families interact to generate the expected array of neurons and glial cells.

The astrocytes of the spinal cord arise from multiple progenitor pools. One of the best-characterized of these is the p2 domain immediately dorsal to the previously mentioned pMN (fig. 8.10). Expression of the HD transcription factor Pax6 marks three progenitor domains dorsal to pMN: p0, p1, and p2. The p2 domain gives rise primarily to the V2 population of

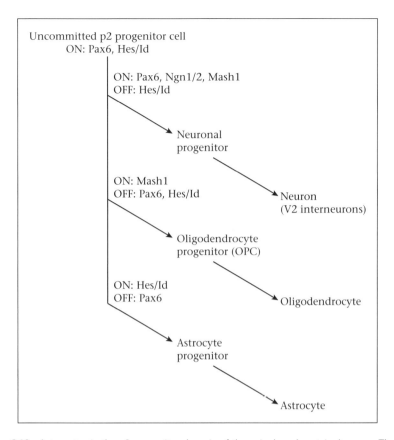

Figure 8.10. Astrocytes in the p2 progenitor domain of the spinal cord ventricular zone. The p2 domain of the spinal cord ventricular zone produces interneurons, oligodendrocytes, and astrocytes. A simplified overview of the profile of gene expression associated with the different cell types is shown. The homeodomain protein Pax6 is expressed in uncommitted progenitors and neuronal progenitors but not in oligodendrocyte and astrocyte progenitors. Expression of the inhibitory bHLH proteins Hes/Id together with Pax6 blocks neurogenesis. When Hes/Id are downregulated, expression of the proneural proteins Ngn1/2 and Mash1 together with Pax6 generates neurons. Expression of Mash1 in the absence of Pax6 promotes production of oligodendrocytes. Downregulation of Pax6 with continued expression of antineural Hes/Id promotes production of astrocytes. Drawing adapted from Guillemot (2007).

interneurons and astrocytes, although oligodendrocytes are also produced. The induction of Ngn1, Ngn2, and Mash1 in the presence of Pax6 expression commits the progenitor cells of p2 to produce the V2 interneurons. Later in the lineage, downregulation of Pax6 expression permits the production of OPCs. Downregulation of Pax6 and Mash1 establishes the conditions required for astrogenesis.

The end of neurogenesis in the p2 domain is signaled by expression of another bHLH transcription factor called stem cell leukemia or SCL. (As the name suggests, SCL is also expressed in blood cell progenitors.) One might think of SCL as the Olig2 of astrocytes. Starting at 10.5 dpc, SCL is expressed in the p2 domain of the spinal cord. The astrocyte marker S100B is expressed in this same region at 14.5 dpc, and expression of S100B and SCL colocalize. GFAP expression is not evident until 15.5 dpc. Knockout of SCL in the developing spinal cord resulted in a decrease in S100B-expressing cells in these animals. Regulatory interactions between astrocyte-promoting SCL and oligodendrocyte-promoting Olig2 are implied by the finding that SCL knockout mice have more than the usual number of oligodendrocytes, whereas Olig2 knockout mice have a larger than expected number of astrocytes. SCL has also been demonstrated to repress Olig2 expression.[21]

Results from these and other studies of knockout mice reinforce themes developed in Chapters 5 and 6, including roles for bHLH transcription factors, Notch signaling, and morphogen gradients. Current research on gliogenesis in the mouse is exciting because it allows researchers to address a key question in developmental biology: how can a single progenitor cell generate multiple types of daughter cells? An understanding of how the neuronal or glial switch is flipped is a first step toward harnessing the power of progenitor cells to repair damaged nervous systems.

The final topic we cover before moving to gliogenesis in humans is the newest population of cells to be discovered in the mammalian central nervous system, the NG2 cells (also sometimes referred to as O-2A cells). The NG2 for which these cells are named is a membrane-spanning chondroitin sulfate proteoglycan with a large extracellular domain and a short cytoplasmic tail. Because these cells depend for their survival on a protein called platelet-derived growth factor (PDGF), antibodies that recognize the PDGF receptor are excellent markers for NG2 cells, as are antibodies directed against NG2 itself.

Like oligodendrocytes and astrocytes, NG2 cells are conspicuously and ubiquitously present in the brain. They originate in the ventricular zones of the brain and spinal cord. In morphology they resemble astrocytes, and, as in the case of astrocytes, the processes of neighboring NG2 cells rarely overlap. NG2 cells generated intense interest among neuroscientists when they were first discovered because in cultures of rat optic nerve they appeared to produce diverse progeny: oligodendrocytes, astrocytes, and neurons. In the intact adult nervous system, however, NG2 cells produce oligodendrocytes, not neurons. After injury to the nervous system, NG2 cells proliferate, producing more NG2 cells before producing oligodendrocytes and an occasional astrocyte. How to harness the neurogenic potential displayed by cultured NG2 cells in vivo is currently an unanswered question.

The processes of NG2 glia are associated with synapses and with axons, and, in some cases, NG2 glia form and receive inhibitory and excitatory synapses. For example, NG2 glia in the corpus callosum receive glutamatergic synapses from unmyelinated axons and respond to glutamate with receptor-mediated changes in membrane potential.[22] These findings support a view that signals produced by neurons recruit myelinating oligodendrocytes.

The discovery of the NG2 population of cells poses a challenge to a simple dichotomous neuron/glia classification scheme. Several new names have been proposed for NG2 cells—polydendrocytes, syntanocytes, glial progenitor cells—but so far none has been universally adopted.

Glia in Humans

One of the hallmarks of primate brains is an abundance of glia. It is, however, difficult to provide hard numbers to support the view that primate brains in general and human brains in particular are special in this regard, at least relative to the brains of other mammals. A major source of uncertainty regarding glial numbers and densities is the finding that glial cell distribution is not uniform throughout the brain. For example, one study reported that the human thalamus contains 17 glial cells for every neuron, whereas an equally authoritative study reported that neurons outnumber glial cells in the human cerebellum by a factor of 25 to 1.[23] Another source of uncertainty is that any count is only as good as the marker used. Any particular marker may overcount or underestimate the extent of the cell population being studied. In particular, the NG2 cells of the human brain are difficult to count using immunolabeling of the NG2 antigen. This is because many standard histological procedures apparently cause loss of the NG2 antigen from brain tissue. Another confounding factor for counters is that anti-NG2 antibodies also label certain types of vascular cells. Accurate counts require optimization of cell-marking protocols and the use of multiple markers. At present it is not possible to say with certainty how many glial cells are found in the human brain. It is, however, probably safe to say that the widespread claim that the brain contains 10 times more glial cells than neurons is an inaccurate rough guess.

Studies of the development of glial cell populations in human embryos rely heavily on interpretation of immunolabeling profiles in autopsy specimens. For example, an analysis of embryonic human spinal cord using antibody markers for oligodendrocytes revealed that oligodendrocyte precursors are first seen in ventral progenitor cell domains, in the region of the neural tube immediately dorsal to the floor plate. This pattern, evident at 45 dpc, is consistent with the morphogen-derived pattern of progenitor domains known from studies in mice. Myelin proteolipid protein (typically referred to as PLP) was the earliest myelin marker to be detected (30 dpc); myelin basic protein was detected at 9 weeks of gestation, and the first myelin (associated with the nerves of the spinal cord) was evident at 10–11 weeks. The formation of myelin begins later in the brain than in the spinal cord, but by 18 weeks both oligodendrocyte precursors and NG2 cells were present in forebrain white matter and the cortical mantle. GFAP immunolabeling is first evident in radial glial cells associated with the ventricular zone of the forebrain by week 16 of gestation; by weeks 28–29, GFAP-positive cells are present in other locations, including the marginal layer of the developing cortex. The GFAP-positive cells associated with the ventricular zone are assumed to represent radial glia, whereas the dispersed GFAP-positive cells that appear later are thought to be differentiated astrocytes. GFAP remains a marker for astrocytes in the adult human as well as in the embryo.

To recap, as is so often the case, most of what is known about gliogenesis in the developing human brain is an extrapolation of the mouse model, supported by analysis of a limited number of human autopsy specimens. Two aspects of glial cells in humans, however, warrant closer attention. The first is the protracted time course of the myelination of the human nervous system. The second is the response of glial cells to brain injury, a phenomenon called reactive gliosis.

As noted, substantial populations of glial cells, including myelinating oligodendrocytes, are present prenatally in the central nervous system of humans. Other oligodendrocytes are generated during postnatal life, perhaps by the NG2 cells. There are regional differences in the onset and extent of myelination, but in general the caudal regions of the central nervous system initiate myelination earlier and complete myelination sooner. Several studies have indicated that the spinal cord and brain stem complete most myelination by a child's first birthday. The human forebrain, however, continues to myelinate throughout the first five decades of life.

New information on the time course of brain myelination in older children and adults is based on the use of noninvasive neuroimaging techniques to study development in healthy brains. Sensitive techniques such as diffusion tensor imaging (DTI) permit the estimation of gray and white matter volumes and the assessment of tissue water content. Tissue water content provides a crude index of myelination because myelination decreases water content whereas myelin breakdown and loss increase water content. These techniques have been used to suggest that white matter volume increases linearly in humans between the ages of 4 and 20 and that a higher degree of myelination is associated with improved performance on specific cognitive tasks. In another study, the volume of the gray and white matter subcompartments in the frontal and temporal lobes of the cortex was determined in 70 healthy men 19–76 years of age.[24] The investigators reported that there was an age-related linear loss in gray matter volume but also found that white matter volume increased through age 44 in the frontal lobes and age 47 in the temporal lobes. These studies have led to speculation that later age-related declines in cognitive processing reflect later age-related declines in myelination. Perhaps more pertinent for the developmental neuroscientist is the question of how similar the mechanisms controlling oligodendrocyte production and maturation in adults are to the molecular regulatory mechanisms defined in embryos.

Reactive gliosis has been described as the universal reaction to brain injury.[25] In response to an experimentally inflicted clean stab wound, pre-existing astrocytes in the affected area enlarge and upregulate their expression of GFAP. Some astrocytes proliferate (fig. 8.11). The result is formation of a glial scar in the region of the injury. The glial scar is beneficial in that it fills in a damaged brain region but also potentially harmful in that its presence inhibits axonal growth that could lead to recovery of function after a brain injury. In vivo, astrocytes that participate in reactive gliosis produce only astrocytes, but when they are removed from the brain and cultured in small clusters of cells called neurospheres, they can also produce oligodendrocytes and neurons. These observations illustrate the potential interactions between studies of nervous system development and studies of nervous system pathology. For example, the observation that astrocytes proliferate in the adult central nervous system forces us to ask why oligodendrocytes and neurons are doggedly postmitotic, whereas differentiated astrocytes have the capacity to reenter the cell cycle. Another question raised by these observations is why such astrocytes and NG2 cells behave as multipotent stem cells in the culture dish but produce only glial cells in the living brain. Defining the differences between development and healing at the molecular level may lead to a better understanding of how to treat abnormalities in both processes.

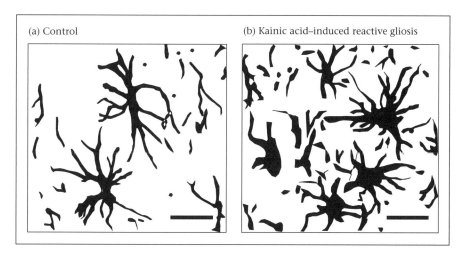

(a) Control

(b) Kainic acid–induced reactive gliosis

Figure 8.11. Reactive gliosis. An increase in glial fibrillary acid phosphatase (GFAP) immuno-labeling (indicated in black in this drawing) is evident in 10-micrometer-thick sections through the CA1 region of the rat hippocampus five days after an intraperitoneal injection of kainic acid. Control rats received no injection (a); treated rats experienced seizures and death of more than half of the hippocampal pyramidal cells in addition to an increase in GFAP-positive astrocytes and processes (b). Scale bars, 25 micrometers. Redrawn from photomicrographs in Jabs et al. (1997).

Fruit Flies and Glioblastoma

Malignant brain tumor is a feared diagnosis, and rightly so. The term *tumor* simply means an abnormal mass of tissue resulting from overactive cell proliferation. Some tumors are benign, which means that they may grow larger but will not spread beyond the initial mass. Although benign brain tumors may cause symptoms as they grow and push against normal brain tissues, once they are surgically removed they typically do not recur. The contrasting term *malignant* refers to the uncontrollable proliferation of abnormal cells capable of invading nearby normal tissues or even spreading to other parts of the body. An untreated malignant brain tumor will eventually overwhelm the normal cells of the brain. Treatment for malignant brain tumors typically combines surgery with radiation or chemotherapy in an effort to reduce abnormal cell proliferation. Despite these interventions, survival of patients diagnosed with malignant brain tumors is often poor.

Glioblastoma is the most common primary malignant brain tumor diagnosed in adults. The term *primary* is used to designate a tumor that has its origin in the cell populations of the brain.[26] A glioblastoma (also called glioblastoma multifore) is a specific tumor type within the broader glioma category of brain tumors. Gliomas may result from transformation of either astrocytes or oligodendrocytes. A glioblastoma is a fast-growing, aggressive type of astrocytic glioma. It is often untreatable, and the 5-year survival rate post diagnosis is less than 10 percent.[27]

Research on glioblastomas removed from human patients has revealed that these tumors originate from glial progenitor cells, not neural stem cells. Many genetic mutations and abnormalities of gene expression have been identified in these tumor cells, indicating that different patients have different molecular subtypes of this cancer. But molecular biologists have identified two abnormalities shared by the majority of glioblastomas. They are mutations in the epidermal growth factor receptor tyrosine kinase and mutations in the phosphatidylinositol 3-kinase pathway. Note that both of these pathways are vitally important during development, when regulated cell proliferation is normal and necessary. Malignancies arise, however, when cell proliferation escapes its normal control mechanisms.

The epidermal growth factor receptor tyrosine kinase (EGFR) is a transmembrane receptor protein. EGFR belongs to a large family of receptor protein–tyrosine kinases that includes many receptors for neuronal and glial trophic factors (a trophic factor is a ligand that blocks the activation of intrinsic cell suicide pathways and thereby promotes cell survival; see Chapter 9). A family of epidermal growth factors (EGFs) provides the ligands for EGFRs. The EGFs generally support survival and proliferation of their target cells.[28]

Normal EGFRs are activated when they bind their EGF ligands. Activated receptor tyrosine kinases turn on intracellular signaling pathways by catalyzing the transfer of a phosphate from ATP to tyrosine residues on specific target proteins. One target of activated EGFR is an intracellular protein called Ras. Ras is often described as a switch, meaning that it has the ability to activate or inactivate intracellular signaling networks. A common mutation of EGFR found in glioblastoma cells results in constitutive activation of Ras. This means that all of the downstream processes regulated by Ras are always turned on. Constitutive activation of Ras has been associated with cancers in many tissues, not just those of the brain.[29]

A single mutation, however, does not result in a malignant tumor. Instead, mutations in multiple genes are required. One of the other mutations often found in glioblastoma cells involves, as noted, the gene that encodes phosphatidylinositol 3-kinase (PI3K). PI3Ks phosphorylate a membrane phospholipid called phosphatidylinositol. PI3Ks sit on top of a multistep pathway that, when activated, promotes cell growth and proliferation. Mutations in PI3Ks or other pathway components that lead to overactivation of this pathway are, like mutations involving EGFRs and Ras, often associated with malignant tumors.

The *Drosophila* genome contains homologs for genes that encode EGFR, Ras, and PI3K pathway components. In many cases, only a single counterpart of the mammalian gene is present, which simplifies experimental analysis of mutations in these pathways in flies. Researchers can drive overexpression of specific genes or combinations of genes in specific cell types using carefully selected transcriptional drivers. As you might have already guessed, *repo* provides a useful transcriptional driver because it is specific for *Drosophila* glial cells.

The effects of coexpression of a constitutively activated EGFR and PI3K in *Drosophila* are striking.[30] The larval brain contains up to 100 times the normal number of glial cells. Normal brain structure is disrupted by multi-layered packets of abnormal glia distributed throughout the brain. These abnormal glial cells also passed another test for tumorigenicity: when tissue from the combined EGFR–PI3K mutants was transplanted into the abdomen of wild-type host flies, the transplanted cells (which could be tracked because they expressed GFP) proliferated and formed large tumors that filled the host abdomens, leading to death. By contrast, single mutations in either EGFR or PI3K led to much smaller increases in the number of glial cells or to no increase at all. Taken together, these results show that potential new drugs might be efficiently screened for their ability to slow tumor growth by feeding them to fruit fly larvae and that other genes involved in tumor progression can be identified using modern variations on traditional forward genetic screens. Such studies, which begin with a basic understanding of development, have the potential to repay the favor by revealing new details of intracellular signaling pathways important in both normal development and tumorigenesis.

1. Oscar is the informal name given to awards bestowed annually for achievement in the motion picture industry by the American Academy of Motion Picture Arts and Sciences.

2. I first encountered the idea that it is hard to study how glial cells contribute to neuronal function if neurons die when you ablate them in the concluding remarks of an article by Oikonomou and Shaham (2011).

3. The movement of calcium into, out of, and between compartments in cells can be tracked by using fluorescent dyes that change their spectral properties when they bind calcium.

4. The toxin used in these studies is diphtheria toxin. In nature, this toxin is secreted by the bacterium that causes the human upper respiratory tract infection diphtheria. In the research laboratory, the gene encoding this toxin can be inserted into an animal's genome under the control of a cell type–specific promoter. When the toxin gene is expressed in a cell, the cell dies because the toxin blocks the translation of RNA into protein.

5. See Bacaj et al. (2008).

6. To humans, isoamyl alcohol has a mild, fruity odor; benzaldehyde smells like almonds.

7. See Hilgetag and Barbas (2009) for a thoughtful discussion of the myth that this ratio is significantly higher in the brains of humans than in the brains of other animals.

8. Dextran is a branched polysaccharide of glucose molecules. Dextran chains can be short or long and can range in mass from 3 to hundreds of kilodaltons.

9. See Stork et al. (2008).

10. See Xiong et al. (1994).

11. The dye DiI, a lipophilic cyanine dye, is the fluorescent dye that was used for lineage tracing of longitudinal glia (see Chapter 1).

12. The term *endothelium* is used to describe the thin layer of epithelial cells that lines the inner surface of blood vessels in vertebrates.

13. Neuromast cells are similar to the hair cells of the vertebrate ear.

14. *Placode* is a general term for an ectodermal thickening from which a sensory structure develops. As is so often the case during development, a placode is a transient structure. The eyes, ears, and noses of humans and other mammals develop from the optic, otic, and olfactory placodes. The anterior pituitary develops from the adenohypophyseal placode, and the trigeminal ganglion complex of cranial nerve V develops from the trigeminal placode. Other vertebrates have additional placodes, including the lateral line placodes of fish discussed in this text. Placodes are often visible on the surface of the embryo. Placodes provide valuable examples of the models that can be found in our existing models. See Schlosser (2006).

15. Neural crest cells form adjacent to each side of the neural tube at the time of neural tube closure. Neural crest cells migrate away from their site of origin to form the entire peripheral nervous system and key parts of other structures, including the teeth and the heart. For the most part, we are ignoring the neural crest in this introductory text because it deserves a textbook of its own.

16. See Gilmour et al. (2002).

17. See Cahoy et al. (2008).

18. See Rabadán et al. (2012).

19. See Zheng et al. (2010).

20. Intermediate filaments are protein filaments that are part of the cellular cytoskeleton. They are not as thick as myosin filaments and not as thin as actin filaments —hence they are designated intermediate.

21. See Muroyama et al. (2005).

22. See Ziskin et al. (2007).

23. See Herculano-Houzel (2009).

24. See Bartzokis et al. (2001).

25. See Buffo et al. (2008).

26. Malignant brain tumors can also result from migration of abnormal cells from other organs such as the lungs or kidneys into the brain via the circulatory or lymphatic system. Such tumors are referred to as secondary or metastatic brain tumors.

27. This information is from the M. D. Anderson Cancer Center's OncoLog, a publication designed to update physicians on advances in cancer research. See http://www2.mdanderson.org/depts/oncolog/articles/11/3-mar/3-11-1.html.

28. The 1986 Nobel Prize in Physiology or Medicine was awarded jointly to Stanley Cohen and Rita Levi-Montalcini for their discoveries of growth factors. See http://www.nobelprize.org/nobel_prizes/medicine/laureates/1986/.

29. The *ras* gene and the Ras protein are named for rat sarcoma, a type of connective tissue cancer.

30. See Read (2011), especially figure 1.

Investigative Reading

1. The role of *gcm* in regulation of the development of the appropriate number of lateral glial cells in the *Drosophila* central nervous system is well established. In *Drosophila,* the 3′ untranslated region (3′ UTR) contains an AUUUA motif. This pentamer is referred to as AU-rich elements or ARE. It has been hypothesized that AREs regulate mRNA stability. If the presence of this ARE reduces the stability of *gcm* mRNA, predict the effect on the development of the *Drosophila* central nervous system of mutating the wild-type ARE to the nonfunctional mutant element AGGUA if the mutation increases the stability of *gcm* mRNA. What marker will you use to assess glial populations in your fly embryos?

> Soustelle, Laurent, Nivedita Roy, Gianluca Ragone, and Angela Giangrande. 2008. *Molecular and Cellular Neuroscience* 37: 65–62.

2. Exposure to radiation, even at low doses, reduces cell proliferation. Given this fact, predict the results of exposing the brains of aged rats to low-dose radiation after production of a brain lesion by middle cerebral artery occlusion for a period of 50 minutes, a procedure designed to mimic the damage and impairment of motor function caused by some types of stroke in humans What marker will you use to assess glial populations in histological sections of your rat brains?

Titova, Elena, Robert P. Ostrowski, Arash Adami, Jerome Badaut, Serafin Lalas, Nirmalya Ghosh, Roman Vlkolinsky, John H. Zhang, and Abdre Obenaus. 2011. *Journal of the Neurological Sciences* 306: 143–53.

3. Recovery from acute spinal cord injury is poor in both mice and humans. Your research on the development of the nervous system has convinced you that a good way to promote recovery from spinal cord injury would be to restore the signaling environment present during the formation of the spinal cord. You have developed nontoxic, biodegradable microspheres that can be implanted in a mouse near the site of an experimentally induced transection of the spinal cord. These microspheres slowly release their contents over a period of 7 days. What developmental signal do you load into your microspheres, and why? Describe at least three different methods (functional, histological, and molecular) that you will use to assess the outcome of your treatment. Do you think this (or similar) treatments might one day be used in human patients? Why or why not?

Lowry, Natalia, Susan K. Gocerie, Patricia Lederman, Carol Charniga, Michael R. Gooch, Kristina D. Gracey, Akhilesh Banerjee, Supriya Punyani, Jerry Silver, Ravi S. Kane, Jeffrey H. Sterna, and Sally Temple. 2012. *Biomaterials* 33: 2892–901.

NINE

Maturation

Growing Up

Studies of the development of the nervous system typically focus on the earliest events, in part because the results are amazing (for example, the brain of a newborn baby) and in part because so many disorders of the human nervous system have their origins during prenatal life. At the opposite end of life, new methods for studying the human brain have revealed that even fundamental aspects of development such as myelination and neurogenesis persist for decades after birth. In between, the fundamental plasticity of the nervous system—the capacity to refine synaptic connections and even gene expression on the basis of experience—means that the normal status of the brain is *under construction*.

In many animals, however, the transition from the intensity of the embryonic period of nervous system development to the *everyday plasticity* of adult life is punctuated by dramatic transitions in morphology and behavior. In arthropods and amphibians, this transition is known as metamorphosis. In humans, the interval between childhood and adulthood is referred to as adolescence. The goal of this chapter is to review the events in the nervous system that correspond to metamorphosis and adolescence. The term *neurometamorphosis* is already in use to describe the transformation of the larval insect nervous system into that of the adult. This chapter introduces the corresponding term *neuroadolescence* to describe the changes that occur in the human adolescent's brain. One question that is commonly asked about neurometamorphosis can also be asked about neuroadolescence: are the same signals that control the initial formation of the nervous system also important for the later events? That is, does metamorphosis recapitulate embryogenesis?

Metamorphosis

Several familiar groups of insects, including the Lepidoptera (moths and butterflies), Diptera (flies), Hymenoptera (ants, bees, and wasps), and Coleoptera (beetles), display a pattern of postembryonic development referred to as complete metamorphosis or holometaboly (Chapter 3). Holometabolous insects have distinct larval and adult stages, with an intervening pupal stage. Unraveling the hormonal regulation of insect metamorphosis was one of the great achievements of twentieth-century insect physiology. Contemporary studies, aided by the availability of sequenced and annotated insect genomes, have begun to delineate specific molecular pathways controlled by hormones during metamorphosis.

The changes in body morphology characteristic of insect metamorphosis (here a mental image of caterpillars and butterflies is extremely helpful) result from modification and replacement of embryonic structures. Some persisting tissues display sequential polymorphism. For example, the same epidermal cells secrete, in turn, larval, pupal, and adult cuticle. Other larval cells also persist but are respecified for new functions, as in the case of thoracic body wall muscles in fly larvae that form the scaffold of the flight muscles of the adult. Some adult-specific tissues, such as wings, genitalia, and compound eyes, are produced by proliferation of populations of cells called imaginal disks. These islands of embryonic tissue lie beneath the epidermis and remain undifferentiated until the end of larval life. Internal organs such as the gut and excretory system contain proliferation centers that are inactive during larval life but produce new cells at the metamorphic transition. The death of specific larval cells and tissues provides a backdrop of creative destruction for the process of adult development.

Metamorphosis of the body and development of adult sensory organs such as compound eyes are accompanied by changes in behavior. Although the behavioral repertoires of larvae and adults may overlap, the typical larva is specialized for crawling and feeding, whereas the typical adult is specialized for flight and reproduction. These changes in behavior require structural and functional modifications at all possible levels of behavioral control. As noted, metamorphosis is therefore accompanied by neurometamorphosis. Neurometamorphosis involves changes in the brain and ventral nerve cord, in the peripheral nervous system, and in the motoneurons that control the muscles of the body wall and the appendages. As in the case of the other body tissues, neurometamorphosis blends persisting larval neural elements with new, adult-specific elements.

Studies of two model insects—a large sphinx moth (*Manduca sexta*) and our old friend *Drosophila*—have provided numerous examples of neurometamorphosis and have also provided evidence for the direct actions of the hormones that regulate metamorphosis on neurons, glial cells, and muscle fibers. The following examples from *Drosophila* highlight the phenomena of postembryonic neurogenesis, programmed cell death, and respecification (fig. 9.1). Note that the period of adult development via metamorphosis is often said to begin with pupariation—the darkening and hardening of the larval cuticle at the end of the third larval stage—which is preceded by changes in

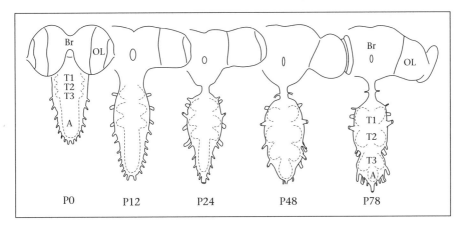

Figure 9.1. Metamorphosis of the central nervous system of *Drosophila*. The changes that occur from the formation of the puparium (P0) through 78 hours after pupariation (P78) are depicted. Dashed lines indicate the outlines of the central neuropil in the ventral nerve cord. A, abdominal neuromeres; Br, brain; OL, optic lobe region; T1, T2, T3, thoracic neuromeres. Drawings adapted from Truman et al. (1993).

hormone profiles, behavior, and bodywide regulation of gene expression of tissue-specific gene expression profiles. At 25° C, the time from pupariation to the emergence of the fully formed adult fly is approximately 100 hours.

Postembryonic Neurogenesis in Drosophila

Persisting embryonic neuroblasts account for postembryonic neurogenesis in *Drosophila* and other insects. These neuroblasts have been mapped in the brain and ventral nerve cord of larval fruit flies using markers for DNA synthesis (e.g., ^{3}H-thymidine, BrdU) and a dye called toluidine blue, which is applied to living tissue.[1] Toluidine blue is not a specific label for neuroblasts, but under controlled conditions other cells in the nervous system can be destained while the larger neuroblasts retain a dark blue color. Larvae can be exposed to markers such as BrdU by incorporating the marker into their food; flies that have already pupariated can be partially dissected and immersed for a short time in BrdU-containing medium. These methods have revealed that the number and distribution of postembryonically active neuroblasts in *Drosophila* are highly stereotyped.

Neuroblasts in the brain and ventral nerve cord divide throughout larval life. As was the case in the embryo, most neuroblasts undergo repeated rounds of asymmetrical divisions that result in a ganglion mother cell and a regenerated neuroblast. The ganglion mother cell then undergoes a symmetrical division to produce two postmitotic neurons. An unusual feature of postembryonic neurogenesis is that the neurons to be rest a bit before differentiating. The neurons born during larval life do not complete development until the onset of metamorphosis. They therefore play no role in larval life. The pairing of repeated rounds of asymmetrical division with delayed differentiation results in the formation of clusters of small neurons around each neuroblast. These clusters are referred to informally as nests (fig. 9.2). The neuroblasts of the ventral nerve cord terminate their lineages by dying 12–18 hours after pupariation, but some neuroblasts in the brain continue dividing throughout metamorphosis. An expansion of the neuropils of the ventral nerve cord occurs between 18 and 36 hours after pupariation, an event that reflects the differentiation of the adult-specific neurons.

The ventral nerve cord contains the neurons that innervate the thoracic and abdominal segments of the fly. These structures are notably unspectacular in vermiform (wormlike) larvae, in which both thoracic and abdominal segments are easily controlled by small populations of neurons in the corresponding ganglia. During metamorphosis, however, the thoracic segments are embellished with legs, wings, and club-shaped sensory organs called halteres. Controlling these new structures requires a significant expansion of the neuronal populations in the thoracic neuromeres. Each thoracic neuromere contains at least 47 neuroblasts, which will collectively generate more than 4,000 adult-specific neurons per segment. By contrast, the abdominal neuromeres, which do not add appendages, each contain only 6 neuroblasts and produce a correspondingly smaller number of adult-specific neurons.

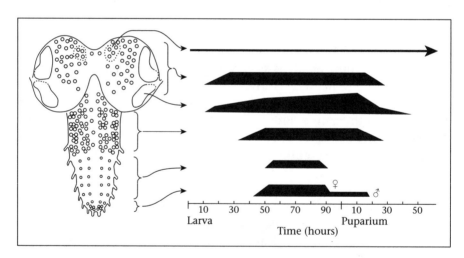

Figure 9.2. Nests of neuroblasts in the metamorphosing *Drosophila* central nervous system. The location of neuroblasts is shown on the left; a summary of the time course of neurogenesis (based on studies of BrdU incorporation) is shown on the right. Unlike the other neuroblasts, the mushroom-body neuroblasts shown at the top of the brain enclosed by a dashed outline are active from hatching through the end of metamorphosis. Note the subtle sex difference in duration of neurogenesis in the terminal ganglion of the ventral nerve cord. Drawings adapted from Truman et al. (1993).

What is the function of the adult-specific neurons? In the central nervous system, there is strong evidence that most larval motoneurons and neurosecretory cells persist through metamorphosis without any significant changes in number. The same is true for neuromodulatory cells that contain biogenic amines such as serotonin. This means that all, or nearly all, of the adult-specific neurons in the brain and ventral nerve cord must be interneurons. It is not hard to imagine how addition of new interneurons to existing neural circuits leads to new behaviors, but at present almost nothing is known of the mechanisms by which adult-specific interneurons in *Drosophila* become polarized, grow processes, find their targets, and form synapses in the context of a preexisting nervous system.

Neuronal Death during Neurometamorphosis

Some larval neurons do not survive metamorphosis. In some cases, the reason for the death of a neuron is easily explained in terms of changes in morphology or physiology. For example, the death of a muscle may leave a motoneuron without a target, or a peptide secreted from a neurosecretory cell may control a response useful only during larval life. Two periods of neuronal death have been described during *Drosophila* metamorphosis. The first occurs shortly after puparium formation; the second occurs shortly after the emergence of the adult fly from its puparial case.

Before discussing specific examples of neuronal death during metamorphosis, it is important to note that neuronal death is a normal and widespread phenomenon in developing nervous systems. Both differentiated neurons and neural progenitor cells die in predictable patterns in vertebrates and invertebrates. These neuronal deaths are not the result of injury but reflect responses to the appearance of specific signals or the lack of specific signals. The relevant signals are chemically diverse, ranging from circulating steroid hormones that act via nuclear receptors to neurotrophic proteins that bind to receptor tyrosine kinases. The molecular mechanisms of neuronal death are not unique to neurons. One commonly used pathway is apoptosis, a form of cellular suicide defined by fragmentation of nuclear DNA and formation of membrane-bounded particles that allow the cell to die without damage to its neighbors. Apoptosis can be detected in tissues by use of the TUNEL method for detecting DNA fragmentation.[2] The upstream details of apoptotic pathways are different in different groups of animals, but the final event is the activation of cysteine-dependent aspartate-directed proteases, typically referred to as caspases. The life of a cell can be thought of as depending on the balance between caspase-inhibiting and caspase-activating proteins. Developmental signals can tip the balance in one direction or the other. Another type of cell death that has been described in the developing nervous system is enzymatic self-digestion (autophagy).

Many molecular tools are now available to help researchers distinguish different forms of programmed cell death.[3] It was possible, however, to detect neuronal death in the developing nervous system prior to the introduction of molecular markers using standard histological stains and cell counting.[4] Another useful category of stains are those excluded by healthy cells with intact plasma membranes. For example, propidium iodide is a fluorescent molecule that binds to DNA and enters only dying or damaged cells. Another method that can be used to detect neuronal death is to focus on neurons that produce an uncommon gene product, such as a neuropeptide, that can be detected by immunolabeling or in situ hybridization (fig. 9.3). The disappearance of a previously expressed gene product from the nervous system is not a guarantee that the neuron that produced it has died, but it is a vital clue. For example, corazonin is a cardioacceleratory neuropeptide produced by insects that is involved in the regulation of molting behavior. In *Drosophila* larvae, a small number of neurosecretory cells in the brain and the ventral nerve cord synthesize corazonin. They can be identified by immunolabeling the nervous system with an anticorazonin antibody. The corazonin-immunoreactive neurons can be identified in the brain at all stages of life, but all corazonin-immunoreactive neurons disappear from the ventral nerve cord between 2 and 6 hours after puparium formation.[5]

Another method for tracing the fate of specific neurons during metamorphosis is to use a genetically-introduced label such as GFP. The RP2 moto-

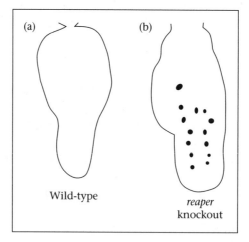

Figure 9.3. Death of identified neurons in the ventral nerve cord of *Drosophila*. Eight pairs of corazonin-expressing neurons can be identified in the ventral nerve cord of third-instar larvae. Between 12 and 24 hours after puparium formation, however, all corazonin expression disappears from the ventral nerve cord (a). If developmental programmed cell death is blocked by a genetic manipulation that eliminates expression of the *reaper* gene (*reaper* knockout) (b), many corazonin-expressing neurons (shown in black) are still present 12–16 hours after puparium formation. Drawings based on photomicrographs from Choi et al. (2006).

neurons of the larval ventral nerve cord have been labeled with membrane-bound GFP using a standard Gal4/UAS system combined with FLP/*FRT*-mediated expression.[6] The RP2 motoneurons are located in segmentally repeated, bilaterally symmetrical pairs in the abdominal neuromeres of the ventral nerve cord. They form neuromuscular junctions on Muscle 2 of the dorsal muscle groups of the abdominal segments. At 8 hours after puparium formation, abdominal neuromeres A1–A7 each contained a pair of GFP-labeled RP2 motoneurons. A day later (30 hours after puparium formation), the RP2 motoneurons were gone from every segment but A1. The investigators performing the study were able to use commercially available cell-death reagents and hormone manipulations to establish that the RP2s undergo apoptosis in response to an increase in circulating levels of a steroid molting hormone.

Regression and Regrowth

Persisting larval neurons can change their position in neural circuits during metamorphosis by altering the placement of their axon terminals and/or the shape of their dendritic arborization. A substantial number of studies based on reconstructions of identified motoneurons in the large moth *Manduca sexta* revealed that these alterations can be spectacular. For example, an abdominal motoneuron can have an ipsilateral dendritic arborization during larval life but ipsilateral *and* contralateral dendritic arborizations in the adult. In another example, visceral motoneurons found in the terminal ganglion (posterior end of the ventral nerve cord) innervate the cryptonephridial complex of the larval hindgut, but when that structure disappears during metamorphosis, innervate the oviduct instead.[7] The cryptonephridial neurons provide an interesting window into the sex-specific nature of some aspects of development: male moths have the larval cryptonephridial motoneurons but of course lack an oviduct; in males, these visceral motoneurons do not survive the metamorphic transition.[8]

These and many other studies of the metamorphosing ventral nerve cord of the moth relied heavily on the retrograde (toward the soma) and antero-grade (away from the soma) diffusion of cobalt chloride ions applied to the cut ends of nerves. After the cobalt has filled the entire cell, it can be precipitated to give a black color to the entire neuron, including fine processes.[9] The cobalt-filling procedure is technically challenging in the moth but even more so in the tiny fruit fly. Yet the information gained from cobalt fills has been essential for documenting the changes in neural circuitry that accompany metamorphosis. A modern variant of this classic technique combines intracellular injection of fluorescent dyes into selected motoneurons with the filling of cut nerves using fluorescent dyes.[10] It was discovered that motoneurons born during embryogenesis (named MN1, MN2, MN3, and MN4) innervate body wall muscles during larval life but switch to muscles associated with wing movements in the adult fly. This switch in targets is accompanied by massive restructuring of their central dendritic arborizations by a process of regression followed by regrowth. Regression of the larval dendrites of these motoneurons was detectable two hours after puparium formation. The regressed dendritic arbors did not begin to reexpand until 19–20 hours after puparium formation, with growth of the adult-specific dendritic arborizations continuing until almost the time of emergence of the adult from the puparial case.

A motoneuron named MN5 innervates the same population of flight muscles in the adult as do MN1–MN4 but, unlike these target-switching neurons, has no target during larval life (fig. 9.4). Studies using BrdU birth-dating in which female fruit flies laid their eggs in food containing BrdU revealed that the MN5 neurons are also born during embryogenesis but that during the larval stages they have no dendrites and possess only a short axon that does not innervate a muscle target. MN5 is therefore the exception that proves the rule—it is typical for dendritic growth during insect metamorphosis to be preceded by dendritic regression; the only exception is when there is nothing to regress!

Dendritic regression has also been documented in multipolar sensory neurons associated with the body wall described in Chapter 7. In *Drosophila*, some of the multipolar sensory neurons present in the larva survive metamorphosis. The dendrites of these peripheral sensory neurons can be studied

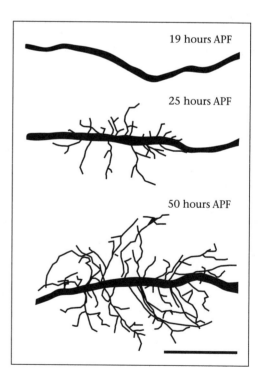

Figure 9.4. Dendritic outgrowth of *Drosophila* motoneuron MN5 during metamorphosis. The MN5 motoneuron is born during embryogenesis of the *Drosophila* nervous system but forms no dendrites until metamorphosis begins. Drawings are based on confocal images of MN5 neurons labeled by intracellular injections of fluorescent dye at three stages between the early pupa (19 hours after puparium formation, APF) and the late pupa (50 hours APF). Scale bar, 50 micrometers. Drawings adapted from Consoulas et al. (2002).

19 hours APF

25 hours APF

50 hours APF

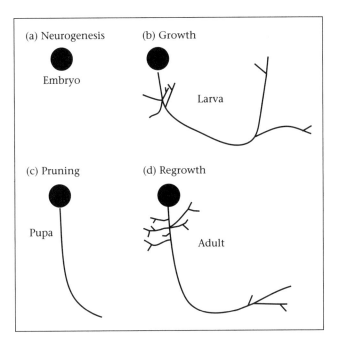

Figure 9.5. Metamorphosis of brain neurons in *Drosophila*. The Kenyon cells of the central brain extend branched axons during larval life (a, b). One axon branch enters a structure called the medial lobe; the other axon branch enters a structure called the dorsal lobe. During metamorphosis, a subpopulation of Kenyon cells called the gamma neurons are restructured (c, d), resulting in an adult Kenyon cell with only medial axon branches.

using time-lapse microscopy in living pupae that have had their puparial cases removed (fig. 9.5).[11] One sensory neuron that can be readily identified across individuals is named ddaE. ddaE has a larval dendritic arbor that regresses between 10 and 25 hours after puparium formation. The process of regression is highly stereotyped: the higher-order branches disappear first, followed by the second order branches. The primary branches retract but survive. The regrowth of the dendritic arborization begins approximately 40 hours after puparium formation. At this time, growth cones form on the tips of the dendrites. By 54 hours, the basic scaffold of the adult arbor is evident. This is followed by a hormone-sensitive period of continued growth and fine-tuning during which branches come and branches go.

These observations provide evidence that dendritic regression—sometimes referred to using the gardener's term *pruning*—is an orderly, regulated process, but they do not reveal the underlying molecular mechanisms by which dendrites are destroyed. A study of the metamorphosis of a population of neurons in a brain structure called the mushroom bodies of the *Drosophila* brain suggests some possible mechanisms. The gamma neurons of the mushroom bodies are present in larvae but regress and regrow their dendrites and axons during metamorphosis. Careful observations of the changes in these neurons prior to regression indicated that a period of "blebbiness" (apparent fragmentation) precedes the disappearance of dendrites and axons. A comparison of the regressing and nonregressing portions of axons revealed that expression of tubulin was lost from the microtubule cytoskeleton, but only where regression was occurring. Genetic perturbations of the ubiquitin system for proteasome-mediated protein destruction blocked pruning of the gamma neuron axons.[12] These findings suggest that regressing axons and dendrites are not recycled by being absorbed back into their parent neuron but are instead disposed of locally via the ordinary, ubiquitin-dependent mechanisms for cellular protein turnover. As you will soon learn, pruning is definitely not unique to insect neurometamorphosis.

Adolescence

Insect metamorphosis is a process, not an event, but it is not difficult to define the beginning and the end of the process. It is trickier to define the boundaries of human adolescence. One widely used starting point is the onset of puberty, although defining this event is also challenging given the many changes in primary and secondary sex characteristics that could be used, including growth of pubic hair (in males and females), changes in the external genitalia (males) or breasts (females), and onset of menstruation (females).[13] It is even trickier to define the end of adolescence, which is typically said to occur *when an individual becomes an adult*. Age is often used as a surrogate indicator for the state of adolescence, but this is inadequate because there are important individual and sex differences in the timing of puberty (girls typically begin puberty earlier than boys). All definitions of adolescence, however, include the early teenage years.

The scientific study of the years between childhood and adulthood has largely been the province of psychologists and anthropologists. Adolescent humans are generally described in terms of how their behavior differs from that of children and adults. For example, adolescents are said to spend more time with their peers than do children and to be more willing to engage in dangerous behavior than are adults. Adolescence is also a time of increased risk (relative to childhood and adulthood) for an initial diagnosis of a mood disorder, an eating disorder, and schizophrenia.[14] One thing adolescence is definitely *not* is a time for overall brain growth: the overall size of the human brain does not increase after age 5 or 6.

Until the advent of noninvasive brain imaging, neuroscientists had relatively little to contribute to the discussion of adolescence. Most studies of puberty in rats and mice have focused on the maturation of the hypothalmo–pituitary–gonadal axis and the effects of gonadal steroids on brain structure and reproductive behavior.[15] A small number of studies of post mortem human brains, however, has had a large (and continuing) impact on our view of the adolescent brain. The results of these studies indicated that, at the cellular level, one of the major characteristics of the adolescent brain is a reduction in the number of synapses.

One of the first studies to reach this conclusion was a survey of synaptic density in Layer III of the cerebral cortex in 21 human brains ranging in age from newborn to 90 years (fig. 9.6).[16] Small blocks of cortex were obtained at autopsy and prepared for electron microscopy using a method selective for

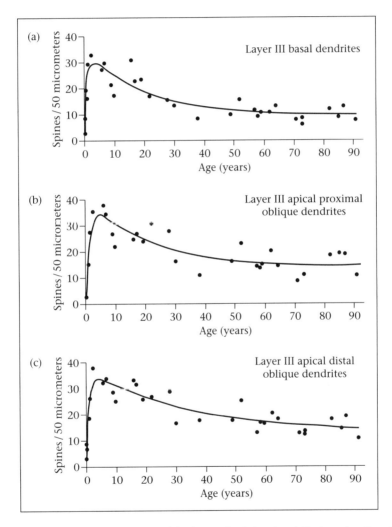

Figure 9.6. Synaptic density in Layer III of the human frontal cortex plotted as a function of age. Human brains were prepared for electron microscopy using a phosphotungstic acid staining method selective for synapses. Graphs represent populations of dendrites with different points of origin from cortical pyramidal neurons. Drawing based on Huttenlocher (1979).

synapses. Synaptic density (synapses per cubic millimeter × 10⁸) was estimated in Layer III of the cerebral cortex on the basis of counts of synaptic profiles in consecutive sections, with a correction to take into account the possibility that a single synapse might be counted in more than one section.

Despite the relatively small sample, the results were clear. Synaptic density in Layer III of the cortex increased after birth, appearing to reach a peak by the second year of postnatal life. Then a decline in synaptic density occurred between ages 2 and 16, followed by a prolonged stable phase. The oldest subject in the sample (age 90) had a synaptic density roughly equivalent to that of the newborn baby.

Adolescent brains were poorly represented in this study, and age is the sole indicator of stage of development provided by the author. A subsequent study from the same laboratory provided data on synaptic density from four additional adolescent brains (aged 12–19 years).[17] Despite the small number of brains sampled, this fundamental observation is oft repeated: human adolescence is a time of synaptic pruning. In fact, this finding has served as the basis of psychological theories of how teen brains function and has even been used to explain why the initial manifestation of schizophrenia typically occurs during adolescence.[18]

The extremely limited number of human brains on which this conclusion is based may (or should) make you uneasy. Nevertheless, a small but growing literature on peripubertal changes in synaptic density in the brains of rodents and nonhuman primates indicates that the phenomenon is real. Synaptic pruning has also been observed in subcortical regions of the brain such as the amygdala and the lateral geniculate nucleus. Studies using the mouse model, in part inspired by the *Drosophila* model, will eventually answer questions regarding the molecular requirements for developmental pruning, but extraordinary creativity on the part of researchers will be required to link these molecular mechanisms to theories of why human teenagers do what they do.

On the other hand—and in this case the other hand is very, very important—the widespread application of noninvasive neuroimaging techniques to the adolescent human brain has already offered new insights into neuroadolescence. Brain imaging techniques do not offer synapse-level resolution, but because they discriminate so well between white and gray matter, they are well suited to studies of myelination.

Neuroanatomical MRI studies consistently report that the volume of white matter (myelinated axons) increases throughout the adolescent years. This trend is evident in childhood prior to puberty and continues into adulthood. Myelination is the backdrop for adolescence. By contrast, consistent with prior descriptions of synaptic pruning, the volume of cortical gray matter decreases during the same time period. The results of such purely anatomical studies can in theory be correlated with different measures of behavior or cognitive ability, but the conclusions that can be drawn from correlative studies are typically limited by small sample sizes. They are also rarely replicated.

As noted in Chapter 8, studies using diffusion tensor imaging (DTI) permit individual myelinated tracts to be examined in the living brain. This powerful tool is already being used to study atypical development. For example, DTI studies of the brains of dyslexic adolescents revealed that the state of myelination of a tract called the right superior longitudinal fasciculus was a better predictor of future improvement in reading ability than were school-type tests that directly measured skills associated with reading.[19] The superior longitudinal fasciculus is a major interhemispheric network of fiber tracts that connects frontal, temporal, and parietal association areas of the cortex. This interesting result is potentially clinically useful. This study exemplifies the type of question best answered by brain imaging techniques: it focuses on a well-defined phenotype in a carefully characterized group of subjects. These conditions are most often met in patient populations. By contrast, the answers that neuroscience can provide to questions such as why teens are more impulsive than their younger siblings (or more peer oriented, more prone to eating disorders, or more prone to send text messages) remain limited. Read the fascinating brain imaging literature as it grows, but read it with the proverbial grain of salt close at hand.[20]

Summary

The juxtaposition of neurometamorphosis in *Drosophila* and neuroadolescence in humans is an artificial construct, but several interesting parallels can be drawn between these interesting biological phenomena. First, both cases speak to the primacy of the embryonic period in establishing the overall organization of the nervous system, which persists throughout life. Second, both examples highlight the role of regressive events in the development of the nervous system. A fully functional adult nervous system cannot be achieved solely through neurogenesis, process outgrowth, and synapse formation. Neuronal death, process retraction, and synaptic pruning are equally relevant.

Further, the importance of ongoing myelination to the postembryonic development of the human nervous system is now widely recognized. In a longitudinal study of the brains of more than 100 healthy human subjects between the ages of 5 and 32, every one of the 10 white matter tracts imaged using DTI increased in volume in repeated measurements separated by about 4–5 years.[21] The connections that link the frontal and temporal cortices— brain regions that are at a minimum responsible for bringing planning, memory, and meaning to our lives—appear to be the last to mature. An upside to this protracted development is the opportunity afforded to the environment and our experiences (including our education) to shape our adult brain. A downside is that certain aspects of human brain structure and function remain vulnerable to harmful exogenous influences for decades. For example, the deleterious effects of overconsumption of ethanol on the brain can be seen in fetuses, children, adolescents, and adults.

Notes

1. This easy-to-use method works for all arthropod nervous systems and probably for other invertebrates as well. The basic method is described by Altman and Bell (1973).

2. The TUNEL procedure is based on terminal deoxynucleotidyl transferase dUTP nick-end labeling. In this procedure a labeled dUTP is added to a fragment of DNA and then detected by visualization of the label. This procedure is popular because it can be applied to DNA in tissue sections. See Gavrieli et al. (1992).

3. Search the web site of any large company that supplies reagents for life sciences research using the term *apoptosis* to get a sense of the broad range of tools available for the study of cell death.

4. Dying neurons can often be identified because their appearance becomes altered in predictable ways. Two terms used to describe dying or dead neurons are *pyknotic* and *apoptotic*. Pyknotic cells are characterized by the appearance of a half-moon of condensed chromatin at the margin of the nucleus. Apoptotic bodies are membrane-bounded packets of cell components. The final removal of the remains of dead neurons is usually accomplished by phagocytic cells.

5. See Choi et al. (2006).

6. See Winbush and Weeks (2011).

7. The cryptonephridium is a specialization of the hindgut present in some larval insects, including the caterpillars of moths and butterflies. In a cryptonephridium, the tips of the insect kidney-equivalents, the Malpighian tubules, are kept in contact with the basal side of the hindgut epithelium by the perinephric membrane. This arrangement facilitates the uptake of water from the hindgut. In some cases the reabsorbed water may come from the atmosphere as well as from the diet.

8. See Thorn and Truman (1989).

9. The cobalt backfill method produces neuronal labeling comparable or superior to that produced by the Golgi method. If you have a relatively large arthropod and lack ready access to a fluorescent microscope, cobalt is still your tracer of choice for analysis of neuronal cytoarchitecture. See Bacon and Altman (1977).

10. See Consoulas et al. (2002).

11. These studies were performed in flies engineered to express GFP in this cell population. The images were acquired with a multiphoton fluorescence microscope. Because multiphoton microscopes use longer wavelengths of light to excite fluorophores in tissue than do laser scanning confocal microscopes, the production of phototoxic reaction products is reduced, and tissue damage is minimized. See Williams and Truman (2005).

12. See Watts et al. (2003).

13. The detailed sequence of externally observable events that occurs during puberty in human boys and girls is often described using Tanner stages, named after Professor James Tanner (1920–2010), a pediatrician and researcher who founded the modern study of human growth. See Marshall and Tanner (1969, 1970).

14. See Spear (2000).

15. See Sisk and Foster (2004).

16. See Huttenlocher (1979).

17. See Huttenlocher and Dabholkar (1997).

18. See Boksa (2012).

19. See Hoeft et al. (2011).

20. The English-language idiom of taking something with a grain of salt (that is, considering that something you learn is possibly but not certainly true) is a translation of the Latin phrase *cum grano salis*. The phrase is ancient and sometimes attributed to Pliny the Elder, but its origins are unclear.

21. See Lebel and Beaulieu (2011).

Investigative Reading

1. Katanin proteins are a family of microtubule-severing proteins. Design a study to test your hypothesis that katanins are required to regulate dendritic regression during neurometamorphosis in *Drosophila*. Design a second study to test the hypothesis that katanins regulate axonal reorganization during neurometamorphosis.

Lee, Hsiu-Hsiang, Lily Yeh Jan, and Yuh-Nung Jan. 2009. *Proceedings of the National Academy of Sciences U.S.A.* 106: 6363–68.

2. Human adolescents have been reported to engage in more novelty-seeking behavior than do younger children. You wish to study the neuroanatomical correlates of this behavior in a rodent model. Devise a behavioral method that you can use to study novelty seeking in juvenile rats and mice.

Cyrenne, De-Laine M., and Gillian R. Brown. 2011. *Developmental Psychobiology* 52: 670–76.

3. You are studying the development of the cerebral cortex in humans. You know that human adolescence is characterized by a loss of gray matter. Your goal is to determine which cortical areas lose gray matter first. What methods can you use to answer this question?

Gogtay, Nitin, Jay N. Giedd, Leslie Lusk, Kiralee M. Hayashi, Deanna Greenstein, A. Catherine Vaituzis, Tom F. Nugent III, David H. Herman, Liv, S. Clasen, Arthur W. Toga, Judith L. Rapoport, and Paul M. Thompson. 2004. *Proceedings of the National Academy of Sciences U.S.A.* 101: 8174–79.

4. *Challenge Question:* Spinophilin is an actin-binding protein that is enriched in dendritic spines. An antibody to spinophilin can be used to identify dendritic spines in the hippocampus of rodents. Design a study using this antibody to investigate whether dendritic spine density changes

during adolescence in female rodents and whether estradiol secreted by the ovaries at the time of puberty has any effect on hippocampal spine density. Do you expect to find an increase or a decrease? Is your study a light microscopy or electron microscopy study?

Yildirim, Murat, Oni M. Mapp, William G. M. Janssen, Weiling Yin, John H. Morrison, and Andrea C. Gore. 2008. *Experimental Neurology* 210: 339–48.

Thinking about Intellectual Disability in the Context of Development

Neuroscience and Intellectual Disability

Millions of children in the United States have received a diagnosis of intellectual disability. The estimated prevalence of what was once commonly referred to as mental retardation is at least 1 percent.[1] The defining feature of intellectual disability is the coexistence of limitations in performance of the tasks of everyday life (often referred to as adaptive behavior) and below-average intellectual function. Impairment is evident before age 18 and may or may not be associated with other physical or cognitive disabilities. The *Diagnostic and Statistical Manual of Mental Disorders* (DSM-IV-TR) defines subcategories of mild, moderate, severe, and profound intellectual disability based on IQ test scores.

The term *intellectual disability* is used to describe symptoms of heterogeneous etiology (a medical term that means "causes"). Known causes of intellectual disability include heritable genetic mutations, chromosomal abnormalities, neonatal hypothyroidism, intrauterine infections, and prenatal exposure to ethanol. In individual cases, particularly when the disability is relatively mild, the etiology is never resolved.

The neuroscientist's perspective offers guidance to our thinking about intellectual disability, a perspective that can significantly extend the clinical perspective. The differences in cognitive and adaptive development displayed by individuals with intellectual disability must, from this perspective, occur because of abnormal brain structure leading to abnormal brain function. This view has consequences. The first is that, despite myriad etiologies, the resulting common set of symptoms suggests the causal agents are converging on a relatively small number of brain abnormalities. The second is (again, despite myriad etiologies) that, once specific brain abnormalities are identified, it may be possible to devise ameliorative therapeutic interventions.

Signaling pathways active during normal development of the brain offer a source of candidate therapeutic drugs for this purpose. An analogy might be found in current drug treatments for high blood pressure: the causes of hypertension are numerous, and in most patients, especially those with mild hypertension, the origin of the condition will never be determined. Regardless, a relatively small number of drug types—mainly diuretics, ACE inhibitors, beta blockers, and calcium channel blockers—are widely and successfully used to relieve high blood pressure in many patients. An individual patient diagnosed with hypertension can feel optimistic that, with some trial and error, he or she can find an effective treatment for the symptom without having to cure the underlying condition. Might intellectual disability be viewed in the same light one day?[2]

The goal of this chapter is to summarize current understanding of the neurobiological basis of intellectual disability in humans and animal models. This summary is followed by five examples that link abnormal brain development and intellectual disability: neonatal hypothyroidism, fetal alcohol syndrome, Rett syndrome, Down syndrome, and fragile X syndrome. These examples were selected to highlight the increasingly dynamic interplay between developmental neuroscience and intellectual disability research.

Obviously, birth defects or injuries that cause large parts of the central nervous system to fail to develop or to develop inappropriately will be devastating for intellectual and adaptive function, and, as described in previous chapters, severe defects such as holoprosencephaly are associated with early death. But gross brain malformations are not typical of most cases of intellectual disability. It is necessary to study the details of neuronal structure and connectivity to understand what is different in the brains of these individuals. Our knowledge of the abnormalities associated with intellectual disability come from study of a relatively small number of human brains post mortem. Early investigations were limited to parameters that can be measured in histological sections prepared from autopsy specimens, primarily estimates of neuronal number and density and analysis of lamination in layered structures such as the cortex. In the 1970s, basic histology was supplemented by application of the Golgi method to human brains. This method randomly but completely labels a tiny proportion of neurons in a block of brain tissue with a microcrystalline form of silver nitrate that appears black when viewed with an ordinary light microscope. This permits detailed analysis of dendritic branching and dendritic spine morphology.[3] Studies using the Golgi method drew attention to dendrites as the possible locus of pathology in intellectual disability. Two other methods occasionally applied to autopsy specimens are electron microscopy of synaptic ultrastructure and immunolabeling to examine the distribution of specific proteins in the brain. Information is also available from neuroanatomical and functional brain imaging studies of persons with intellectual disability. For example, a recent study compared neural activation in the brain during an object recognition task in typically developing young adults and in persons with Down syndrome, a condition that results from having an extra copy of Chromosome 21.[4]

These sparse studies on human brains have inspired animal models in which intellectual disability–like symptoms are induced by experimental manipulations such as malnutrition, hypothyroidism, and prenatal alcohol exposure. Most such studies are performed using rodent models. The ability to create transgenic mice has extended animal model studies to the effects of mutation of candidate intellectual disability genes. Yet other genetic models, as has been amply documented in the preceding chapters, also inform our understanding of development of the human nervous system. Such reasoning has led to the recognition that *Drosophila* can be used not only as tools for analysis of the development deficits that lead to mental retardation but also as cost-effective tools for drug discovery. For example, primary cultures of *Drosophila* neurons bearing mutations associated with abnormal morphology can be used to screen libraries of candidate therapeutic compounds.[5]

Several decades of studies on human brains and animal models have identified two categories of brain deficits as primary causal candidates for the symptoms of developmental disability. The first category is a perturbation of the neuronal migration that normally occurs during development of the cerebral and cerebellar cortices. The second category comprises quantitative changes in dendritic branching and dendritic spine morphology. Both abnormal migration and abnormal dendrites are likely to result in changes in neural network connectivity (some of which can now be revealed using fMRI). At present, in part because neuronal migration occurs prenatally, it is difficult to envision therapeutic interventions designed to restore normal migration. Dendrites, however, offer enticing therapeutic targets.

Perturbations of Neuronal Migration

The term *neuronal migration* refers to the directed movement of neurons away from their sites of origin to their final location. Examples of neuronal migration were noted in Chapter 5 (rostral migratory stream) and Chapter 6 (formation of the cerebral cortex). Brain structures that have clearly demarcated layers are easy to assay for perturbations of neuronal migration because the absence of usual layers is so readily apparent. Like the cerebral cortex, the cerebellar cortex is also a layered structure, albeit simpler. The adult cerebellar cortex contains only three layers: a superficial molecular layer, a middle Purkinje layer, and, closest to the underlying white matter, the granule cell layer. The molecular layer contains two types of inhibitory interneurons—stellate and basket cells—plus the dendrites of the Purkinje cells and the axon terminals of the granule cells. The Purkinje layer contains the large somata of the output neurons of the cerebellar cortex, the Purkinje cells. The inner granule layer contains the densely packed small granule cells. The large, neatly arrayed Purkinje cells and the tiny but extremely numerous and densely packed granule cells define the unique structure of the cerebellum (fig. 10.1).

The cerebellar cortex has a different appearance prior to differentiation of the mature Purkinje cell and granule cell phenotypes. There is no internal

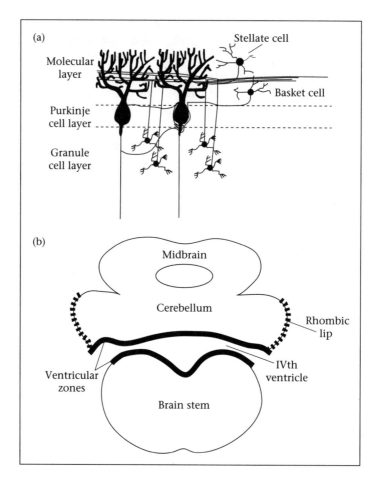

Figure 10.1. Structure and development of the cerebellar cortex. The mature cerebellar cortex comprises three distinct layers (a). Some of the characteristic cerebellar circuitry is hinted at in this schematic diagram. The large and beautiful Purkinje cells are the output neurons of the cerebellum; they send their axons to the deep cerebellar nuclei. The ventricular zone of the IVth ventricle is the major source of cerebellar neurons (and glial cells), including the Purkinje cells (b). A secondary neurogenic zone called the rhombic lip, however, is the source of the granule cell progenitors. The Purkinje cells migrate away from the ventricular zone, much as the neurons of the cerebral cortex migrate radially away from the ventricular zone. The granule cell progenitors migrate away from the rhombic lip along a path tangential to the developing cerebellum to form a transient proliferative center (the external germinal layer). Drawings adapted from Hatten and Roussel (2011).

granule cell layer, and, instead of a superficial molecular layer, the young cortex is capped by an external germinal layer (EGL).[6] The first cells in the EGL arise from a neuroepithelium of a transient metencephalic structure called the rhombic lip. The rhombic lip defines the border of the cerebellum and produces granule cell precursors that migrate onto the surface of the forming cortex. Eventually, the cells born in the rhombic lip cover the cerebellar cortex with a thin layer of proliferating cells. Continued proliferation within the EGL results in formation of a large pool of granule cell precursors.

In the mouse cerebellum, the initial monolayer of the EGL is present by E16. After the pup is born, the precursor cells of the EGL continue to divide and the EGL becomes thicker. Its days, however, are numbered.[7] A shift from proliferation to migration during the second week of postnatal life results in the disappearance of the EGL. Once postmitotic granule cells begin to differentiate, they move into the cortex and migrate radially to their final position *below* the Purkinje cell layer. In the mouse, this process is complete by the end of the second week of postnatal life. In humans, histological examination of brains post mortem demonstrated that the equivalent of the mouse EGL was visible in the cerebellar cortex by 24 weeks of gestation.[8] The EGL maintained a roughly constant thickness until the second month of postnatal life, when it began to thin. Sometime before the end of the first year (and sometimes by as early as the end of the fourth month), the EGL disappeared.

The disappearance of the EGL in both mouse and human reflects migration of the granule cells into the cerebellar cortex. Persistence of the EGL is therefore an indicator of a *failure of migration*. One well-known cause of intellectual disability in humans is hypothyroidism. Lack of sufficient thyroid hormone during the prenatal and neonatal periods alters development in many ways, resulting in the intellectual disability syndrome described later in this chapter. One of the signature effects of hypothyroidism, however, is delay of granule cell migration. In mice nursed by mothers given an anti-thyroid drug in their drinking water, a clearly defined EGL was still visible on postnatal Day 21.[9]

Another cause of intellectual disability is prenatal exposure to ethanol. As in the case of hypothyroidism, ethanol alters brain development in many different ways. However, in affected infants and children, a frequent autopsy finding of leptomeningeal neuroglia heterotopia—a sheet of misplaced neurons and glia on the surface of the cerebrum, cerebellum, and brain stem—suggested that one of the mechanisms by which ethanol alters development is interference with neuronal migration.[10] This finding has been confirmed by studies using the cerebellar cortex of rodents as a model. Ingestion of ethanol by pups during the early postnatal period—roughly equivalent to the third trimester of prenatal development in humans—resulted in a significant reduction in the number of postmigratory granule cells in the internal granular layer. The suspected impact of ethanol on granule cell migration was subsequently demonstrated directly by real-time observation of granule cell movement in cerebellar slices prepared from the brains of mice on postnatal Day 10. In this study cerebellar granule cells were labeled immediately after preparation of the slice by brief immersion in medium containing the lipophilic carbocyanine dye DiI. Migratory progress could then be tracked over the next several hours using a confocal microscope. Treatment with ethanol slowed granule cell migration significantly (fig. 10.2).[11]

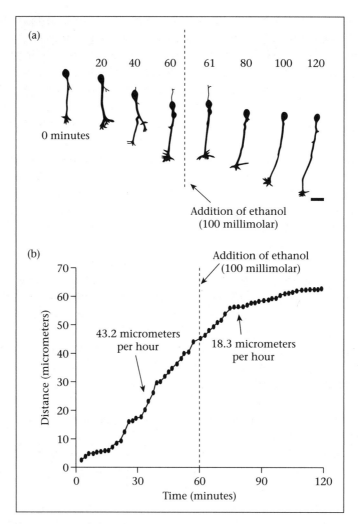

Figure 10.2. Direct effects of ethanol on granule cell migration in the developing cerebellar cortex. Isolated granule cells migrate considerable distances from microexplant cultures prepared from the cerebellum of P0–P3 mice; this migration is inhibited in a dose-dependent fashion by ethanol. The upper panel (a) shows time-lapse images of the effects of an application of 100-millimolar ethanol to a microexplant culture of P1 cerebellum. Scale bar, 12 micrometers. The lower panel (b) shows the distance traveled by the cell shown in (a) before and after the application of ethanol. The average rate of migration was 43.2 micrometers per hour before ethanol and 18.3 micrometers per hour after. Migrating granule cells in control cultures not exposed to ethanol maintained an average migration rate of approximately 50 micrometers per hour over a 2-hour observation. Data redrawn from Kumada et al. (2006).

The simple organization and relatively late migration of the cerebellar granule cells make the cerebellar cortex a good assay for studies of how factors that cause intellectual disability alter brain development. Some of the features associated with these syndromes—delayed motor development, poor fine motor skills, lack of balance and coordination—reflect abnormal functioning of the cerebellum. But the effects of hypothyroidism and ethanol on neuronal migration are unlikely to be limited to the cerebellar granule cells: they are just most easily assessed in this population of neurons. Perturbation of neuronal migration should always be considered a possible cause of impaired intellectual disability and adaptive function.

Dendritic Abnormalities

Mispositioned neurons may not be able to take their customary places in neural circuits. For example, cerebellar granule cells that do not enter the internal granule layer are unlikely to make connections with mossy fiber terminals. Disruption of neural circuit function can also be caused by abnormalities in neuronal dendrites. Two categories of dendritic abnormalities have been described. The first involves a change in the extent of dendritic branching. The second involves changes in the structure and density of dendritic spines. Given the role of dendritic spines as the major postsynaptic partners in excitatory central synapses (Chapter 7), it is not difficult to understand why investigators studying intellectual disability study dendrites.

Several articles published in the early 1970s provided evidence of altered dendritic branching and reduced dendritic spine density in the brains of children with intellectual disability. One study used the Golgi method to examine the cytoarchitecture of pyramidal neurons in samples of cerebral cortex from patients with *moderate to severe* intellectual disability.[12] The photomicrographs in this article are shocking to anyone familiar with the appearance of normal pyramidal neurons. Cortical neurons from the brains of several of the individuals studied displayed truncated dendritic arborizations with very few branches. Although other abnormalities were present in several cases involving young children with profound intellectual disability, the abnormal dendrites were the sole or major pathology described.

The first study to document dendritic spine abnormalities was based on examination of Golgi-impregnated cortical neurons in two children with Down syndrome (fig. 10.3).[13] A subsequent study expanded this finding to children with intellectual disability who had normal chromosome counts.[14] Both studies described an increase in the proportion of spines with a long, threadlike morphology. Such spines are regarded as immature, because they are rarely seen in Golgi material prepared from adult brains. Golgi studies of animal models of intellectual disability have subsequently reported comparable abnormalities. In addition to cortical neurons, hippocampal pyramidal neurons and cerebellar Purkinje cells have been reported to be affected by

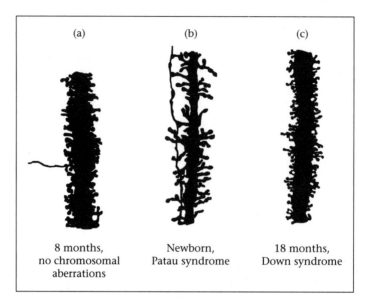

| (a) | (b) | (c) |

8 months,
no chromosomal
aberrations

Newborn,
Patau syndrome

18 months,
Down syndrome

Figure 10.3. Golgi studies of apical dendritic spines of cortical pyramidal neurons. Shown are studies of an infant with no chromosomal aberrations (a), a newborn infant with Patau syndrome (trisomy 13) (b), and an 18-month-old with Down syndrome (c). Drawings based on camera lucida (drawing tube) drawings of the visible spines of representative neurons. Redrawn from Martin-Padilla (1972).

manipulations that induce intellectual disability, including experimental hypothyroidism.

Evidence for an association of abnormal dendrites with intellectual disability in human patients and animal models accords with our understanding that the synapse is the locus of all forms of neuronal plasticity, including learning and memory. These findings help explain why the many different causes (genetic and environmental) of intellectual disability produce roughly comparable phenotypes. The data from the Golgi studies imply that many of the differences between normal brains and brains from individuals with intellectual disability are quantitative rather than qualitative in nature. They raise the tantalizing possibility that treatments targeted to dendrites might reduce some of the symptoms of intellectual disability.

The five examples that follow are currently considered incurable, although the effects of neonatal hypothyroidism can be blunted if treatment begins sufficiently early. Each example is a challenge to present and future neuroscientists. These examples also illustrate one of the most beautiful attributes of biomedical research: the study of pathology can lead to breakthroughs in our understanding of normal function, which in turn leads to the development of rational therapeutic interventions.

Neonatal Hypothyroidism

The thyroid hormones thyroxine (T_4) and triiodothyronine (T_3) are synthesized in the thyroid glands. These hormones are unique in that they are amino acid derivatives that contain the element iodine. The source of iodine is the diet. Human populations that lived in iodine-poor regions prior to the recognition of the requirement for dietary iodine suffered many deleterious consequences of hypothyroidism, including cretinism. Individuals with cretinism experience stunted physical growth and intellectual disability. The term *cretinism* is now rarely used because it is recognized that this syndrome is caused by congenital deficiency of thyroid hormones, reflecting either maternal or fetal hypothyroidism or both. Dietary supplementation (e.g., the use of iodized salt) has reduced the incidence of congenital hypothyroidism in the developed world. In many countries, newborns are screened for this condition by immunoassay for both T_4 and thyroid-stimulating hormones so that hormone replacement can be initiated if necessary. This condition is not common, but neither is it rare. It is estimated that 1 in every 4,000 infants born in the United States is affected by hypothyroidism. Although standard medical practice is to test newborns and treat them if needed, the use of tests for maternal hypothyroidism is growing, given evidence that reduced maternal thyroid hormone levels early in pregnancy predict intellectual disability in the children of such mothers. (Because the fetal thyroid gland does not begin to secrete thyroid hormones until Week 16 of gestation, early pregnancy is a time of particular vulnerability to maternal hypothyroidism).

In part because hypothyroidism is not a fatal condition and in part because it is typically diagnosed in neonates, few studies have examined the brains of hypothyroid human fetuses. The current model of thyroid hormone action during development depends heavily on the use of rodent models. Studies in rats and mice have focused on the first 2–3 weeks of postnatal life. During this time abnormalities in the cerebellum, including delayed granule cell migration, are among the most prominent features of hypothyroidism. As adults, developmentally hypothyroid rats and mice show impaired learning in standard laboratory tests such as the Morris water maze and appear to be more anxious than their euthyroid (with a functioning thyroid gland) counterparts.

The access of thyroid hormones to targets in the developing brain is regulated by binding proteins and cell membrane transporters. Evidence from rodent studies suggests that T_4 enters astrocytes via a cell membrane protein called OATp1c1 that acts as a specific transporter. Inside astrocytes T_4 is deiodinated to T_3. T_3 then enters adjacent neurons via another transporter, MCT8. The requirement for a transporter to move T_3 into neurons is indicated by the phenotype that results from an inherited mutation in the MCT8 transporter of humans. This mutation causes a rare X-linked recessive disorder characterized by severe intellectual disability called Allan–Herndon–Dudley syndrome.

Inside neurons, T_3 binds to its receptor, which is a member of the family of animal nuclear receptors that mediate the cellular actions of steroid hormones such as androgens and estrogens. These receptors function as dimerizing transcription factors that bind to specific DNA sequences in the regulatory regions of target genes. The presence of T_3 controls the binding of the thyroid receptor to other nuclear proteins that function as either corepressors or coactivators; the presence of the hormone therefore controls whether transcription is repressed or activated.

Two isoforms of the thyroid receptor (TRα and TRβ) are expressed in many brain regions, including those that develop abnormally as a result of hypothyroidism. Although the presence of two receptors complicates experimental analysis of thyroid hormone action in the brain, the effects of TRα and TRβ knockouts on development of the cerebellum have been carefully analyzed. In general, TRα appears to regulate granule cell migration, whereas TRβ regulates differentiation of Purkinje cells. One surprise arising from the study of TR knockout mice was the finding that not all thyroid hormone receptor knockouts mimic the effects of hypothyroidism. For example, as expected, knockout of the TRβ receptor impaired the postnatal maturation of Purkinje cells, with effects similar to those of hypothyroidism. But knockout of the TRα receptor did not prevent the disappearance of the EGL. This finding is interpreted as evidence that the *unliganded* receptor, rather than being inactive, binds to DNA and acts a transcriptional repressor. Both knocking out the receptor via a genetic manipulation and binding the hormone to the receptor relieve the transcriptional repression.

Rett Syndrome

Rett syndrome is a complex disorder that affects approximately 1 out of every 10,000 female births. Individuals with Rett syndrome appear normal at birth but display developmental reversals (loss of previously acquired skills) between 6 and 18 months of age. Intellectual disability is one of a long list of symptoms that includes poor speech, loss of purposeful hand movements, and irregular breathing. A patient's condition typically becomes progressively worse until the teenage years, at which time symptoms may stabilize (but not improve). Rett syndrome occurs primarily in girls. This finding suggested that an X-linked dominant mutation was at fault and that boys with this mutation did not survive. This was confirmed in a 1999 study showing that all individuals in a sample that included sporadic and familial cases of Rett syndrome shared mutations in a single gene on the X chromosome.[15] Since then, more than 600 Rett-associated mutations in this gene have been reported, including missense, nonsense, frameshift, and deletion mutations. All of these mutations alter the function of a gene that encodes a protein called methyl CpG–binding protein 2 (MeCP2). A diagnosis of Rett syndrome is based on a combination of observed symptoms and the results of tests for mutations in this gene.

MeCP2 is a protein that binds to methylated 5'-CpG-3' dinucleotides in nuclear DNA. MeCP2 is present in many cell types but is abundant in neurons. The protein contains a transcriptional repressor domain in addition to the methyl CpG–binding domain. Its functional role appears to be the suppression of transcription, but at present its specific role in the regulation of gene expression is unclear. Knockout mice that lack MeCP2 protein in all tissues display a lack of coordination, stereotyped paw movements, and breathing abnormalities reminiscent of but less severe than those observed in human patients. Studies comparing mice with neuron-specific knockouts and mice with neuron-only expression of MeCP2 have implicated lack of neuronal MeCP2 as the cause of Rettlike symptoms. Golgi studies of cortical pyramidal neurons in brains of Rett patients have revealed subtle but significant dendritic alterations in widespread regions of frontal cortex, motor cortex, and temporal lobes (fig. 10.4). Overall, Layers III and V pyramidal neurons from Rett brains have a smaller number of shorter apical and basal dendrites compared with the same populations sampled from non-Rett brains.

Golgi studies of Layers II or III cortical pyramidal neurons in the brains of knockout mice also found a reduction in the complexity of dendritic arborizations. No differences were detected in the density or distribution of dendritic spines. These findings were confirmed in a subsequent study in which dissociated cortical neuronal precursors from the brains of GFP-expressing wild-type and MeCP2 knockout mice were transplanted into the brains of non-GFP-expressing newborn pups.[16] After an 8-week posttransplantation survival period, images of the surviving fluorescing cortical neurons were captured and analyzed. Once again, the neurons that lacked MeCP2 expression had reduced dendritic arbors. As in human patients, both apical and basal dendrites were affected. The transplantation technique allowed dendritic growth to be assessed in MeCP2-null neurons growing in a wild-type cortex and in wild-type neurons growing in the MeCP2 knockout brain. The impact on dendritic development appeared to be primarily cell autonomous, for the effects were strongest in neurons lacking their own expression of MeCP2.

Despite its complex phenotype, Rett syndrome is an attractive target for therapeutic intervention. Could some of the symptoms be relieved if normal dendritic arbors could be restored to the affected cortical pyramidal neurons? Traditional gene therapy—that is, replacement of the mutant gene with a normal copy, typically by means of a viral vector—faces formidable obstacles in terms of both delivery (right place, right time) and dose (there is evi-

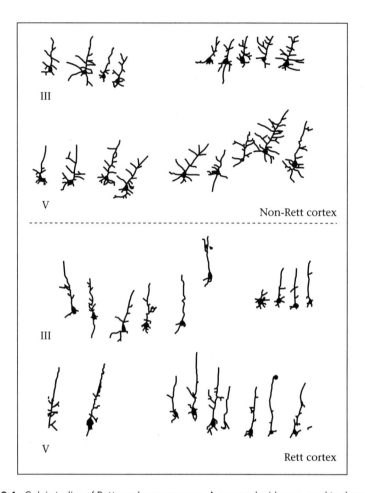

III

V

Non-Rett cortex

III

V

Rett cortex

Figure 10.4. Golgi studies of Rett syndrome neurons. A camera lucida was used to draw representative cortical Layer III and Layer V pyramidal neurons from a Rett and a non-Rett brain. A photograph of the camera lucida drawing was published as Figure 1 in Armstrong et al. (1995), and I adapted it for this book.

dence that too much MeCP2 is as harmful to neuronal function as too little). It is formally possible that drugs that mimic the regulatory effects of MeCP2 on transcription might be developed, but this approach is also challenging in terms of delivery and dose. One possibly beneficial approach could be to treat with nongenetic, nondrug interventions shown in other contexts to promote dendritic branching and cortical growth. Environmental enrichment is an elegant example of such an approach.

To the nonneuroscientist, the phrase *enriched environment* probably brings to mind a high-quality daycare center. To a neuroscientist, however, this phrase has a surprisingly specific meaning. Environmental enrichment refers to an experimental manipulation in which laboratory animals (typically rats or mice) live in cages that provide more sensory, cognitive, and motor stimulation than does standard laboratory housing.[17] Environmental enrichment produces many changes in the brains and behavior of rodents. One well-known change is an increase in thickness of the cerebral cortex. Other studies, some of which used the Golgi method, have demonstrated that neurons, including cortical pyramidal cells, respond to environmental enrichment with dendritic growth.

It should be straightforward to assess the impact of environmental enrichment on the brains of MeCP2-null mice. Several studies have demonstrated that environmental enrichment reduces anxiety and improves motor coordination and performance in the Morris water maze in MeCP2 knockout mice, but to date no study has investigated environmentally regulated dendritic growth in this mouse model. Positive results in the laboratory, however, would raise the interesting question of what specific housing and educational conditions might constitute effective environmental enrichment for human children with Rett syndrome. Such enrichment might resemble that provided to all children by the aforementioned high-quality daycare.

Fragile X Syndrome

Intellectual disability is a hallmark of fragile X syndrome, an inherited condition often accompanied by anxiety, speech disorders, and hypersensitivity to visual, auditory, and tactile stimuli. Symptoms may be mild or severe. Fragile X syndrome is probably the most common inherited intellectual disability. It is estimated to affect as many as 1 in 1,250 males and 1 in 2,500 females. Symptoms are more severe in boys than in girls. Most diagnoses are made in preschoolers. The name of the syndrome is not derived from a symptom displayed by individuals with the condition. Instead, the name refers to the appearance of the X chromosome of these individuals. A portion of the X chromosome looks as if it could easily break off (fig. 10.5).

The cause of fragile X syndrome was discovered in 1991. Symptoms arise from mutations in a gene located on the X chromosome called FMR1. The FMR1 gene encodes an RNA-binding protein called FMRP (fragile X mental retardation protein). This protein is expressed most abundantly in neurons and in testes. Its major function appears to be regulation of protein synthesis by means of its ability to bind and transport mRNAs within neurons. An association of this protein with ribosomes is consistent with this proposed function.

The most common mutation that causes fragile X syndrome is a trinucleotide repeat, a type of insertion that, when it achieves a certain length, results in loss of expression of FMRP. The most common repeated sequence is CGG (a cytosine followed by two guanines), and the insertion is found in the 5′ untranslated region of the gene that encodes FMRP. Having 40 or fewer CGG repeats in this gene is normal. A person with more than 200 repeats displays fragile X symptoms. An intermediate number of repeats is termed a pre-

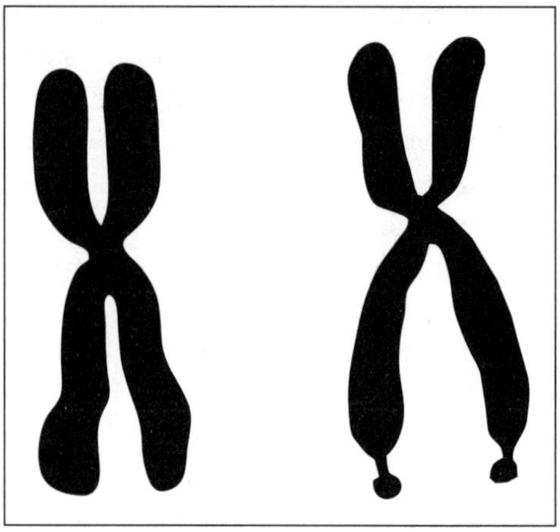

Figure 10.5. Fragile X chromosome. A normal human X chromosome is on the left. A fragile X human chromosome is on the right. These chromosomes are depicted at the metaphase stage of mitosis of the cell cycle.

mutation because the length of the repeat can increase during meiosis. This means that a parent with a premutation (and no symptoms) can produce children with a full mutation. The loss of FMRP expression in patients with 200 or more repeats occurs because of a change in the methylation pattern of the promoter of *FMR1*. Fragile X syndrome is definitively diagnosed by testing an individual's DNA. The DNA is typically obtained from a blood sample.

As in the case of Rett syndrome, the identification of a mutation in a single gene led quickly to the development of the corresponding knockout mouse. The mouse homolog of human *FMR1* is *Fmr1*. A consortium of scientists based in Belgium and the Netherlands developed the first knockout model for fragile X syndrome in mice. These mice lack any Fmr1 protein. Initial behavior tests revealed that the knockout mice were overall surprisingly similar to wild-type mice. But the knockouts displayed subtle deficits in Morris water maze tests: although they were capable of learning and remembering the location of a hidden platform, they took longer than their littermates to swim to the platform.

Despite the mild nature of their cognitive impairments, Fmr1 knockout mice have made extraordinary contributions to understanding of the normal role of FMRP in development. They have also proved to be valuable tools in the search for treatments.

Application of the Golgi method to brains obtained post mortem from several patients diagnosed with fragile X syndrome revealed a striking abnormality: the dendrites of their cortical pyramidal cells had thin, elongated dendritic spines that resembled spines normally found only in the very young cortex. A subsequent Golgi study of Layer V pyramidal cell dendrites directly compared the dendritic spines of fragile X patients with the dendritic spines of age-matched controls. The fragile X patients had more long, immature spines than did the controls. This study also reported that fragile X patients had a higher than normal density of spines on both apical and basal dendrites, a result that was interpreted as a failure of normal synaptic pruning. Is the same defect found in the brains of Fmr1 knockout mice? Could this be a clue to the function of FMRP in brain development?

Careful Golgi studies of cortical pyramidal cells in Fmr1 knockout mice revealed defects matching those seen in human patients. The knockout mice had more long dendritic spines and more immature dendritic spines than did wild-type controls. The knockouts also displayed a trend toward slightly higher dendritic spine density. These studies revealed that the overall shape of the dendritic tree characteristic of cortical pyramidal neurons was not altered by the absence of FMRP. Taken together, the human and mouse findings directed the focus of research to the postsynaptic side of the synapse.

As previously noted, the cell organelles most commonly associated with FMRP are ribosomes. Dendritic spines and ribosomes at first glance appear to represent two fundamentally different aspects of neuronal life: communication and protein synthesis. The apparent conundrum was resolved by the finding that both FMRP and mRNA are expressed in dendritic spines and the demonstration that localized protein synthesis can occur far from the neuronal soma, even in dendritic spines. Postsynaptic protein synthesis is now the basis of a model in which protein synthesis driven by activation of metabotropic glutamate receptors (mGluR) is responsible for certain forms of synaptic plasticity including long term potentiation and long term depression. Studies using preparations of isolated synapses called synaptoneurosomes revealed that one of the proteins synthesized in dendritic spines in response to activation of this category of glutamate receptors is FMRP.[18] The finding that *Fmr1* knockout mice displayed enhanced long-term depression relative to wild-type mice led to the development of a novel hypothesis for the cause of intellectual disability in patients with fragile X syndrome. The hypothesis, which is based on the assumption that FMRP functions primarily as a repressor of protein synthesis, predicts that the symptoms of fragile X syndrome reflect out-of-control protein synthesis in dendritic spines experiencing mGluR activation.

Many experimental results have supported this hypothesis since its formulation in the early 2000s. This has led to attempts to bypass the genetic deficit in fragile X syndrome through treatments targeting mGluR. Both genetic and drug-based manipulations of mGluRs have successfully treated the fragile X syndrome–like symptoms displayed by Fmr1 knockout mice. Small-scale studies in patients with drugs that alter mGluR activity (already approved for use in humans for other disorders) produced encouraging results consistent with the results obtained in studies using Fmr1 knockout mice. For example, young adults treated with an mGluR antagonist showed improvement in language skills. The mood should be optimistic as trials of other drugs targeting mGluR and downstream effects of mGluR activation proceed both in the mouse model and in human patients.

Down Syndrome

Triplication—trisomy—of Chromosome 21 in humans results in a complex phenotype called Down syndrome (fig. 10.6).[19] All patients with Down syndrome have some degree of intellectual disability. Because Down syndrome is diagnosed in approximately 1 in 750 live births, it is one of the most common and widely recognized causes of intellectual disability. In addition to the central nervous system, many other organ systems and aspects of growth and development are affected. Two common phenotypes are congenital heart disease and distinctive facial features including a round face, almond-shaped eyes, a small chin, and a protruding tongue. Down syndrome is definitively diagnosed by inspection of the individual's karyotype (appearance of the chromosomes). A newborn may be tested based on a doctor's and parents' assessment of the baby's appearance. Down syndrome can also be diagnosed in prenatal karyotypes by sampling fetal cells in the amnionic fluid (amniocentesis), placenta (chorionic villus sampling), or cord blood (percutaneous umbilical blood sampling). Maternal blood tests can be used as a screen for increased risk of chromosomal abnormalities. Positive screen results can be followed by a diagnostic test.

Trisomy of Chromosome 21 is caused by a meiotic nondisjunction event resulting in an individual with 47 chromosomes instead of the typical 46. Why does having an extra copy of a chromosome result in so many problems, including intellectual disability? The answer is found in the concept of dosage-sensitive genes. *Gene dosage* refers to the number of copies of a particular gene present in a cell. In some but not all cases, a higher gene dosage results in higher levels of the protein encoded by that gene. The resulting imbalance in gene products can disrupt normal cell function, and in humans, the typical outcome of trisomy is death of the embryo. Chromosome 21 is an exception to this rule, likely because it is the smallest human chromosome (47 million base pairs). Even the Y chromosome is larger (60 million base pairs). Yet even a small chromosome contains hundreds of genes. It is estimated that Chromosome 21 contains between 300 and 400 genes. It is unlikely that all of these genes will be found to be dosage sensitive. The task facing the medical geneticist is to identify the gene or subset of genes on Chromosome 21 that are dosage sensitive *and* involved in Down syndrome. The task facing the developmental neuroscientist is to understand how changes in the expression of these genes—both singly and in combination—are linked to structure and function in the nervous system. Given the complexity of the Down syndrome phenotype, it is highly unlikely that a single Down syndrome gene will be identified.

Analysis of rare cases of humans with partial Chromosome 21 trisomy led to the characterization of a Down syndrome critical region (DSCR) on Chromosome 21. Several genes found in the DSCR have been studied in animal models as possible contributors to the behavioral and cognitive aspects

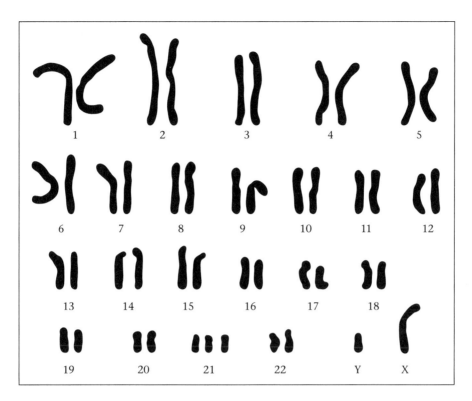

Figure 10.6. Trisomy 21 karyotype. A karyotype is an organized diagram prepared from photo-micrographs of an individual's chromosomes. The chromosomes are photographed during the metaphase of the cell cycle, when the chromosomes are most condensed. Karyotypes are often prepared from white blood cells obtained in a blood sample. The drug colchicine is used to block the progression of the cell cycle so that the chromosomes can be more easily studied. The karyotype depicted is from a boy with trisomy 21.

of the Down syndrome phenotype. Most of the research on the genetic basis of Down syndrome, however, has focused on genetically engineered triso-mic mice. But mice are not simply tiny humans. One of the many obvious differences between mice and humans is that mice do not have a Chromo-some 21. The mouse nuclear genome consists of 19 autosomes and a pair of sex chromosomes. But this does not mean that mice lack mouse versions of genes found on human Chromosome 21. Mapping blocks of orthologous genes—a tool of comparative genomics called synteny mapping—has revealed that the genes located on Chromosome 21 in humans are conserved as blocks of genes found on three mouse chromosomes. The relevant mouse chromosomes are 10, 16, and 17. The result is that, instead of having only one mouse model for Down syndrome, investigators can choose from sev-eral. Although this may appear to be a disadvantage, there is a benefit in being able to work with smaller blocks of genes, for this may make it easier to link a particular gene (or genes) with a particular phenotype.

A widely used mouse model designated Ts65Dn is trisomic for segments of mouse Chromosome 16. The genetic alteration in this mouse strain was randomly induced by irradiation. Approximately 55 percent of the mouse orthologs of Chromosome 21 genes are triplicated in Ts65Dn. One reason that this model is favored is that Ts65Dn mice display more severe learning and memory deficits than other lines of mice trisomic for Chromosome 10 or 17. Another mouse line of interest, Tc1, is *transchromosomic* rather than trisomic. Tc1 mice have the normal complement of mouse chromosomes, but their genome has been engineered to contain an additional chromosome: human Chromosome 21. Tc1 mice are in theory therefore trisomic for all of the genes for which human patients with Down syndrome are trisomic. Although Tc1 mice have been used in some studies, it has been discovered that Tc1 mice are mosaics, with human Chromosome 21 lost from many of their cells. Because none of the existing mouse models exactly matches human Down syndrome, alternatives continue to be actively generated.

The brain anomalies that cause the intellectual disability of Down syndrome resist a simple summary. It is likely that many of the critical events occur during early development. One reason to suspect this is that at birth the total brain volume of humans with Down syndrome is typically only 85 percent of normal. Some brain regions are more affected than others. The volume of the cerebellum, for example, is typically less than 75 percent of normal. Individuals with Down syndrome show reductions in the number and density of cerebellar granule cells.

Histological and high-resolution MRI studies have demonstrated that Ts65Dn mice also have reduced cerebellar volume and display a reduction in the number of granule cells in the internal granule cell layer. The developmental origins of this deficit were explored in a study in which Ts65Dn mice received a single injection of a Shh agonist (SAG 1.1) on the day they were born. SAG 1.1 is a small molecule that has been shown to cross the blood–brain barrier. An agonist is a drug that activates a response. The agonist SAG 1.1 activates the Shh pathway by binding to the mouse homolog of the *Drosophila Smoothened* gene. In Ts65Dn mice, injection of SAG 1.1 restored cerebellar volume and cerebellar granule cell numbers to normal. Studies of the responses of cultured granule cell precursors to activation of Shh pathways revealed that trisomic precursor cells were less responsive to the proliferative effects of Shh signaling than were wild-type precursor cells cultured under the same conditions. Apparently, hyperstimulation (with an injected dose of the agonist SAG 1.1 that was huge: 20 micrograms per gram) of the trisomic precursor cells resulted in a restoration of nearly normal levels of mitosis.[20]

The other major brain phenotypes found in the brains of Ts65Dn mice are abnormal dendritic spines in cortex and hippocampus and reduced long-term potentiation in the CA1 region of the hippocampus. These phenotypes suggest that trisomy alters signal transduction pathways related to synaptic plasticity. A specific hypothesis consistent with these findings is that inhibitory GABAergic synapses are overactive. The development of drug-based therapies that interfere with inhibitory signaling is one focus of current research on Down syndrome. Environmental enrichment strategies have also been pursued in both humans and mouse models. Studies using Ts65Dn mice have produced conflicting results. An enriched early rearing environment that yielded a significant increase in number of dendritic spines on cortical pyramidal neurons in wild-type mice had no effect on dendritic spines of Ts65Dn mice. By contrast, housing adult Ts65Dn mice in an enriched environment improved their performance in the Morris water maze, restored normal long-term potentiation, and reversed Down syndrome–associated alterations in the visual cortex. A single study of Ts65Dn mice that combines neuroanatomical and functional measures is essential. It will also be necessary to test the effects of environmental enrichment in other mouse models of Down syndrome and in humans with Down syndrome. Here it should be noted that, historically speaking, the societal transition from institutional to home-based care of children with Down syndrome may constitute a large-scale experiment in environmental enrichment.

The combination of changes in number of neurons, neuronal morphology, density of dendritic spines, and excessive inhibition likely account for much if not all of the intellectual disability associated with Down syndrome. Another aspect of Down syndrome found in mouse models and humans is early onset of brain pathologies and cognitive impairments associated with Alzheimer's disease, which often diminishes function as individuals with trisomy 21 age. Taken together, the developmental, signal transduction, and, with aging, neuropathological aspects of Down syndrome can seem overwhelming, almost to a point at which research that looks for therapies might seem of little use. Keep in mind, however, that even small changes in brain structure or function have the potential to translate into improvement in intellectual function and that even slight improvements in intellectual function (including those well short of a "cure") have the potential to enhance the lives of individuals and their families, friends, and teachers.

Fetal Alcohol Syndrome

Fetal alcohol syndrome is another common cause of intellectual disability.[21] The estimated incidence depends on the criteria used and the region surveyed. Typical estimates for the United States are 0.5 to 2.0 persons affected per live birth. The cause of fetal alcohol syndrome is prenatal exposure to alcohol (ethanol) as a result of maternal consumption. Many aspects of development are affected. Children with fetal alcohol syndrome often have distinctive facial features (including a thin upper lip, an upturned nose, and a flat philtrum, the vertical groove found above the upper lip) and grow more slowly than children not exposed prenatally to ethanol. In severe cases, frank microcephaly can be present. Intellectual disability can range from mild to profound and is often accompanied by other behavioral problems such as clumsiness, hyperactivity, and anxiety. These problems persist into adulthood. Although fetal alcohol syndrome shares features with other forms of intellectual disability, the syndrome never arises spontaneously. That is, fetal alcohol syndrome in humans is entirely the result of in utero ethanol exposure and therefore entirely preventable. In addition to physical signs, diagnosis is based on maternal history. The concentration of fatty acid ethyl esters in meconium (the green first feces produced by an infant) is considered a biomarker for prenatal ethanol exposure, but this test is almost never performed in clinical settings.

So many areas of the brain have been reported to be affected by prenatal ethanol exposure that it was once thought that the effects of ethanol were nonspecific. Modern neuroimaging studies, however, have identified regions that are particularly vulnerable, such as the parietal lobe of the cerebral cortex. The specific anomalies present in any individual case reflect the combination of maternal blood alcohol level and stage of development at the time of exposure. More severe effects are associated with blood alcohol levels of greater than 200 milligrams per deciliter. Studies comparing the impact of prenatal ethanol exposure in mice and humans have demonstrated that exposure during gastrulation (about the third week of human gestation) results in craniofacial abnormalities and reductions in size of the pool of neural progenitor cells. Exposure between 7 and 20 weeks of gestation reduces neuronal numbers in the cerebral cortex and hippocampus. Exposure to ethanol during the third trimester is associated with reduction in the volume of the cerebellum. Two features of fetal alcohol syndrome uncommon in other intellectual disability syndromes are abnormalities of the corpus callosum and reductions in glial cell number. The impact on glial cell populations is not subtle; the results of applying modern brain imaging methods to individuals diagnosed with fetal alcohol syndrome have highlighted a loss of white matter in the cerebral cortex.

Individuals with normal facial features despite prenatal ethanol exposure —sometimes referred to as nondysmorphic—also can exhibit changes in brain structure and intellectual disability. The term *fetal alcohol spectrum disorder* has been introduced to describe the broad range of disabilities, including nondysmorphic variants. Using this terminology, fetal alcohol syndrome represents one of the extreme ends of a spectrum of disability.

Despite its environmental origin, fetal alcohol syndrome shares two features typical of most forms of intellectual disability: dendritic spine abnormalities and impaired neuronal migration. A significant increase in the number of long, thin spines and a corresponding decrease in the number of stubby, mature spines on the apical dendrites of cortical pyramidal cells have been documented using the Golgi method in human samples post mortem and in rats and mice whose mothers were fed ethanol during pregnancy. The migration of cerebellar granule cells from the EGL to the internal granular layer has been used as a model system to study the impacts of ethanol on neuronal migration. As previously noted, in rodents the early postnatal period is the equivalent of the third trimester in humans. Studies in neonatal rodents have revealed that even a single exposure (via injection or intragastric tube) to ethanol early in life significantly reduces the number of granule cells in the internal granular layer. In contrast to hypothyroid animals, there is no corresponding increase in the thickness of the EGL; instead the granule cells that cannot migrate undergo apoptotic cell death.

What are the cellular and molecular mechanisms by which ethanol perturbs cerebellar development? Clues come from studies of granule cell migration in cerebellar slices acutely exposed to concentrations of ethanol roughly comparable to those that might be experienced by a human baby in utero as a result of maternal binge drinking. A comprehensive in vitro study of granule cell migration in slices prepared from mouse pups on postnatal Days 7, 10, and 13 revealed that exposure to ethanol immediately slows the tangential migration of granule cells in the EGL. The later phases of granule cell migration—radial migration in the molecular layer and final positioning within the internal granular layer—were also affected. All of the effects of acute administration of ethanol, however, could be reduced or even reversed by simultaneous pharmacological manipulation of calcium, cGMP, and cAMP pathways.[22] Investigators were able to demonstrate that the same manipulations were effective in vivo in blocking the effects of acute alcohol exposure on granule cell migration, but that they did not block the effects of acute alcohol on granule cell survival. These studies simultaneously reveal the impact on cerebellar development of a single exposure to ethanol and suggest possible treatments. The efficacy of treatments that improve granule cell migration, however, will be limited unless the death of granule cells stalled in the EGL can also be controlled.

Like many other teratogens (substances that disrupt prenatal development), ethanol has the potential to alter gene expression. Several microarray analyses comparing cultured mouse embryos exposed to ethanol with control embryos sampled during neurulation have revealed that ethanol reduces the expression of neurogenic genes such as members of the neurogenin family and Sox5. The expression of numerous other genes was also affected by ethanol, consistent with the bodywide effects of ethanol on growth and organ development. Another mechanism through which ethanol may alter developmental processes is epigenetic modification of DNA. Studies that demonstrated the potential of ethanol to modify DNA methylation profiles by inhibition of folate-mediated methionine synthesis led investigators to conduct a genomewide survey of gene methylation using microarrays paired with mapping of promoter methylation using methylated DNA immunoprecipitation.[23] Exposure of neurulating cultured embryos to ethanol resulted in significant alterations in methylation profiles. One of the many fascinating findings of this study was that chromosomes containing genes known to be regulated by epigenetic mechanisms were particularly affected by ethanol exposure. One such chromosome is the X chromosome. Recall from your study of introductory genetics that in mammals many genes on one of the X chromosomes of females are normally silenced to achieve dosage compensation. Methylation is one of the known epigenetic mechanisms of gene silencing. Ethanol treatment resulted in a 15-fold increase in the number of methylated genes on the X chromosome of young embryos bearing neural tube defects as a consequence of ethanol exposure.

Nonmammalian Models

Can a fruit fly display intellectual disability? The answer is no if the focus is on verbal skills, school-type tasks, and logical reasoning, but possibly yes if we ask a fly to perform fly-appropriate tasks. We do not know if intellectually disabled fruit flies occur spontaneously in natural populations. In the laboratory, however, fruit flies have been the focus of intensive research on the molecular mechanisms of learning and memory for several decades. Many mutations that affect learning have already been identified, and laboratory assays have been developed for the purpose of studying the ability of flies to learn and to remember what they learned.

The combination of an extensive behavioral literature with efficient and powerful tools for genetic transformation has led to development of several *Drosophila* models of Down syndrome and fragile X syndrome. These models are valuable because they allow investigators to study disrupted neural circuits from the electrophysiological, morphological, and behavioral perspectives in large numbers of individuals. It is also possible to study the brain and behavior in the same individuals. *Drosophila* also provides cost-effective models for proof-of-concept tests of potential therapeutic interventions. Three examples from a large and growing literature are introduced here. The first two are based on the analysis of mutants created specifically for the study of intellectual disability. The third is a model for fetal alcohol syndrome based on larval ingestion of ethanol.

Overexpression of Calcipressin in Drosophila

One of the genes located in the Down syndrome–critical region of Chromosome 21 has been named the Down syndrome critical region 1 gene (*DSCR1*). This gene encodes a protein that is a member of a family of proteins called calcipressins. Calcipressins inhibit the activity of calcineurins, a family of serine or threonine protein phosphatases. The *Drosophila* homolog of human *DSCR1* is named *sarah*.[24] Although loss-of-function mutations provide information on the normal function of the *sarah* gene product, the hypothesis that the deleterious effects of trisomy 21 are caused by the high dosage of genes in the Down syndrome–critical region can be tested directly by over-expression of *sarah*. This can be achieved by developing lines of flies that constantly (constitutively) overexpress *sarah*. By placing expression of *sarah* under control of the appropriate drivers, calcipressin can be overexpressed in all tissues or only in neurons. Another approach is to use an inducible transgene, which permits the effects of overexpression during development to be dissociated from the effects of overexpression at the time of testing. For example, in one type of inducible system, a chimeric regulator protein activates transcription only when it is bound to synthetic antiprogestin drugs such as RU486 (mifepristone).[25] Because wild-type fruit flies have nei-ther progesterone nor the associated progestin receptors, transcription can be activated only when the researcher provides RU486. Fruit flies can be fed RU486 as adults or larvae, then tested for their ability to remember an asso-ciation between an odor and an electric shock. In this assay, flies indicate that they remember the learned association by avoiding locations where that odor is present.

In contrast to wild-type *Drosophila*, flies that constitutively overexpressed *sarah* could not learn to avoid odors paired with electric shock.[26] Transient overexpression induced by feeding RU486 to adult flies with the inducible transgene also impaired learning, but tests of flies fed RU486 to drive *sarah* expression as larvae but then switched to RU486-free medium as adults yielded wild-type levels of learning. Such results are grounds for cautious optimism that treatments—in this case treatments that restore calcineurin-mediated signaling—might relieve some of the intellectual disability associ-ated with Down syndrome in humans, even in adults.

A Drosophila *Fragile X Syndrome Model*

In addition to the *FMR1* gene that encodes FMRP, the human genome con-tains two other members of the fragile X gene family: *FXR1* and *FXR2*. The mouse genome similarly encodes three related genes: *Fmr1, Fxr1,* and *Fxr2.* To date, almost all of the phenotype associated with fragile X syndrome has been attributed to *FMR1* and *Fmr1,* but the presence of the additional genes complicates analysis of the relationship between genotype and phenotype. By contrast, the *Drosophila* genome encodes a single homolog of *FMR1* named *dFmr1*. Like human FMRP, dFMRP also binds RNA. Null mutants of *dFMR1* are viable, and their phenotypes have been extensively analyzed. Morpho-logical analyses reveal that the nervous system of mutants is subtly ab-normal, with both axon terminals and dendritic arborizations displaying more branching than seen in wild-type flies. Behavior observations have revealed that, relative to wild-type flies, mutants cannot maintain circadian rhythms of locomotor activity when housed under conditions of constant darkness, and they have abnormally low levels of interest in courtship. Do these flies display any evidence of intellectual disability?

The results of studies probing the ability to learn an association between an odor and electric shock are instructive. Mutant flies were defective in their ability to remember a pairing one day after training relative to wild-type flies, although the deficit was evident only when using a specific training procedure.[27] Further studies of dFmr1 flies revealed that this effect of dFMRP is acute rather than developmental and that dFMRP interacts with a protein called Argonaute1, an essential component of the RNAi pathway. Most striking, this mutant permits a direct test of the hypothesis that the intellectual disability of fragile X syndrome patients reflects loss of activity-dependent control of protein synthesis at synapses. As this model predicts, mutant flies fed protein synthesis inhibitors showed improvements in their ability to remember the pairing of an odor and an electric shock.

Larval Alcohol Syndrome

Our final example of the use of fruit flies to study intellectual disability is not a genetic model. This model is instead based on the effects of exposure to ethanol during larval life. *Drosophila* larvae feed during each of the three larval instars. Larvae reared on medium containing ethanol conspicuously failed to thrive.[28] Many did not make it through metamorphosis, and those that did were visibly smaller than controls. The brains of larvae that consumed ethanol were smaller than normal. Studies using the BrdU and TUNEL techniques clearly showed that the reason for the shrunken brains was a reduction in the number of replicating cells (as opposed to an increased incidence of neuronal death). Flies in this study were not tested using learning and memory tasks, but those that survived to adulthood showed changes in their responses to ethanol in vapor form: they took longer to lose their righting reflex (the ability to return to a normal position after a fall) and took longer to shift from ethanol-induced hyperactivity to ethanol-induced hypoactivity than did control flies.

Why develop a *Drosophila* model for fetal alcohol syndrome? The avoidance of any maternal-mediated effects seems, in the general scheme of things, a rather small benefit, especially given that each batch of affected flies must be carefully reared on the ethanol diet. The advantage lies in the power of combining forward and reverse genetics approaches. The effects of the environmentally induced phenotype can be studied in terms of the resulting changes in gene expression; the results of gene expression studies can be used to guide the creation of transgenic fly lines. In these initial studies, exposure of larvae to ethanol reduced the expression of insulin-like peptides and insulin-like receptors in the brain. Some aspects of the ethanol phenotype were rescued by transgenic expression of insulin-like peptides, suggesting that a reduction in insulin signaling may cause some of the symptoms associated with fetal alcohol syndrome. These results are consistent with results showing ethanol-induced inhibition of insulin-like growth factors and receptors in rodent brains. The performance of flies exposed to ethanol during larval life on tests of learning and memory will be of the greatest interest.

Reality Check

Application of the tools of modern neuroscience to the challenge of human intellectual disability makes this an exciting time for investigators interested in these syndromes. The relatively small number of morphological phenotypes—problems with neurogenesis and gliogenesis, problems with migration, problems with dendritic branching, problems with dendritic spines—associated with intellectual disability can make it seem as if effective therapies are within reach. The insight that, as in the cases of many other diseases, it is not necessary to cure the underlying defect if one can restore near-normal or at least closer-to-normal function is a genuine cause for optimism.

Recent progress should promote reflection on what researchers and clinicians can realistically expect to accomplish. It is possible that the underlying deficits in some forms of intellectual disability are purely functional rather than developmental. In these cases, any treatments that are developed should be effective any time they are given, and treatment of intellectual disability might then be viewed as similar to management of a serious chronic disease such as diabetes. It is also possible that many of the deficits are developmental and that once a particular stage (sometimes referred to as a "critical period") of development has passed, treatments will no longer be effective. In these cases, early diagnosis will be essential so that intervention can begin as soon as possible.

Many challenging questions await young neuroscientists interested in a career in research related to human intellectual disability. For example, what molecular mechanisms account for the effects of environmental enrichment? Why does environmental enrichment work in some settings but not in others? Do populations of neurons beyond the ever-popular trio of cortical pyramidal cells, hippocampal pyramidal cells, and cerebellar Purkinje cells also show intellectual disability–associated defects in dendritic branching and dendritic spine maturation? Can stimulation of adult neurogenesis (or stem cell–based therapies) compensate for failure to produce sufficient neurons early in development? What side effects will be acceptable to patients in exchange for amelioration of intellectual disability? Another category of questions that will occupy not only future developmental neuroscientists but also bioethicists relates to the fine line between therapy and enhancement. Will treatments that improve cognitive function in individuals with intellectual disability also enhance cognitive function in individuals of so-called normal intellect? Will such treatments eventually define a *new normal* for brain development in our species?

Notes

1. The older term *mental retardation* is rapidly being supplanted by the term *intellectual disability*. As an example of the ongoing search for accurate and nonpejorative terminology, consider the American Association on Intellectual and Developmental Disabilities (AAIDD), a nonprofit professional organization that advocates on behalf of persons with intellectual disabilities. AAIDD was founded in 1876 as the Association of Medical Officers of American Institutions for Idiotic and Feebleminded Persons. It was later renamed the American Association on Mental Deficiency, in turn followed by the American Association on Mental Retardation. The present name was adopted in 2006. A discussion of terms used to describe mental retardation or intellectual disability can be found in Panek and Smith (2005).

2. I first heard this type of question raised by Dr. Linda L. Restifo in a seminar in the Department of Biology at Wake Forest University in 2007. See Restifo (2005).

3. The Nobel Prize in Physiology or Medicine for 1906 was jointly awarded to the inventor of the Golgi method, the Italian scientist Camillo Golgi, and to the Spanish neuroanatomist who perfected the method, Santiago Ramón y Cajal. See the Online Resources for a link to the always-informative Nobel web site.

4. The persons with Down syndrome were found to have an unusual pattern of neural activation compared with age-matched typically developing individuals. See Jacola et al. (2011).

5. See Restifo (2005) for a description of this and other approaches.

6. The external germinal layer is also called the external granule layer by some authors. Fortunately, the widely used designation EGL can be used for both.

7. To say that a thing's days are numbered means that its existence is coming to an end. This phrase can also be applied to people or institutions. The meaning conveyed is that the days remaining are so few that they can be easily counted (numbered). The origin of the phrase is uncertain. One possibility is that it was introduced in an English translation of a verse from the Old Testament Book of Daniel: "God hath numbered thy reign and finished it" (Chapter 5, Verse 26).This verse is the famous "writing on the wall" that David translates for King Belshazzar.

8. See Friede (1973).

9. An antithyroid drug commonly used in animal studies is 2-mercapto-1-methylimidazole. See Morte et al. (2002).

10. The meninges are the three membranes that cover the brain and spinal cord: the dura mater, the arachnoid membrane, and the pia mater). The term *leptomeninges* refers to the two delicate membranes closest to the surface of the brain, the arachnoid membrane and the pia mater.

11. See Kumada et al. (2006).

12. See Huttenlocher et al. (1974).

13. See Martin-Padilla (1972).

14. See Purpura (1974).

15. See Amir et al. (1999).

16. The transplant method, in which GFP-expressing neurons are transplanted to

a nonfluorescing host brain, permits Golgi-style analyses to be performed on the scattered fluorescing neurons because the background provided by the nonfluorescing host brain is dark. See Kishi and Macklis (2010).

17. A discussion of different aspects of environmental enrichment and a reference to the earliest documented enrichment study performed by a neuroscientist (D. O. Hebb) are provided by Nithianantharajah and Hannan (2006).

18. Synaptoneurosomes are components of a fractionated brain homogenate, typically prepared from a region of a rat or mouse brain. An individual synaptoneurosome consists of a synapse with attached presynaptic (axon terminal) and postsynaptic (dendritic spine) sacs. The sheared plasma membrane reseals itself so that synapses can be studied in isolation from neuronal somata. A synaptoneurosome is similar but not identical to a synaptosome, which is an isolated axon terminal (also prepared by fractionating a brain homogenate). The components of synaptoneurosomes can be confirmed by electron microscopy, biochemical analysis, and immunoblotting.

19. Earlier terms for this condition included *Mongolian idiocy* and *mongolism*. These terms have been for the most part supplanted by the interchangeable terms *Down's syndrome* and *Down syndrome*. The latter term (without the possessive) is the current preference of professional and advocacy groups in the United States. John Langdon Down was the British physician who published a description of the syndrome in 1866. The chromosomal basis of Down syndrome was discovered in 1959 by French researcher Jérôme Lejeune.

20. See Roper et al. (2006).

21. The term *fetal alcohol syndrome* was first used in an article published in *The Lancet*, a British medical journal, in 1973. See Jones and Smith (1973).

22. See Kumada et al. (2006).

23. Methylation refers to the addition of a methyl group to cytosines in DNA (in mammals mainly at the C5 position of CpG dinucleotides). Because repressors are recruited to the methylated site, the effect of methylation is to block the initiation of transcription. One procedure for identification of methylated regions of DNA is methylated DNA immunoprecipitation (MeDIP). Fragmented genomic DNA is immunoprecipitated using antibodies specific to 5-methylcytosine. The methylated DNA can then be characterized by hybridization to DNA arrays (MeDIP-chip) or by sequencing (MeDIP-seq).

24. A synonym for *sarah,* the *Drosophila* calcipressin gene that is a homolog of human DSCR1, is *nebula. sarah* is a reference to the Sarah, wife of Abraham, the biblical patriarch. Sarah was a very, very old woman when she gave birth to her only son (Isaac). *sarah* mutants are infertile.

25. See http://www.geneswitch.com/index.html.

26. See Chang et al. (2003).

27. See Bolduc et al. (2008).

28. See McClure et al. (2011).

Investigative Reading

1. You produce an *Fmr1* knockout zebrafish. How do you validate that your knockout is indeed a true knockout? What phenotypes do you expect will result?

den Broeder, Marjo J., Herma van der Linde, Judith R. Bouwer, Bem A. Oostra, Rob Willemsen, and René F. Ketting. 2009. *PLoS ONE* 4: e7910.

2. In humans, intellectual disability is more common in males than in females. Generate a hypothesis to account for this observation. How do you test your hypothesis?

Raymond, F. L. 2006. *Journal of Medical Genetics* 43: 193–200.

3. You want to study altered profiles of gene expression in the brains of patients with Rett syndrome. What methods can be used to achieve your goal?

Gibson, Joanne H., Barry Slobedman, Harikrishnan KN, Sarah L. Williamson, Dimitri Minchenko, Assam El-Osta, Joshua L. Stern, and John Christodoulou. 2010. *BMC Neuroscience* 11: 53.

4. The intellectual disability characteristic of fragile X syndrome is caused by more than 200 CGG repeats in the 5′ untranslated portion of the fragile X mental retardation 1 gene. Individuals with 55–200 CGG repeats are typically described as having a premutation repeat expansion. Do such individuals display any symptoms despite being capable of producing FMRP? What tests would you perform to study these individuals?

Hagerman, Randy, Jacku Au, and Paul Hagerman. 2011. *Journal of Neurodevelopmental Disorders* 3: 211–24.

Many of the abbreviations listed are the names of genes and their associated symbols. (A gene symbol is a unique abbreviation of a gene name.) Different species have different gene and gene symbol naming conventions, and I apologize in advance for any deviations from preferred usage, which varies across different research communities. It is typical that the names of genes and their associated gene symbols are given in italics, the names of gene products (proteins) in regular type. For some species, such as *Drosophila melanogaster,* names of genes known to be dominant begin with an uppercase letter, whereas recessive genes are written in all lowercase letters. The symbols for human genes are typically italicized, but the full names of the genes are not. The names of human proteins are often written using uppercase letters.

abdA	*abdominal A* gene of *Drosophila,* a member of the *bithorax* complex
AbdB	*Abdominal B* gene of *Drosophila,* a member of the *bithorax* complex
AMPA	α-amino-3-hydroxyl-5-methyl-4-isoxazole-propionate, defining ligand for the AMPA glutamate receptor
ANS	autonomic nervous system
APC, APC	adenomatous polyposis coli gene and protein of humans; the protein acts as an antagonist of the Wnt signaling pathway
APF	after puparium formation, used to describe timing of events that occur during metamorphosis in *Drosophila*; for example, 10 APF describes an event observed 10 hours after puparium formation
ARE	adenylate-uridylate-rich elements (AU-rich elements), feature of the 3′ untranslated region in many mRNAs
ART	assisted reproductive technology

AS-C	*Achaete-scute* gene complex of *Drosophila*; contains four genes (*achaete, scute, lethal of scute,* and *asense*) that encode bHLH proteins
ASPM	abnormal spindlelike microcephaly-associated gene of humans; encodes the putative ortholog (ASPM) of the *Drosophila* abnormal spindle protein (asp)
AVE	anterior visceral endoderm, structure found in very young mouse embryos
BALB/c	inbred strain of albino mice frequently used in biology research
bHLH	basic helix-loop-helix protein motif (or a transcription factor that contains this motif)
Bmp	one of the bone morphogenetic proteins, a member of the transforming growth factor-β (TGF-β) superfamily of proteins
BOLD	blood oxygenation level–dependent fMRI
BrdU	5-bromo-2′-deoxyuridine, a thymidine analog
CAM	cell adhesion molecule
cAMP	cyclic adenosine monophosphate
CAT	computerized axial tomography, also called CT
CDK	cyclin-dependent kinase
cDNA	complementary deoxyribonucleic acid
cGMP	cyclic guanosine monophosphate
ChIP	chromatin immunoprecipitation
CNS	central nervous system
CpG	any site in DNA where a guanine nucleotide immediately follows a cytosine nucleotide; important sites of methylation in mammalian DNA
Cre	Cre recombinase
cRNA	complementary ribonucleic acid
Cyp26	a member of the superfamily of cytochrome P450 enzymes responsible for degrading the morphogen retinoic acid
DAPT	*N*-[(3,5-difluorophenyl)acetyl]-L-alanyl-2-phenylglycine-1, 1-dimethylethyl ester, a Notch signaling inhibitor
DCC, DCC	human deleted in colorectal cancer gene and protein; a receptor for netrin axon guidance cues, mutated in a subset of patients with colorectal carcinoma
DCX, DCX	doublecortin gene and protein of humans, primarily expressed by neural progenitor cells and immature neurons

ddaE	multipolar sensory neuron found in the peripheral nervous system of *Drosophila*
dFmr1	*Drosophila* fragile X mental retardation gene
dFMRP	*Drosophila* homolog of human fragile X mental retardation protein
dhtt or *htt*	gene that encodes the Huntingtin protein of *Drosophila*
DIC	differential interference contrast microscopy
DiI	a fluorescent lipophilic carbocyanine dye used to trace connections within the nervous system
Dll1, Dll1	human Delta-like 1 gene and protein, a homolog of the *Drosophila* ligand for Notch called Delta
DNA	deoxyribonucleic acid
dpc	days postconception or days postcoitum
DSCR1, DSCR1	human Down syndrome critical region 1 gene and protein
dsRNA	double-stranded ribonucleic acid
DTI	diffusion tensor imaging
E2F1	transcription factor that interacts with other proteins at the E2 recognition site
E-box	sequence that lies upstream of a gene in a promoter region; a binding site for bHLH transcription factors
EGF	epidermal growth factor
EGFR	epidermal growth factor receptor, a cell surface receptor for epidermal growth factors
EGL	external germinal layer of the developing cerebellar cortex, also called the external granule layer
ENU	*N*-ethyl-*N*-nitrosourea, a mutagen
ERK	extracellular signal-related kinase
ES cells	embryonic stem cells
EVL	enveloping layer of the blastoderm in vertebrate embryos
FasI	fasciclinI, a cell adhesion molecule of *Drosophila* typically expressed by commissural axons, also known as Fas1, Fas I, Fas-1
FasII	fasciclinII, a cell adhesion molecule of *Drosophila* typically expressed by longitudinal axons, also known as Fas2, Fas II, Fas-2
FGF	fibroblast growth factors, a family of growth factors involved in many aspects of development and wound repair
FLP-*FRT*	site-directed recombination technology involving recombination of sequences between short flippase recognition target by the flippase recombination enzyme

FMR1	fragile X mental retardation 1, human gene that codes for the fragile X mental retardation protein (FMRP)
fMRI	functional magnetic resonance imaging
FMRP	human fragile X mental retardation protein
FP	floor plate, the ventral midline of the vertebrate neural tube
FXR1, FXR2	human genes that code for FXR1 and FXR2, proteins similar to FMRP in structure and possibly function
G0	resting phase of the eukaryotic cell cycle, pronounced G-zero
G1	first phase of interphase of the eukaryotic cell cycle prior to the S phase
G2	phase between the S phase and mitosis of the eukaryotic cell cycle
GABA	gamma-aminobutyric acid, an inhibitory neurotransmitter
GABA$_A$	specific subtype of receptor for gamma-aminobutyric acid
Gal4/UAS	a method used for targeted gene expression in *Drosophila* and other insects; Gal4 refers to the yeast transcription activator protein Gal4, UAS refers to the Upstream Activation Sequence, the enhancer to which Gal4 binds to activate transcription
GAPDH	glyceraldehyde-3-phosphate dehydrogenase
gcm, gcm	the *glial cells missing* gene and protein of *Drosophila*
GFAP	glial fibrillary acidic protein, a marker for astrocytes
GFP	green fluorescent protein
GSK3	glycogen synthase kinase 3, a component of the canonical Wnt signaling pathway
GTP	guanosine triphosphate
HD	homeodomain transcription factor
Hes	a *Drosophila* bHLH protein important for the development of the nervous system; known by many names, including Hairy/enhancer of split; encoded by the gene *enhancer of split m8*
HES1, HES1	human Notch-activated gene and the bHLH protein encoded by that gene; a member of the same family of proteins that includes the *Drosophila* Hes protein
HGPPS	horizontal gaze palsy with progressive scoliosis, a disorder of humans that affects vision and causes abnormal curvature of the spine
hh, hh	the *hedgehog* gene and protein of *Drosophila*
HMG	high-mobility group chromosomal proteins

Hox	homeobox
hpf	hours postfertilization, often used to describe the stages of zebrafish embryonic development
HRE	Hox response element
HTT, HTT	human Huntingtin gene and protein
IACUC	institutional animal care and use committee; monitors research on animal subjects
Id	a member of the inhibitor of DNA-binding family of proteins; forms heterodimers with bHLH proteins, thereby blocking their ability to function as transcription factors
IHC	immunohistochemistry
IQ	intelligence quotient
IRB	institutional review board; monitors research on human subjects
Ki-67	protein expressed in proliferating cells in humans and other mammals
KO	a genetically engineered mouse in which a specific gene has been inactivated; the term *knockout* is more widely used than the abbreviation
KOMP	Knockout Mouse Project, a project of the United States National Institutes of Health designed to produce a repository of mouse embryonic stem (ES) cells that contain a null mutation in every gene in the mouse genome
L1–L4	postembryonic larval stages of *C. elegans*
lacZ	one of the structural genes of the lac operon of *E. coli*
LDL	low-density lipoproteins ("bad cholesterol")
LDLR	low-density lipoprotein receptors, required for cellular uptake of LDL
LEF	lymphoid enhancer factor; transcription factor that binds to DNA through a high mobility group domain; dimerizes with β-catenin as a component of the Wnt signaling pathway
lin-22, lin-32	in *C. elegans*, genes encoding for abnormal cell lineages; the corresponding proteins are lin-22 and lin-32
Lnfg, Lnfg	*lunatic fringe* gene and protein of *Drosophila;* member of the fringe family of glycosyltransferases that affect signaling in the Notch pathway
LoxP	locus of chromosomal crossover in the bacteriophage P1
Lrp4	low-density lipoprotein receptor-related protein, an agrin receptor expressed by vertebrate muscle fibers

M	mitotic phase of the eukaryotic cell cycle
MAP	microtubule-associated protein
Mash1, Mash1	mammalian achaete-scute homolog 1 gene and protein; the protein has proneural activity
miRNA	microribonucleic acid
MCPH	one of the genetic loci for human autosomal recessive primary microcephaly
MCT8	a thyroid hormone transporter; facilitates entry of thyroid hormones into cells
MeCP2, MeCP2	methyl CpG binding protein 2, human gene and protein associated with Rett syndrome
MeDIP	methylation analysis by DNA immunoprecipitation
mGluR	metabotropic glutamate receptor
MMS	medial migratory stream, one of two human brain pathways defined by migrating neurons born in the subventricular zone
MN1–MN5	motoneurons that innervate the flight muscles of insects, including *Drosophila*
MRI	magnetic resonance imaging
mRNA	messenger ribonucleic acid
MuSK	muscle-specific kinase of vertebrates, essential for the formation of the neuromuscular junction
ncRNA	noncoding ribonucleic acid
NG2	glial cells in the mammalian central nervous system that express NG2 proteoglycan
ngn, ngn	neurogenin genes and proteins, a family of bHLH proteins essential for neuronal differentiation; specific members of the family are indicated by the prefix ngn followed by a number
Nkx, Nkx	a family of homeodomain transcription factors required for the normal development of the nervous system and other organs; specific members of the family are indicated by the prefix Nkx followed by a number
NMDA	*N*-methyl-D-aspartic acid, the ligand for the NMDAR glutamate receptor
OATp1c1	a thyroid hormone transporter, facilitates entry of thyroid hormones into cells
Olig, Olig	oligodendrocyte transcription factors, human bHLH proteins important for the development of the oligo-dendrocytes; specific Oligs are designated by number; Olig2 is expressed during development and by glial tumors

OPC	oligodendrocyte progenitor cell
oRG	outer subventricular zone radial glialike cells
OSVZ	outer subventricular zone of the human brain
P1	a bacteriophage that encodes a site-specific recombinase
Par, Par	genes and proteins required for normal cytoplasmic partitioning during mitosis
Pax, Pax	members of the highly conserved family of "paired box" transcription factors; specific genes and their corresponding proteins are designated by Pax followed by a number; several Pax proteins are essential for normal development of the nervous system
PC12	a cell line derived from a tumor of the rat adrenal medulla; used to study neuronal differentiation in vivo
PCNA	proliferating cell nuclear antigen, a marker for cell proliferation
PCR	polymerase chain reaction
PDE	posterior dereid neurons of *C. elegans*, used to study neuronal outgrowth
PDGF	platelet-derived growth factor
PET	positron emission tomography
Phr1	mouse gene that encodes a ubiquitin ligase; when mutated, causes a phenotype called Magellan, which is characterized by defective growth of axons
PI3K	phosphatidylinositol 3-kinase
pLL	posterior lateral line sensory system of fish
PLP	proteolipid protein, an important component of myelin
PMC	PubMed Central
pMN	motoneuron progenitor domain of the vertebrate neural tube
PNS	peripheral nervous system
PSA-NCAM	polysialylated neural cell adhesion molecule, a marker for immature neurons
qRT-PCR	quantitative reverse-transcription polymerase chain reaction
RALDH	retinaldehyde dehydrogenase, enzyme that generates the morphogen retinoic acid in vertebrate embryos
Rb, RB	retinoblastoma tumor gene and protein
RCR	responsible conduct of research
repo, repo	*Drosophila repolarized* gene and gene product, a glial-specific homeodomain protein

RG	radial glial cells
RMS	rostral migratory stream, one of two human brain pathways defined by migrating neurons born in the subventricular zone
Rn	in male *C. elegans,* a hypodermal ray precursor cell
RnA, RnB	in male *C. elegans,* specific ray neurons
RNA	ribonucleic acid
RNA-seq	whole-transcriptome shotgun sequencing
RNAi	ribonucleic acid interference
RNase	ribonuclease
Rnst	in male *C. elegans,* a structural cell of the ray
robo, robo	*roundabout* gene and gene product
RP	roof plate, the dorsal most structure of the vertebrate neural tube
RP2	larval motoneuron of *Drosophila*
rRNA	ribosomal RNA
S	DNA synthesis phase of the eukaryotic cell cycle
S100B	calcium-binding protein expressed by mammalian astrocytes
SCL, SCL	mammalian stem cell leukemia gene, encodes a bHLH protein important for production of blood cells; also expressed in the developing nervous system
SGZ	subgranular zone, region of the vertebrate hippocampus where adult neurogenesis occurs
Shh, Shh	*Sonic hedgehog* gene and protein of vertebrates
SMA	spinal muscular atrophy, an inherited disease of humans that causes muscle damage and weakness
SVZ	subventricular zone, cell layers associated with the lateral ventricles of the vertebrate brain; a site of neurogenesis and gliogenesis
T_3, T_4	triiodothyronine and thyroxine, the two forms of thyroid hormone
TCF	T-cell factor, a transcription factor that binds to DNA through a high mobility group domain; dimerizes with β-catenin as a component of the Wnt signaling pathway
TGF-β1	defining member of the transforming growth factor-β superfamily of proteins
TRα, TRβ	the two mammalian genes that encode thyroid hormone receptors

TUNEL	terminal deoxynucleotidyl transferase dUTP nick end labeling, an assay a used to detect the DNA fragmentation associated with apoptotic cell death
Ubx	*Ultrabithorax* gene of *Drosophila,* a member of the *bithorax* complex
unc	prefix given to the name of genetic loci that produce uncoordinated phenotypes in *C. elegans*
UTR	untranslated region of a mRNA; can be 3′ or 5′
UV	ultraviolet
V4, V5	specific blast cells found in the *C. elegans* embryo
VLDLR	very low-density lipoprotein receptor; binds the reelin protein; required for normal development of the cerebellum
VZ	ventricular zone, origin of the neurons that form the vertebrate cortex
wit	whitetail, a zebrafish mutation that results in the absence of melanocytes in the tail
Wnt	a family of highly conserved secreted signaling molecules
YSL	yolk syncytial layer of vertebrate embryos

REFERENCES

Preface

Anderson, Kathryn V., and Philip W. Ingham. 2003. "The transformation of the model organism: A decade of developmental genetics." Supplement, *Nature Genetics* 33 (2003): 285–93.

Chapter 1   Introduction

Casey, B. J., Jonathan D. Cohen, Peter Jezzard, Robert Turner, Douglas C. Noll, Rolf H. Trainor, Jay Giedd, Debra Kaysen, Lucy Hertz-Pannier, and Judith L. Rapoport. 1995. "Activation of prefrontal cortex in children during a nonspatial working memory task with functional MRI." *NeuroImage* 2: 221–29.

Chalfie, Martin. 2009. "GFP: Lighting Up life." Nobel Lecture. Published in *Proceedings of the National Academy of Sciences U.S.A.* 106: 10073–80.

Chalfie, Martin, Yuan Tu, Ghia Euskirchen, William W. Ward, and Douglas C. Prasher. 1994. "Green fluorescent protein as a marker for gene expression." *Science* 263: 802–5.

David, Della C., Frederic Hoerndli, and Jürgen Götz. 2005. "Functional genomics meets neurodegenerative disorders, Part 1: Transcriptomic and proteomic technology." *Progress in Neurobiology* 76: 153–68.

Dhawale, Ashesh and Upinder S. Bhalla. 2008. "The network and the synapse: 100 years after Cajal." *HFSP Journal* 2: 12–6.

Eisen, Judith S., and James C. Smith. 2008. "Controlling morpholino experiments: Don't stop making antisense." *Development* 135: 1735–43.

Fujimori, Toshihiko, Yoko Kurotaki, Jun-ichi Miyazaki, and Yo-ichi Nabeshima. 2003. "Analysis of cell lineage in two- and four-cell mouse embryos." *Development* 130: 5113–22.

Gratzner, Howard G. 1982. "Monoclonal antibody to 5-bromo- and 5-iododeoxyuridine: A new reagent for detection of DNA replication." *Science* 218: 474–75.

Hayes, Peter C., C. Roland Wolf, and John D. Hayes. 1989. "Blotting techniques for the study of DNA, RNA, and proteins." *British Medical Journal* 299: 965–68.

Heisenberg, Martin, Monika Heusipp, and Christiane Wanke. 1995. "Structural plasticity in the *Drosophila* brain." *Journal of Neuroscience* 15: 1951–60.

Holen, T., and C. V. Mobbs. 2004. "Lobotomy of genes: Use of RNA interference in neuroscience." *Neuroscience* 126: 1–7.

Kastanenka, Ksenia V., and Lynn T. Landmesser. 2010. "*In vivo* activation of channelrhodopsin-2 reveals that normal patterns of spontaneous activity are required for motoneuron guidance and maintenance of guidance molecules." *Journal of Neuroscience* 30: 10575–85.

Leif, Robert C., Jeanne H. Stein, and Robert M. Zucker. 2004. "A short history of the initial application of anti-5-BrdU to the detection and measurement of S phase." *Cytometry Part A* 58A: 45–52.

Luskin, Marla B. 1994. "Neuronal cell lineage in the vertebrate central nervous system." *FASEB Journal* 8: 722–30.

Mukherjee, Pratik, and Robert C. McKinstry. 2006. "Diffusion tensor imaging and tractography of human brain development." *Neuroimaging Clinics* 16: 19–43.

Norris, David G. "Principles of magnetic resonance assessment of brain function." 2006. *Journal of Magnetic Resonance Imaging* 23: 794–807.

Poirier, Colline, Tiny Boumans, Marleen Verhoye, Jacques Balthazart, and Annemie Van der Linden. 2009. "Own-song recognition in the songbird auditory pathway: Selectivity and lateralization." *Journal of Neuroscience* 29: 2252–58.

Sanes, Joshua R. 1989. "Analysing cell lineage with a recombinant retrovirus." *Trends in Neurosciences* 12: 21–28.

Shaw, Philip, Jason Lerch, Deanna Greenstein, Wendy Sharp, Liv Clasen, Alan Evans, Jay Giedd, F. Xavier Castellanos, and Judith Rapoport. 2006. "Longitudinal mapping of cortical thickness and clinical outcome in children and adolescents with attention-deficit/hyperactivity disorder." *Archives of General Psychiatry* 63: 540–49.

Silk, Timothy J., and Amanda G. Wood. 2011. "Lessons about neurodevelopment from anatomical magnetic resonance imaging." *Journal of Developmental and Behavioral Pediatrics* 32: 158–68.

Van der Linden, Annemie, Vincent Van Meir, Tiny Boumans, Colline Poirier, and Jacques Balthazart. 2009. "MRI in brains displaying extensive plasticity." *Trends in Neurosciences* 32: 257–66.

Zhang, Feng, Li-Ping Wang, Edward S. Boyden, and Karl Deisseroth. 2006. "Channelrhodopsin-2 and optical control of excitable cells." *Nature Methods* 3: 785–92.

Chapter 2   Overview of Nervous System Development in Humans

Centers for Disease Control and Prevention. 2004. "Spina bifida and anencephaly before and after folic acid mandate—United States, 1995–1996 and 1999–2000." *MMWR Morbidity and Mortality Weekly Report* 53: 362–65.

Chapman, Teresa, Manuela Matesan, Ed Weinberger, and Dorothy I. Bulas. 2010 "Digital atlas of fetal brain MRI." *Pediatric Radiology* 40: 153–62.

Copp, Andrew J., and Nicholas D. E. Greene. 2010. "Genetics and development of neural tube defects." *Journal of Pathology* 220: 217–30.

Dekaban, Anatole S. 1978. "Changes in brain weights during the span of human life: Relation of brain weights to body heights and body weights." *Annals of Neurology* 4: 345–56.

Garel, Catherine, Emmanuel Chantrel, Hervé Brisse, Monique Elmaleh, Dominique Luton, Jean-Françoise Oury, Guy Sebag, and Max Hassan. 2001. "Fetal cerebral cortex: Normal gestational landmarks identified using prenatal MR imaging." *American Journal of Neuroradiology* 22: 184–89.

Lumsden, Andrew, and Roger Keynes. 1989. "Segmental patterns of neuronal development in the chick hindbrain." *Nature* 337: 424–28.

Prayer, Daniela, Gregor Kasprian, Elisabeth Krampl, Barbara Ulm, Linde Witzani, Lucas Prayer, and Peter C. Brugger. 2006. "MRI of normal fetal brain development." *European Journal of Radiology* 57: 199–216.

Wellner, Karen. 2010. "Carnegie Institution of Washington Department of Embryology." Embryo Project Encyclopedia. ISSN: 1940–5030.

Chapter 3   Animal Models

Acevedo-Arozena, Abraham, Sara Wells, Paul Potter, Michelle Kelly, Roger D. Cox, and Steve D. M. Brown. 2008. "ENU mutagenesis, a way forward to understand gene function." *Annual Review of Genomics and Human Genetics* 9: 49–69.

Adouette, André, Guillaume Balavoine, Nicolas, Lartillot, Olivier Lespinet, Benjamin Prud'homme, and Renaud de Rosa. 2000. "The new animal phylogeny: Reliability and implications." 97: 4453–56.

Amacher, Sharon L. 2001. "Zebrafish embryo as a developmental system." *Encyclopedia of Life Sciences,* Wiley Online Library.

Andrews, Paul L. R. 2011. "Laboratory invertebrates: Only spineless, or spineless and painless?" *ILAR Journal* 52: 121–25.

Ardiel, Evan L., and Catharine H. Rankin. 2010. "An elegant mind: Learning and memory in *Caenorhabditis elegans." Learning and Memory* 17: 191–201.

Behringer, Richard R., Maki Wakamiya, Tania E. Tsang, and Patrick P. L. Tam. 2000. "A flattened mouse embryo: Leveling the playing field." *Genesis* 28: 23–30.

Bellen, Hugo J., Chao Tong, and Hiroshi Tsuda. 2010. "100 years of *Drosophila* research and its impact on vertebrate neuroscience: A history lesson for the future." *Nature Reviews Neuroscience*: 514–22.

Blaxter, Mark. 2011. "Nematodes: The worm and its relatives." *PLoS Biology* 9: e1001050. doi:10.1371/journalpbio.1001050.

Bownes, Mary. 1975. "A photographic study of development in the living embryo of *Drosophila melanogaster." Journal of Embryology and Experimental Morphology* 33 : 789–801.

C. elegans Sequencing Consortium. 1998. "Genome sequence of the nematode *C. elegans:* A platform for investigating biology." *Science* 282: 2012–18.

Campos-Ortega, José A., and Volker Hartenstein. 1985. *The embryonic development of Drosophila melanogaster.* Berlin: Springer.

Clarke, Jon. 2009. "Role of polarized cell divisions in zebrafish neural tube formation." *Current Opinion in Neurobiology* 19: 134–38.

Downs, Karen M., and Tim Davies. 1993. "Staging of gastrulating mouse embryos by morphological landmarks in the dissecting microscope." *Development* 118: 1255–66.

Geurts, Aron M., and Carol Moreno. 2010. "Zinc-finger nucleases: New strategies to target the rat genome." *Clinical Science* 119: 303–11.

Grunwald, David Jonah, and Judith S. Eisen. 2002. "Headwaters of the zebrafish—Emergence of a new model vertebrate." *Nature Reviews Genetics* 3: 717–24.

Haffter, Pascal, Michael Granato, Michael Brand, Mary C. Mullins, Matthias Hammerschmidt, Donald A. Kane, Jörg Odenthal, Fredericus J. M. van Eeden, Yun-Jin Jiang, Carl-Philipp Heisenberg, Robert N. Kelsh, Makoto Furutani-Seiki,

Elisabeth Vogelsang, Dirk Beuchle, Ursula Schach, Cosima Fabian, and Christiane Nüsslein-Volhard. 1996. "The identification of genes with unique and essential functions in the development of the zebrafish, *Danio rerio.*" *Development* 123: 1–36.

Hartenstein, Volker. 1993. *Atlas of Drosophila Development.* Cold Spring Harbor Laboratory Press.

Jekosch, Kerstin. 2004. The zebrafish genome project: Sequence analysis and annotation. *Methods in Cell Biology* 77 (): 225–39.

Kimmel, Charles B., William W. Ballard, Seth R. Kimmel, Bonnie Ullmann, and Thomas F. Schilling. 1995. "Stages of embryonic development of the zebrafish." *Developmental Dynamics* 203: 253–310.

Lewcock, Joseph W., Nicolas Genoud, Karen Lettieri, and Samuel L. Pfaff. 2007. "The ubiquitin ligase Phr1 regulates axon outgrowth through modulation of microtubule dynamics." *Neuron* 56: 604–20.

Mori, Ikue. 1999. "Genetics of chemotaxis and thermotaxis in the nematode *Caenorhabditis elegans.*" *Annual Review of Genetics* 33: 399–422.

Roots, Betty I. 2008. "The phylogeny of invertebrates and the evolution of myelin." *Neuron and Glial Biology* 4: 101–9.

Rossant, Janet, and Patrick P. L. Tam. 2009. "Blastocyst lineage formation, early embryonic asymmetries and axis patterning in the mouse." *Development* 136: 701–13.

Schier, Alexander F., Stephan C. F. Neuhauss, Michele Harvey, Jarema Malicki, Liliana Solnica-Krezel, Didier Y. R. Stainier, Fried Zwartkruis, Salim Abdelilah, Derek L. Stemple, Zehava Rangini, Hong Yang, and Wolfgang Driever. 1996. "Mutations affecting the development of the embryonic zebrafish brain." *Development* 123: 165–78.

Schneider, Judsen, Rachel L. Skelton, Stephen E. Von Stetina, Teije C. Middelkoop, Alexander van Oudenaarden, Hendrik C. Korswagen, and David M. Miller III. 2012. "UNC-4 antagonizes Wnt signaling to regulate synaptic choice in the *C. elegans* motor circuit." *Development* 139: 2234–45.

Snyder, Jason S., Jessica S. Choe, Meredith A. Clifford, Sara I. Jeurling, Patrick Hurley, Ashly Brown, J. Frances Kamhi, and Heather A. Cameron. 2009. "Adult-born hippocampal neurons are more numerous, faster maturing, and more involved in behavior in rats than in mice." *Journal of Neuroscience* 29: 14484–95.

Technau, Gerhard M. 2008. *Brain development in* Drosophila melanogaster. *Advances in Experimental Medicine and Biology,* vol. 628. New York: Springer. 160 pp.

Truman, James W. 1990. "Metamorphosis of the central nervous system of *Drosophila.*" *Journal of Neurobiology* 21: 1072–84.

Wadsworth, William G., and Edward M. Hedgecock. 1996. "Hierarchical guidance cues in the developing nervous system of *C. elegans.*" *BioEssays* 18: 355–62.

Ware, Randle W., David Clark, Kathryn Crossland, and Richard L. Russell. 1975. "The nerve ring of the nematode *Caenorhabditis elegans*: Sensory input and motor output." *Journal of Comparative Neurology* 162: 71–110.

Westerfield, Monte. 2007. *The Zebrafish Book: A Guide for the Laboratory Use of Zebrafish* (Danio rerio), 5th edition. Eugene: University of Oregon Press.

Wood, William B., ed. 1988. *The Nematode* Caenorhabditis elegans. Cold Spring Harbor, NY: Cold Spring Harbor Laboratory Press. 667 pp.

Anderson, Ryan M., Alison R. Lawrence, Rolf W. Stottmann, Daniel Bachiller, and John Klingensmith. 2002. "Chordin and noggin promote organizing centers of forebrain development in the mouse." *Development* 129: 4975–87.

Becalska, Agata N., YoungJung R. Kim, Nicolette G. Belletier, Dorothy A. Lerit, Kristina S. Sinsimer, and Elizabeth R. Gavis. 2011. "Aubergine is a component of a nanos mRNA localization complex." *Developmental Biology* 349: 46–52.

Cheeks, Rebecca J., Julie C. Canman, Willow N. Gabriel, Nicole Meyer, Susan Strome, and Bob Goldstein. 2004. "*C. elegans* PAR proteins function by mobilizing and stabilizing asymmetrically localized protein complexes." *Current Biology* 14: 851–62.

Chitnis, Ajay B. 1999. "Control of neurogenesis—Lessons from frogs, fish, and flies." *Current Opinion in Neurobiology* 9: 18–25.

Collins, Michael D., and Gloria E. Mao. 1999. "Teratology of retinoids." *Annual Review of Pharmacology and Toxicology* 39: 399–430.

Driever, Wolfgang, Vivian Siegel, and Christiane Nüsslein-Volhard. 1990. "Autonomous determination of anterior structures in the early *Drosophila* embryo by the *bicoid* morphogen. *Development* 109: 811–20.

Gilbert, Scott F. 2010. *Developmental Biology,* 9th edition. Sunderland, MA: Sinauer Associates.

Glinka, Andrei, Wei Wu, Hajo Delius, A. Paula Monaghan, Claudia Blumenstock, and Christof Niehrs. 1998. "Dickkopf-1 is a member of a new family of secreted proteins and functions in head induction." *Nature* 391: 357–62.

Kiecker, Clemens, and Andrew Lumsden. 2009. "Recent advances in neural development." *F1000 Biology Reports* 1: 1. doi: 10.3410/B1-1.

Klingensmith, John, Maiko Matsui, Yu-Ping Yang, and Ryan M. Anderson. 2010. "Roles of bone morphogenetic protein signaling and its antagonism in holoprosencephaly." *American Journal of Medical Genetics Part C* (*Seminars in Medical Genetics*) 154C: 43–51.

Levine, Ariel J., and Ali H. Brivanlou. 2007. "Proposal of a model of mammalian neural induction." *Developmental Biology* 308: 247–56.

Liao, Xiaoyan, and Michael D. Collins. 2008. "All-trans retinoic acid–induced ectopic limb and caudal structures: Murine strain sensitivities and pathogenesis." *Developmental Dynamics* 237: 1553–64.

Lowenstein, Elie B., and Eve J. Lowenstein. 2011. "Isotretinoin systemic therapy and the shadow cast upon dermatology's downtrodden hero." *Clinics in Dermatology* 29: 652–61.

Moon, Randall T., Aimee D. Kohn, Giancarlo V. De Ferrari, and Ajamete Kaykas. 2004. "WNT and β-catenin signaling: Diseases and therapies." *Nature Reviews Genetics* 5: 691–701.

Rivera-Perez, Jaime A. 2007. "Axial specification in mice: Ten years of advances and controversies." *Journal of Cellular Physiology* 213: 654–60.

Rowland, James W., Jason J. Lee, Ryan P. Salewski, Eftekhar Eftekharpour, Derek van der Kooy, and Michael G. Fehlings. 2011. "Generation of neural stem cells from embryonic stem cells using the default mechanism: In vitro and in vivo characterization." *Stem Cells and Development* 20. doi: 10.1089/scd.2011.0214.

Schier, Alexander F. 2001. "Axis formation and patterning in zebrafish." *Current Opinion in Genetics and Development* 11: 393–404.

Schier, Alexander F., and William S. Talbot. 2005. "Molecular genetics of axis formation in zebrafish." *Annual Review of Genetics* 39: 561–613.

Srinivas, Shankar, Tristan Rodriguez, Melanie Clements, James C. Smith, and Rosa S. P. Beddington. 2004. "Active cell migration drives the unilateral movements of the anterior visceral endoderm." *Development* 131: 1157–64.

Stern, Claudio D. 2005. "Neural induction: Old problem, new findings, yet more questions." *Development* 132: 2007–21.

———. 2006. "Neural induction: 10 years on since the 'default model.'" *Current Opinion in Cell Biology* 18: 692–97.

Weill, Timothy T., Despina Xanthakis, Richard Parton, Ian Dobbie, Catherine Rabouille, Elizabeth R. Gavis, and Ilan Davis. 2010. "Distinguishing direct from indirect roles for *bicoid* mRNA localization factors." *Development* 137: 169–76.

Wessely, Oliver, James I. Kim, Douglas Geissert, Uyen Tran, and E. M. De Robertis. 2004. "Analysis of Spemann organizer formation in Xenopus embryos by cDNA macroarrays." *Developmental Biology* 269: 552–66.

White, Richard J., Qing Nie, Arthur D. Lander, and Thomas F. Schilling. 2007. "Complex regulation of cyp26a1 creates a robust retinoic acid gradient in the zebrafish embryo." *PLoS Biology* 5: e304. doi: 10.1371/journal.pbio.0050304.

Chapter 5 Neurogenesis

Amrein, Irmgard, Dina K. N. Dechmann, York Winter, and Hans-Peter Lipp. 2007. "Absent or low rate of adult neurogenesis in the hippocampus of bats (Chiroptera)." *PLoS ONE* 2: e455. doi: 10.1371/journal.pone.0000455.

Bielas, Stepanie, Holden Higginbotham, Hiroyuki Koizumi, Teruyuki Tanaka, and Joseph G. Gleason. 2004. "Cortical migration mutants suggest separate but intersecting pathways." *Annual Review of Cell and Developmental Biology* 20: 593–618.

Blader, Patrick, Nadine Fischer, Gerard Gradwohl, François Guillemot, and Uwe Strähle. 1997. "The activity of neurogenin1 is controlled by local cues in the zebrafish embryo." *Development* 124: 4557–69.

Bystron, Irina, Pasko Rakic, Zoltán Molnár, and Colin Blakemore. 2006. "The first neurons of the human cerebral cortex." *Nature Neuroscience* 9: 880–86.

Cayre, Myriam, Jordane Malaterre, Sophie Scotto-Lomassese, Colette Strambi, and Alain Strambi. 2002. "The common properties of neurogenesis in the adult brain: From invertebrates to vertebrates." *Comparative Biochemistry and Physiology Part B* 132: 1–15.

Chapouton, Prisca, Paulina Skupien, Birgit Hesl, Marion Coolen, John C. Moore, Romain Madelaine, Elizabeth Kremmer, Theresa Faus-Kessler, Patrick Blader, Nathan D. Lawson, and Laure Bally-Cuif. 2010. "Notch activity levels control the balance between quiescence and recruitment of adult neural stem cells." *Journal of Neuroscience* 30: 7961–74.

Curtis, Maurice A., Monica Kam, and Richard L. M. Faull. 2011. "Neurogenesis in humans." *European Journal of Neuroscience* 33: 1170–74.

Eriksson, Peter S., Ekaterina Perfilieva, Thomas Björk-Eriksson, Ann-Marie Alborn, Claes Nordborg, Daniel A. Peterson, and Fred H. Gage. 1998. "Neurogenesis in the adult human hippocampus." *Nature Medicine* 11: 1313–17.

García-Bellido, Antonio, and José F. de Celis. 2009. "The complex tale of the achaete-scute complex: A paradigmatic case in the analysis of gene organization and function during development." *Genetics* 182: 631–39.

Gohlke, Julia M., Olivier Armant, Frederick M. Parham, Marjolein V. Smith, Celine Zimmer, Diogo S. Castro, Laurent Nguyen, Joel S. Parker, Gerard Gradwohl, Christopher J. Portier, and François Guillemot. 2008. "Characterization of the proneural gene regulatory network during mouse telencephalon development." *BMC Biology* 6: 15.

Gönczy, Pierre. 2008. "Mechanisms of asymmetric cell division: Flies and worms pave the way." *Nature Reviews Molecular Cell Biology* 9: 355–66.

Götz, Magdalena, and Wieland B. Huttner. 2005. "The cell biology of neurogenesis." *Nature Reviews Molecular Cell Biology* 6: 777–88.

Hansen, David V., Jan H. Lui, Philip R. L. Parker, and Arnold R. Kreigstein. 2010. "Neurogenic radial glia in the outer subventricular zone of human neocortex." *Nature* 464: 554–61.

Herrup, Karl, and Yan Yang. 2007. "Cell cycle regulation in the postmitotic neuron: Oxymoron or new biology?" *Nature Reviews Neuroscience* 8: 368–78.

Hobert, Oliver. 2010. "Neurogenesis in the nematode *Caenorhabditis elegans*." In *WormBook*, ed. The *C. elegans* Research Community. WormBook. doi: 10.1895/wormbook.1.12.2, http://www.wormbook.org.

Hui, Subrhra Prakash, Anindita Dutta, and Sukla Ghosh. 2010. "Cellular response after crush injury in adult zebrafish spinal cord." *Developmental Dynamics* 239: 2962–79.

Kempermann, Gerd. 2011. "Seven principles in the regulation of adult neurogenesis." *European Journal of Neuroscience* 33: 1018–24.

Kizil, Caghan, Jan Kaslin, Voolker Kroehne, and Michael Brand. 2012. "Adult neurogenesis and brain regeneration in zebrafish." *Developmental Neurobiology* 72: 429–61.

Knoth, Rolf, Ilyas Singec, Margarethe Ditter, Georgios Pantazis, Philipp Capetian, Ralf P. Meyer, Volker Horvat, Benedikt Volk, and Gerd Kempermann. 2010. "Murine features of neurogenesis in the human hippocampus across the lifespan from 0 to 100 years." *PLoS ONE* 5: e8809.

Kriegstein, Arnold, and Alvarez-Buyulla, Arturo. 2009. "The glial nature of embryonic and adult neural stem cells." *Annual Review of Neuroscience* 32: 149–84.

Lafenêtre, Pauline, Oliver Leske, Zhanlu Ma-Högemeie, Aiden Haghikia, Zoe Bichler, Petra Wahle, and Rolf Heumann. 2010. "Exercise can rescue recognition memory impairment in a model with reduced adult hippocampal neurogenesis." *Frontiers in Behavioral Neuroscience* 34: doi: 10.3389/neuro.08.034.2009.

Lu, Bingwei, Lily Jan, and Yuh-Nung Jan. 2000. "Control of cell divisions in the nervous system: Symmetry and asymmetry." *Annual Review of Neuroscience* 23: 531–56.

Lyons, David A., Adam T. Guy, and Jonathan D. W. Clarke. 2003. "Monitoring neural progenitor fate through multiple rounds of division in an intact vertebrate brain." *Development* 130: 3427–36.

Molnár, Zoltan, Navneet A. Vasistha, and Fernando Garcia-Moreno. 2011. "Hanging by the tail: Progenitor populations proliferate." *Nature Neuroscience* 14: 538–40.

Portman, Douglas S., and Scott W. Emmons. 2000. "The basic helix-loop-helix transcription factors LIN-32 and HLH-2 function together in multiple steps of a *C. elegans* neuronal sublineage." *Development* 127: 5415–26.

Roegiers, Fabrice, and Yuh Nung Jan. 2004. "Asymmetric cell division." *Current Opinion in Cell Biology* 16: 195–205.

Sanai, Nader, Thuhien Nguyen, Rebecca A. Ihre, Zaman Mirzadeh, Hui-Hsin Tsai, Michael Wong, Nalin Gupta, Mitchel S. Berger, Eric Huang, Jose-Manuel Garcia-Verdugo, David H. Rowitch, and Arturo Alvarez-Buylla. 2011. "Corridors of migrating neurons in human brain and their decline during infancy." *Nature* 478: 382–86.

Sawa, Hitoshi. 2010. "Specification of neurons through asymmetric cell divisions." *Current Opinion in Neurobiology* 20: 44–49.

Shimojo, Hiromi, Toshiyuki Ohtsuka, and Ryoichiro Kageyama. 2011. "Dynamic expression of Notch signaling genes in neural stem/progenitor cells." *Frontiers in Neuroscience* 5, article 78. doi: 10.3389/fnins.2011.00078.

Simpson, Pat, and Cathie Carteret. 1990. "Proneural clusters: Equivalence groups in the epithelium of *Drosophila*." Development 110: 927–32.

Taghert, Paul H., Chris Q. Doe, and Corey S. Goodman. 1984. "Cell determination and regulation during development of neuroblasts and neurons in grasshopper embryo." *Nature* 307: 163–65.

Thornton, Gemma K., and C. Geoffrey Woods. 2008. "Primary microcephaly: Do all roads lead to Rome?" *Trends in Genetics* 25: 501–10.

Woods, C. Geoffrey, Jacquelyn Bond, and Wolfgang Enard. 2005. "Autosomal recessive primary microcephaly (MCPH): A review of clinical, molecular, and evolutionary findings." *American Journal of Human Genetics* 76: 717–28.

Chapter 6   Later Events

Akam, Michael. 1987. "The molecular basis for metameric pattern in the *Drosophila* embryo." *Development* 101: 1–22.

Alexander, Tara, Christof Nolte, and Robb Krumlauf. 2009. "*Hox* genes and segmentation of the hindbrain and axial skeleton." *Annual Review of Cell and Developmental Biology* 25: 431–56.

Carroll, Sean B. 1990. "Zebra patterns in fly embryos: Activation of stripes or repression of interstripes?" *Cell* 60: 9–16.

Chamberlain, Chester E., Juhee Jeong, Chaoshe Guo, Benjamin L. Allen, and Andrew P. McMahon. 2008. "Notochord-derived Shh concentrates in close association with the apically positioned basal body in neural target cells and forms a dynamic gradient during neural patterning." *Development* 135: 1097–1106.

Chiang, Chin, Ying Litingtung, Eric Lee, Keith E. Young, Jeffrey L. Corden, Heiner Westphal, and Philip A. Beachy. 1996. "Cyclopia and defective axial patterning in mice lacking *Sonic hedgehog* gene function." *Nature* 383: 407–13.

Choo, Siew Who, Robert White, and Steven Russell. 2011. "Genome-wide analysis of the binding of the Hbox protein Ultrabithorax and Hox cofactor homothorax in *Drosophila*." *PloS ONE* 6: e14778.

Cowing, Deborah, and Cynthia Kenyon. 1996. "Correct *Hox* gene expression established independently of position in *Caenorhabditis elegans*." *Nature* 382: 353–56.

D'Arcangelo, Gabriella, Ramin Homayouni, Lakhu Keshvara, Dennis S. Rice, Michael Sheldon, and Tom Curran. 1999. "Reelin is a ligand for lipoprotein receptors." *Neuron* 24: 471–79.

Ekker, Stephen C., Anne R. Ungar, Penny Greenstein, Doris P. von Kessler, Jeffery A. Porter, Randall T. Moon, and Philip A. Beachy. 1995. "Patterning activities of vertebrate *hedgehog* proteins in the developing eye and brain." *Current Biology* 5: 944–55.

Frasch, Manfred, Rahul Warrior, Jonathan Tugwood, and Michael Levine. 1988. "Molecular analysis of *even-skipped* mutants in *Drosophila* development." *Genes and Development* 2: 1824–38.

Gerhart, J. 1998. "Johannes Holtfreter." In *Biographical Memoirs*. Washington, DC: National Academies Press, 1–19.

Ghysen, A., L. Y. Jan, and Y. N. Jan. 1985. "Segmental determination in *Drosophila* central nervous system." *Cell* 40: 943–48.

Gilbert, Scott F. 2010. *Developmental Biology,* 9th edition. Sunderland, MA: Sinauer.

Hattori, Mitsuharu. 2011. "What does Reelin actually do in the developing brain?" *Nihon Shinkei Seishin Yakurigaku Zasshi*: 267–71.

Hidalgo, Alicia. 1996. "The roles of engrailed." *Trends in Genetics* 12: 1–4.

Hong, Susan E., Yin Yao Shugart, David T. Huang, Saad Al Shahwan, P. Ellen Grant, Jonathan O'B. Hourihane, Neil D. T. Martin, and Christopher A. Walsh. 2000. "Autosomal recessive lissencephaly with cerebellar hypoplasia is associated with human RELN mutations." *Nature Genetics* 26: 93–96.

Hueber, Stefanie D., and Ingrid Lohmann. 2008. "Shaping segments: *Hox* gene function in the genomic age." *BioEssays* 30: 965–79.

Ingham, P. W., and A. Martinez Arias. 1992. "Boundaries and fields in early embryos." *Cell* 68: 221–35.

Letinic, Kresimir, Roberto Zoncu, and Pasko Rakic. 2002. "Origin of GABAergic neurons in the human neocortex." *Nature* 417: 645–49.

Luque, Juan M. 2007. "Puzzling out the reeler brainteaser: Does reelin signal to unique neural lineages?" *Brain Research* 1140: 41–50.

Martinez-Arias, Alfonso, and Peter A. Lawrence. 1985. "Parasegments and compartments in the *Drosophila* embryo." *Nature* 313: 639–42.

Maves, Lisa, and Charles B. Kimmel. 2005. "Dynamic and sequential patterning of the zebrafish posterior hindbrain by retinoic acid." *Developmental Biology* 285: 593–605.

Meyer, Gundela. 2010. "Building a human cortex: The evolutionary differentiation of Cajal-Retzius cells and the cortical hem. *Journal of Anatomy* 217: 334–43.

Mohler, Jym. 1988. "Requirements for *hedgehog*, a segmental polarity gene, in patterning larval and adult cuticle of Drosophila." *Genetics* 120: 1061–72.

Nüsslein-Volhard, Christiane, and Eric Wieschaus. 1980. "Mutations affecting segment and polarity in Drosophila." *Nature* 287: 795–801.

Prince, Victoria E. 2002. "The Hox paradox: More complexes than imagined." *Developmental Biology* 249 (): 1–15.

Roelink, H., A. Augsburger, J. Heemskerk, V. Korzh, S. Nortin, A. Ruiz I Altaba, Y. Tanabe, M. Placzek, T. Edlund, T. M. Jessell, and J. Dodd. 1994. "Floor plate and motor neuron induction by *vhh-1*, a vertebrate homolog of *hedgehog* expressed by the notochord." *Cell* 76: 761–75.

Small, Stephen, and Michael Levine. 1991. "The initiation of pair-rule stripes in the *Drosophila* blastoderm." *Current Opinion in Genetics and Development* 1: 255–60.

Soriano, Eduardo, and José Antonio del Río. 2005. "The cells of Cajal-Retzius: Still a mystery one century after." *Neuron* 46: 389–94.

Varjosalo, Markku, and Jussi Taipale. 2008. "Hedgehog: Functions and mechanisms." *Genes and Development* 22: 2454–72.

Wieschaus, Eric, Christiane Nüsslein-Volhard, and Gerd Jürgens. 1984. "Mutations affecting the pattern of the larval cuticle in *Drosophila melanogaster. Roux's Archives of Developmental Biology* 193: 296–307.

Woltering, Joost M., and Antony J. Durston. 2006. "The zebrafish hoxDb cluster has been reduced to a single microRNA." *Nature Genetics* 38: 601–2.

Zhao, Shanting, and Michael Frotscher. 2010. "Go or stop? Divergent roles of Reelin in radial neuronal migration." *Neuroscientist* 16: 421–34.

Chapter 7   Becoming a Neuron

Arimura, Nariko, and Kozo Kaibuchi. 2007. "Neuronal polarity: From extracellular signals to intracellular mechanisms." *Nature Reviews Neuroscience* 8: 194–205.

Banker, Gary, and Kimberly Goslin. 1998. *Culturing Nerve Cells,* 2nd edition. Cambridge, MA: MIT Press. 666 pp.

Barnes, Anthony P., and Franck Polleux. 2009. "Establishment of axon–dendrite polarity in developing neurons." *Annual Review of Neuroscience* 32: 347–81.

Brenner, Sidney. 1974. "The genetics of *Caenorhabditis elegans.*" *Genetics* 77: 71–94.

Brose, Katja, Kimberly S. Bland, Kuan Hong Wang, David Arnott, William Henzel, Coresy S. Goodman, Marc Tessier-Lavigne, and Thomas Kidd. 1999. "Slit proteins bind Robo receptors and have an evolutionary conserved role in repulsive axon guidance." *Cell* 96: 795–806.

Buck, Kenneth B., and James Q. Zheng. 2002. "Growth cone turning induced by direct local modification of microtubule dynamics." *Journal of Neuroscience* 22: 9358–67.

Chan, S. S.-Y., H. Zheng, M.-W. Su, R. Wilk, M. T. Killeen, E. M. Hedgecock, and J. G. Culotti. 1996. "UNC-40, a *C. elegans* homolog of DCC (Deleted in Colorectal Cancer), is required in motile cells responding to UNC-6 netrin cues." *Cell* 87: 187–95.

Conde, Cecilia, and Alfredo Cáceres. 2009. "Microtubule assembly, organization and dynamics in axons and dendrites." *Nature Reviews Neuroscience* 10: 319–32.

Craig, Ann Marie, and Gary Banker. 1994. "Neuronal polarity." *Annual Review of Neuroscience* 17: 267–310.

Farmer, S. F., D. A. Ingram, and J. A. Stephens. 1990. "Mirror movements studied in a patient with Klippel-Feil syndrome." *Journal of Physiology* 428: 467–84.

Faux, Maree C., Janine L. Coates, Nadia J. Kershaw, Meredith J. Layton, and Antony W. Burgess. 2010. "Independent interactions of phosphorylated β-catenin with E-cadherin at cell–cell contacts and APC at cell protrusions." *PLoS ONE* 5: e14127.

Gautam, Medha, Peter G. Noakes, Lisa Moscoso, Fabio Rupp, Richard H. Scheller, John P. Merlie, and Joshua R. Sanes. 1996. "Defective neuromuscular synaptogenesis in agrin-deficient mutant mice." *Cell* 85: 525–35.

Greene, Lloyd A., and Arthur S. Tischler. 1976. "Establishment of a noradrenergic clonal line of rat adrenal pheochromocytoma cells which respond to nerve growth factor." *Proceedings of the National Academy of Sciences U.S.A.* 73: 2424–28.

Grueber, Wesley B., Chung-Hui Yang, Bing Ye, and Yuh-Nung Jan. 2005. "The development of neuronal morphology in insects." *Current Biology* 15: R730–R738.

Harrelson, Allan L., and Corey S. Goodman. 1988. "Growth cone guidance in insects: Fasciclin II is a member of the immunoglobulin superfamily." *Science* 242: 700–708.

Hedgecock, Edward M., Joseph G. Culotti, J. Nichol Thomson, and Lizabeth A. Perkins. 1985. "Axonal guidance mutants of *Caenorhabditis elegans* identified by filling sensory neurons with fluorescein dyes." *Developmental Biology* 111: 158–70.

Henkemeyer, Mark, Olga S. Itkis, Michelle Ngo, Peter W. Hickmott, and Iryna M. Ethell. 2003. "Multiple EphB receptor tyrosine kinases shape dendritic spines in the hippocampus." *Journal of Cell Biology* 163: 1313–26.

Izzi, L., and F. Charron. 2011. "Midline axon guidance and human genetic disorders." *Clinical Genetics* 80: 226–34.

Janke, Carsten and Jeannette Chloë Bulinski. 2011. "Post-translational regulation of the microtubule cytoskeleton: Mechanisms and functions." *Nature Reviews Molecular Cell Biology* 12: 773–86.

Jen, Joanna C., Wai-Man Chan, Thomas M. Bosley, Jijun Wan, Janai R. Carr, Udo Rüb, David Shattuck, Georges Salamon, Lili C. Kudo, Jing Ou, Doris D. M. Lin, Mustafa A. M. Salih, Tülay Kansu, Hesham al Dhalaan, Zayed al Zayed, David B. MacDonald, Bent Stigsby, Andreas Plaitakis, Emmanuel K. Dretakis, Irene Gottlob, Christina Pieh, Elias I. Traboulsi, Qing Wang, Lejin Wang, Caroline Andrews, Koki Yamada, Joseph L. Demer, Shaheen Karim, Jeffry R. Alger, Daniel H. Geschwind, Thomas Deller, Nancy L. Sicotte, Stanley F. Nelson, Robert W. Baloh, and Elizabeth C. Engle. 2004. "Mutations in a human *ROBO* gene disrupt hindbrain axon pathway crossing and morphogenesis." *Science* 304: 1509–13.

Kaech, Stefanie, and Gary Banker. 2006. "Culturing hippocampal neurons." *Nature Protocols* 1: 1406–15.

Kidd, Thomas, Kimberly S. Bland, and Corey S. Goodman. 1999. "Slit is the midline repellent for the Robo receptor in *Drosophila*." *Cell* 96: 785–94.

Long, Hua, Christelle Sabatier, Le Ma, Andrew Plump, Wenlin Yuan, David M Ornitz, Atsushi Tamada, Fujio Murakami, Corey S Goodman, and Marc Tessier-Lavigne. 2004. "Conserved roles for Slit and Robo proteins in midline commissural axon guidance." *Neuron* 42: 213–23.

Lovell, Peter, and Leonid L. Moroz. 2006. "The largest growth cones in the animal kingdom: An illustrated guide to the dynamics of *Aplysia* neuronal growth in cell culture." *Integrative and Comparative Biology* 46: 847–70.

Lowery, Laura Anne, and David Van Vactor. 2009. "The trip of the tip: Understanding the growth cone machinery." *Nature Reviews Molecular Cell Biology* 10: 332–43.

Pollard, Thomas D., and John A. Cooper. 2009. "Actin, a central player in cell shape and movement." *Science* 326: 1208–12.

Rolls, Melissa M. 2011. "Neuronal polarity in *Drosophila*: Sorting out axons and dendrites." *Developmental Neurobiology* 71: 419–29.

Sabatier, Christelle, Andrew S. Plump, Le Ma, Katja Brose, Atsushi Tamada, Fujio Murakami, Eva Y.-H. P. Lee, and Marc Tessier-Lavigne. 2004. "The divergent Robo family protein Rig-1/Robo3 is a negative regulator of Slit responsiveness required for midline crossing by commissural axons." *Cell* 117: 157–69.

Seeger, Mark, Guy Tear, Dolors Ferres-Marco, and Corey S. Goodman. 1993. "Mutations affecting growth cone guidance in *Drosophila:* Genes necessary for guidance toward or away from the midline." *Neuron* 10: 409–26.

Serafini, Tito, Timothy E. Kennedy, Michael J. Galko, Christine Mirzayan, Thomas M. Jessell, and Marc Tessier-Lavigne. 1994. "The netrins define a family of axon outgrowth-promoting proteins homologous to *C. elegans* UNC-6." *Cell* 78: 409–24.

Serafini, Tito, Sophia A. Colamarino, E. David Leonardo, Hao Wang, Rosa Beddington, William C. Skarnes, and Marc Tessier-Lavigne. 1996. "Netrin-1 is required for commissural axon guidance in the developing vertebrate nervous system." *Cell* 87: 1001–14.

Srour, Myriam, Jean-Baptiste Rivière, Jessica M. T. Pham, Maire-Pierre Dubè, Simon Girard, Steves Morin, Patrick A. Dion, Géraldine Asselin, Daniel Rochefort, Pascale Hince, Sabrina Diab, Naser Sharafaddinzadeh, Sylvain Chouinard, Hugo Théoret, Frédéric Charron, and Guy A. Rouleau. 2010. "Mutations in DCC cause congenital mirror movements." *Science* 328: 328.

Stiess, Michael, and Frank Bradke. 2011. "Neuronal polarization: The cytoskeleton leads the way." *Developmental Neurobiology* 71: 430–44.

Tessier-Lavigne, Marc, and Corey S. Goodman. 1996. "The molecular biology of axon guidance." *Science* 274: 1123–33.

Williams, D. W., and J. W. Truman. 2005. "Remodeling dendrites during insect metamorphosis." *Journal of Neurobiology* 64: 24–33.

Witte, Harald, Dorothee Neukirchen, and Frank Bradke. 2008. "Microtubule stabilization specifies initial neuronal polarization." *Journal of Cell Biology* 180: 619–32.

Wu, Haitao, Wen C. Xiong, and Lin Me. 2010. "To build a synapse: Signaling pathways in neuromuscular junction assembly." *Development* 137: 1017–33.

Chapter 8   Glia

Allaman, Igor, Mireille Bélanger, and Pierre J. Magistretti. 2011. "Astrocyte-neuron metabolic relationships: For better and for worse." *Trends in Neurosciences* 34: 76–87.

Bacaj, Taulant, Maya Tevlin, Yun Lu, and Shai Shaham. 2008. "Glia are essential for sensory organ function in *C. elegans.*" *Science* 322: 744–47.

Bartzokis, George, Mace Beckson, Po H. Lu, Keith H. Nuechterlein, Nancy Edwards, and Jim Mintz. 2001. "Age-related changes in frontal and temporal lobe volumes in men." *Archives of General Psychiatry* 58: 461–65.

Beckervordersandforth, Ruth M., Christof Rickert, Benjamin Altenhein, and Gerhard M. Technau. 2008. "Subtypes of glial cells in the *Drosophila* embryonic ventral nerve cord as related to lineage and gene expression." *Mechanisms of Development* 125: 542–57.

Booth, Gwendolen E., Edward F. V. Kinrade, and Alicia Hidalgo. 2000. "Glia maintain follower neuron survival during *Drosophila* CNS development." *Development* 127: 237–44.

Buffo, Annalisa, Immaculada Rite, Pratibha Tripathi, Alexandra Lepier, Dilek Colak, Ana-Paula Horn, Tetsuji Mori, and Magdalena Götz. 2008. "Origin and progeny of reactive gliosis: A source of multipotent cells in the injured brain." *Proceedings of the National Academy of Sciences, U.S.A.* 105: 3581–86.

Cahoy, John D., Ben Emery, Amit Kaushal, Lynette C. Foo, Jennifer L. Zamanian, Karen S. Christopherson, Yi Xing, Jane L. Lubischer, Paul A. Krieg, Sergey A. Krupenko, Wesley J. Thompson, and Ben A. Barres. 2008. "A transcriptome database for astrocytes, neurons, and oligodendrocytes: A new resource for understanding brain development and function." *Journal of Neuroscience* 28: 264–78.

Dambly-Chaudière, Christine, Dora Sapède, Fabien Soubiran, Kelly Decorde, Nicolas Gompel, Alain Ghysen. 2003. "The lateral line of zebrafish: A model system for the analysis of morphogenesis and neural development in vertebrates." *Biology of the Cell* 95: 579–87.

Eroglu, Cagla, and Ben A. Barres. 2010. "Regulation of synaptic connectivity by glia." *Nature* 468: 223–31.

Freeman, Marc R. 2010. "Specification and morphogenesis of astrocytes." *Science* 330: 774–78.

Gilmour, Darren T., Hans-Martin Maischein, and Christiane Nüsslein-Volhard. 2002. "Migration and function of a glial subtype in the vertebrate peripheral nervous system." *Neuron* 34: 577–88.

Guillemot, François. 2007. "Spatial and temporal specification of neural fates by transcription factor codes." *Development* 134: 3771–80.

Hartenstein, Volker. 2011. "Morphological diversity and development of glia in *Drosophila*." *Glia* 59: 1237–52.

Hebbar, Sarita, and Joyce J. Fernandes. 2010. "Glial remodeling during metamorphosis influences the stabilization of motor neuron branches in *Drosophila*." *Developmental Biology* 340: 344–54.

Herculano-Houzel, Suzanna. 2009. "The human brain in numbers: A linearly scaled-up primate brain." *Frontiers in Human Neuroscience* 3: Article 31.

Hilgetag, Claus C., and Helen Barbas. 2009. "Are there ten times more glia than neurons in the brain?" *Brain Structure and Function* 213: 365–66.

Jabs, R., I. A. Pateron, and W. Walz. 1997. "Qualitative analysis of membrane currents in glial cells from normal and gliotic tissue in situ: Down-regulation of Na^+ current and lack of P2 purinergic responses." *Neuroscience* 81: 847–60.

Lundkvist, Johan, and Urban Lendahl. 2001. "Notch and the birth of glial cells." *Trends in Neurosciences* 24: 492–94.

Ma, Eva Y., and Raible, David W. 2009. "Signaling pathways regulating zebrafish lateral line development." *Current Biology* 19: R381–86.

Middeldorp, Jinte, and Elly M. Hol. 2011. "GFAP in health and disease." *Progress in Neurobiology* 93: 421–43.

Miyata, Takaki, Daichi Kawaguchi, Ayano Kawaguchi, and Yukiko Gotol. 2010. "Mechanisms that regulate the number of neurons during mouse neocortical development." *Current Opinion in Neurobiology* 20: 22–28.

Monk, Kelly R., and William S. Talbot. 2009. "Genetic dissection of myelinated axons in zebrafish." *Current Opinion in Neurobiology* 19: 486–90.

Muroyama, Yuko, Yuko Fujiwara, Stuart H. Orkin, and David H. Rowitch. 2005.

"Specification of astrocytes by bHLH protein SCL in a restricted region of the neural tube." *Nature* 438: 360–63.

Oikonomou, Grigorios, and Shai Shaham. 2011. "The glia of *Caenorhabditis elegans.*" *Glia* 59: 1253–63.

Rabadán, M.A., J. Cayuso, G. Le Dréau, C. Cruz, M. Barzi, S. Pons, J. Briscoe, and E. Martí. 2012. "Jagged2 controls the generation of motor neuron and oligodendrocyte progenitors in the ventral spinal cord." *Cell Death and Differentiation* 19: 1–11 (advance online publication).

Read, Renee D. 2011. "*Drosophila melanogaster* as a model system for human brain cancers." *Glia* 59: 1364–76.

Richardson, William D., Kaylene M. Young, Richa B. Tripathi, and Ian McKenzie. 2011. "NG2 glia as multipotent stem cells: Fact or fantasy?" *Neuron* 70: 661–73.

Rowitch, David H. 2004. "Glial specification in the vertebrate neural tube." *Nature Reviews Neuroscience* 5: 409–19.

Rowitch, David H., and Arnold R. Kriegstein. 2010. "Developmental genetics of vertebrate glial cell specification." *Nature* 468: 214–22.

Schlosser, Gerhard. 2006. "Induction and specification of cranial placodes." *Developmental Biology* 294: 303–51.

Shaham, Shai. 2006. "Glia–neuron interactions in the nervous system of *Caenorhabditis elegans.*" *Current Opinion in Neurobiology* 16: 522–28.

Stork, Tobias, Daniel Engelen, Alice Krudewig, Marion Silies, Roland J. Bainton, and Christian Klämbt. 2008. "Organization and function of the blood–brain barrier in *Drosophila.*" *Journal of Neuroscience* 28: 587–97.

Xiong, Wen-Cheng, Hideyuki Okano, Nipam H. Patel, Julie A. Blendy, and Craig Montell. 1994. "*repo* encodes a glial-specific homeo domain protein required in the *Drosophila* nervous system." *Genes and Development* 8: 981–94.

Zhang, Su-Chun. 2001. "Defining glial cells during CNS development." *Nature Reviews Neuroscience* 2: 840–43.

Zheng, Kang, Hong Li, Ying Zhu, Qiang Zhu, and Mengsheng Qiu. 2010. "MicroRNas are essential for the developmental switch from neurogenesis to gliogenesis in the developing spinal cord." *Journal of Neuroscience* 30: 8245–50.

Ziskin, Jennifer L., Akiko Nishiyama, Maria Rubio, Masahiro Fukaya, and Dwight E. Bergles. 2007. "Vesicular release of glutamate from unmyelinated axons in white matter." *Nature Neuroscience* 10: 321–30.

Chapter 9 Maturation

Altman, J. S., and E. M. Bell. 1973. "A rapid method for the demonstration of nerve cell bodies in invertebrate central nervous systems." *Brain Research* 63: 487–89.

Bacon, Jonathan P., and Jennifer S. Altman. 1977. "A silver intensification method for cobalt-filled neurons in wholemount preparations." *Brain Research* 138: 359–63.

Boksa, Patricia. 2012. "Abnormal synaptic pruning in schizophrenia: Urban myth or reality?" *Journal of Psychiatry and Neuroscience* 37: 75–77.

Brenhouse, Heather C., and Susan L. Andersen. 2011. "Developmental trajectories during adolescence in males and females: A cross-species understanding of underlying brain changes." *Neuroscience and Biobehavioral Reviews* 35: 1687–1703.

Budnik, Vivian, and Kalpana White. 1988. "Catecholamine-containing neurons in *Drosophila melanogaster:* Distribution and development. *Journal of Comparative Neurology* 268: 400–413.

Choi, Youn-Jeong, Gyunghee Lee, and Jae H. Park. 2006. "Programmed cell death mechanisms of identifiable peptidergic neurons in *Drosophila melanogaster.*" *Development* 133: 2223–32.

Consoulas, Christos, Carsten Duch, Ronald J. Bayline, and Richard B. Levine. 2000. "Behavioral transformations during metamorphosis: Remodeling of neural and motor systems." *Brain Research Bulletin* 53: 571–83.

Consoulas, Christos, Linda L. Restifo, and Richard B. Levine. 2002. "Dendritic remodeling and growth of motoneurons during metamorphosis of *Drosophila melanogaster.*" *Journal of Neuroscience* 22: 4906–17.

Durston, Sarah, Hilleke E. Hulshoff Pol, B. J. Casey, Jay N. Giedd, Jan K. Buitelaar, and Herman van Engeland. 2001. "Anatomical MRI of the developing human brain: What have we learned?" *Journal of the American Academy of Child and Adolescent Psychiatry* 40: 1012–20.

Fahrbach, Susan E., John R. Nambu, and Lawrence M. Schwartz. 2012. "Programmed Cell Death in Insects." In *Insect Molecular Biology and Biochemistry,* ed. Lawrence I. Gilbert, 419–49. San Diego: Academic Press.

Feixa, Carles. 2011. "Past and present of adolescence in society: The 'teen brain' debate in perspective." *Neuroscience and Biobehavioral Reviews* 35: 1634–43.

Gavrieli, Yael, Yoav Sherman, and Schmuel A. Ben-Sasson. 1992. "Identification of programmed cell death *in situ* via specific labeling of nuclear DNA fragmentation." *Journal of Cell Biology* 119: 493–501.

Hoeft, Fumiko, Bruce D. McCandliss, Jessica M. Black, Alexander Gantman, Nahal Zakerani, Charles Hulme, Heikki Lyytinen, Susan Whitfield-Gabrieli, Gary H. Glover, Allan L. Reiss, and John D. E. Gabrieli. 2011. "Neural systems predicting long-term outcome in dyslexia." *Proceedings of the National Academy of Sciences U.S.A.* 108: 361–66.

Huttenlocher, Peter R. 1979. "Synaptic density in human frontal cortex—developmental changes and effects of aging." *Brain Research* 163: 195–205.

Huttenlocher, Peter R., and Arun S. Dabholkar. 1997. "Regional differences in synaptogenesis in human cerebral cortex." *Journal of Comparative Neurology* 387: 167–78.

Ito, Kei. and Yoshiki Hotta. 1992. "Proliferation pattern of postembryonic neuroblasts in the brain of *Drosophila melanogaster.*" *Developmental Biology* 149: 134–48.

Kimura, Ken-ichi, and James W. Truman. 1990. "Postmetamorphic cell death in the nervous and muscular systems of *Drosophila melanogaster.*" *Journal of Neuroscience* 10: 403–11.

Lebel, Catherine, and Christian Beaulieu. 2011. "Longitudinal development of human wiring continues from childhood into adulthood." *Journal of Neuroscience* 31: 10937–47.

Marshall, W. A., and J. M. Tanner. 1969. "Variation in pattern of pubertal changes in girls." *Archives of Disease in Childhood* 44: 291–303.

———. 1970. "Variations in the pattern of pubertal changes in boys." *Archives of Disease in Childhood* 45: 13–23.

Oppenheim, Ronald W. 1991. "Cell death during development of the nervous system." *Annual Review of Neuroscience* 14: 453–501.

Sisk, Cheryl L., and Douglas L. Foster. 2004. "The neural basis of puberty and adolescence." *Nature Neuroscience* 10: 1040–47.

Spear, L. P. 2000. "The adolescent brain and age-related behavioral manifestations." *Neuroscience and Biobehavioral Reviews* 24: 417–63.

Sturman, David A., and Bita Moghaddam. 2011. "The neurobiology of adolescence: Changes in brain architecture, functional dynamics, and behavioral tendencies." *Neuroscience and Biobehavioral Reviews* 35: 1704–12.

Thorn, Robert S., and James W. Truman. 1989. "Sex-specific neuronal respecification during the metamorphosis of the genital segments of the tobacco hornworm moth *Manduca sexta*." *Journal of Comparative Neurology* 284: 489–503.

Tissot, Madeleine, and Reinhard F. Stocker. 2000. "Metamorphosis in *Drosophila* and other insects: The fate of neurons throughout the stages." *Progress in Neurobiology* 62: 89–111.

Truman, James W. 1990. "Metamorphosis of the central nervous system of *Drosophila*." *Journal of Neurobiology* 21: 1072–84.

Truman, James W., and Michael Bate. 1988. "Spatial and temporal patterns of neurogenesis in the central nervous system of *Drosophila melanogaster*." *Developmental Biology* 125: 145–57.

Truman, James W., Barbara J. Taylor, and Timothy A. Awad. 1993. "Formation of the adult nervous system," 1245–75. In *The Development of* Drosophila melanogaster, vol. 2. Cold Spring Harbor, NY: Cold Spring Harbor Laboratory Press.

Watts, Ryan J., Eric D. Hoopfer, and Liquin Luo. 2003. "Axon pruning during *Drosophila* metamorphosis: Evidence for local degeneration and requirement of the ubiquitin–proteasome system." *Neuron* 38: 871–85.

Williams, D. W. and James W. Truman. 2005. "Remodeling dendrites during insect metamorphosis." *Journal of Neurobiology* 64: 24–33.

Winbush, Ari, and Janis C. Weeks. 2011. "Steroid-triggered, cell-autonomous death of a *Drosophila* motoneuron during metamorphosis." *Neural Development* 6: 15.

Chapter 10 Thinking about Intellectual Disability in the Context of Development

Alfonso-Loeches, Silvia, and Consuelo Guerri. 2011. "Molecular and behavioral aspects of the actions of alcohol on the adult and developing brain." *Critical Reviews in Clinical Laboratory Sciences* 48: 19–47.

Amir, Ruthie E., Ignatia B. Van den Veyver, Mimi Wan, Charles Q. Tran, Uta Francke, and Huda Y. Zoghbi. 1999. "Rett syndrome is caused by mutations in X-linked MECP2, encoding methyl-CpG-binding protein 2." *Nature Genetics* 23: 185–88.

Armstrong, Dawna, J. Kay Dunn, Barbara Antalffy, and Renuka Trivedi. 1995. "Selective alterations in the cortex of Rett syndrome." *Journal of Neuropathology and Experimental Neurology* 54: 195–201.

Baxter, Laura L., Timothy H. Moran, Joan T. Richtsmeier, Juan Troncoso, and Roger H. Reeves. 2000. "Discovery and genetic localization of Down syndrome cerebellar phenotypes using the Ts65Dn mouse." *Human Molecular Genetics* 9: 195–202.

Begenisic, Tatjana, Maria Spolidoro, Chiara Braschi, Laura Baroncelli, Marco Milanese, Gianluca Pietra, Maria E. Fabri, Ciambattista Bonanno, Giovanni Cioni,

Lamberto Maffei, and Alessandro Sale. 2011. "Environmental enrichment decreases GABAergic inhibition and improves cognitive abilities, synaptic plasticity, and visual functions in a mouse model of Down syndrome." *Frontiers in Cellular Neuroscience* 5: Article 29.

Bolduc, Françoise V., Kimberly Bell, Hilary Cox, Kendal S. Broadie, and Tim Tully. 2008. "Excess protein synthesis in *Drosophila* fragile X mutants impairs long-term memory." *Nature Neuroscience* 11: 1143–45.

Chang, Karen T., Yi-Jun Shi, and Kyung-Tai Min. 2003. "The *Drosophila* homolog of Down's syndrome critical region 1 gene regulates learning: Implications for mental retardation." *Proceedings of the National Academy of Sciences, U.S.A.* 100: 15794–99.

Das, Ishita, and Roger H. Reeves. 2011. "The use of mouse models to understand and improve cognitive deficits in Down syndrome." *Disease Models and Mechanisms* 4: 596–606.

Dierssen, Mara, and Ger J. A. Ramakers. 2006. "Dendritic pathology in mental retardation: From molecular genetics to neurobiology." *Genes, Brain and Behavior* 5, Suppl. 2: 48–60.

Dutch–Belgian Fragile X Consortium. 1994. "*Fmr1* knockout mice: A model to study fragile X mental retardation." *Cell* 78: 23–33.

Friede, Reinhard L. 1973. "Dating the development of the human cerebellum." *Acta Neuropathologica* 23: 48–58.

Gadalla, Kamal K. E., Mark E. S. Bailey, and Stuart R. Cobb. 2011. "MeCP2 and Rett syndrome: Reversibility and potential avenues for therapy." *Biochemical Journal* 439: 1–14

Hatten, Mary E. 1999. "Central nervous system neuronal migration." *Annual Review of Neuroscience* 22: 511–39.

Hatten, Mary E., and Nathaniel Heintz. 1995. "Mechanisms of neural patterning and specification in the developing cerebellum." *Annual Review of Neuroscience* 18: 385–408.

Hatten, Mary E., and Martine F. Roussel. 2011. "Development and cancer of the cerebellum." *Trends in Neurosciences* 34: 134–42.

Huttenlocher, Peter R. 1974. "Dendritic development in neocortex of children with mental defect and infantile spasms." *Neurology* 24: 203–10.

Inlow, Jennifer K., and Linda L. Restifo. 2004. "Molecular and comparative genetics of mental retardation." *Genetics* 166: 835–81.

Jacola, Lisa M., Anna W. Byars, Melinda Chalfonte-Evans, Vincent J. Schmithorst, Fran Hickey, Bonie Patterson, Stephanie Hotze, Jennifer Vannest, Chung-Yiu Chiu, Scott K. Holland, and Mark B. Schapiro. 2011. "Functional magnetic resonance imaging of cognitive processing in young adults with Down syndrome." *American Journal of Intellectual and Developmental Disabilities* 116: 344–50.

Jones, Kenneth L., and David W. Smith. 1973. "Recognition of the fetal alcohol syndrome in early infancy." *Lancet* 302: 999–1001.

Kishi, Noriyuki, and Jeffrey D. Macklis. 2010. "MeCP2 functions largely cell-autonomously but also non-cell-autonomously, in neuronal maturation and dendritic arborization of cortical pyramidal neurons." *Experimental Neurology* 222: 51–58.

Krueger, Dilja D., and Mark F. Bear. 2011. "Toward fulfilling the promise of molecular medicine in fragile X syndrome." *Annual Review of Medicine* 62: 411–29.

Kumada, Tatsuro, Madepalli K. Lakshmana, and Hitoshi Komuro. 2006. "Reversal of neuronal migration in a mouse model of fetal alcohol syndrome by controlling second-messenger signaling." *Journal of Neuroscience* 26: 742–56.

Kumada, Tatsuro, Yulan Jiang, D. Bryant Cameron, and Hitoshi Komuro. 2007. "How does alcohol impair neuronal migration?" *Journal of Neuroscience Research* 85: 465–70.

Lana-Elola, Eva, Sheena D. Watson-Scales, Elizabeth M. C. Fisher, and Victor L. J. Tybulewicz. 2011. "Down syndrome: Searching for the genetic culprits." *Disease Models and Mechanisms* 4: 586–95.

Liu, Chunhong, Pavel V. Belichenko, Li Zhang, Dawei Fu, Alexander M. Kleschevnikov, Antonio Baldini, Stylianos E. Antonarakis, William C. Mobley, and Y. Eugene Yu. 2011. "Mouse models for Down syndrome–associated developmental cognitive disabilities." *Developmental Neuroscience* 33: 404–13.

Martin-Padilla, Miguel. 1972. "Structural abnormalities of the cerebral cortex in human chromosomal aberrations: A Golgi study." *Brain Research* 44: 625–29.

McClure, Kimberly D., Rachael L. French, and Ulrike Heberlein. 2011. "A *Drosophila* model for fetal alcohol syndrome disorders: Role for the insulin pathway." *Disease Models and Mechanisms* 4: 335–46.

Morte, Beatriz, Jimena Manzano, Thomas Scanlan, Björn Vennström, and Juan Bernal. 2002. "Deletion of the thyroid hormone receptor α1 prevents the structural alterations of the cerebellum induced by hypothyroidism." *Proceedings of the National Academy of Sciences, U.S.A.* 99: 3985–89.

Murphy, Catherine C., Coleen Boyle, Diana Schendel, Pierre Decouflé, and Marshalyn Yeargin-Allsopp. 1998. "Epidemiology of mental retardation in children." *Mental Retardation and Developmental Disabilities Research Reviews* 4: 6–13.

Nithianantharajah, Jess, and Anthony J. Hannan. 2006. "Enriched environments, experience-dependent plasticity and disorders of the nervous system." *Nature Reviews Neuroscience* 7: 697–709.

Panek, Paul E., and Jessi E. Smith. 2005. "Assessment of terms to describe mental retardation." *Research in Developmental Disabilities* 26: 565–76.

Patel, J., K. Landers, H. Li, R. H. Mortimer, and K. Richard. 2011. "Thyroid hormones and fetal neurological development." *Journal of Endocrinology* 209: 1–8.

Purpura, Dominick P. 1974. "Dendritic spine 'dysgenesis' and mental retardation." *Science* 186: 1126–28.

Restifo, Linda L. 2005. "Mental retardation genes in Drosophila: New approaches to understanding and treating developmental brain disorders." *Mental Retardation and Developmental Disabilities Research Reviews* 11: 286–94.

Roper, Randall J., Laura L. Baxter, Nidhi G. Saran, Donna K. Klinedinst, Philip A. Beachy, and Roger H. Reeves. 2006. "Defective cerebellar response to mitogenic Hedgehog signaling in Down's syndrome mice." *Proceedings of the National Academy of Sciences, U.S.A.* 103: 1452–56.

Roper, Randall J., and Roger H. Reeves. 2006. "Understanding the basis for Down syndrome phenotypes." *PLoS Genetics* 2: e50.

Shea, Sarah E. 2006. "Mental retardation in children ages 6 to 16." *Seminars in Pediatric Neurology* 13: 262–70.

Spadoni, Andrea D., Christie L. McGee, Susanna L. Fryer, and Edward P. Riley. 2007. "Neuroimaging and fetal alcohol spectrum disorders." *Neuroscience and Biobehavioral Reviews* 31: 235–45.

Takashima, Sachio, Laurence E. Becker, Dawna L. Armstrong, and Fu-Wah Chan. 1981. "Abnormal neuronal development in the visual cortex of the human fetus and infant with Down's syndrome: A quantitative and qualitative Golgi study." *Brain Research* 225: 1–21.

ONLINE RESOURCES

Chapter 1   Introduction

http://www.brain-map.org/
In situ hybridization atlases (produced by the Allen Institute for Brain Science) showing gene expression in the mouse and human brains, with cell-level resolution. Full documentation of methods is provided. The online tutorials provide essential orientation.

http://cre.jax.org/strainlist.html
Jackson Laboratory list of mouse strains that express Cre.

http://crin.sfn.org/
Need to review a basic neuroscience concept or see a video related to development of the nervous system? The Society for Neuroscience maintains ERIN (Educational Resources in Neuroscience) as a resource for neuroscience students and instructors.

http://www.microscopyu.com/
Online microscopy education provided by Nikon Instruments Inc.

http://www.ihcworld.com/
Online immunohistochemistry center provided by IHC World LLC, with many contributions from individual research laboratories.

http://www.invitrogen.com/site/us/en/home/References/Molecular-Probes-The-Handbook.html
Online version of a popular comprehensive reference on fluorescent labeling and detection.

Chapter 2   Overview of Nervous System Development in Humans

http://www.visembryo.com/
Online resource on human development illustrating all stages of embryonic and fetal development. Highly recommended.

http://embryology.med.unsw.edu.au/
Comprehensive online resource on embryology of many species, including humans. Hosted by the University of New South Wales, Australia.

Chapter 3 Animal Models

http://flybase.org/
    An invaluable database for *Drosophila* genes, genomes, and genetics.

http://www.cbs.umn.edu/CGC/
    Information about the Caenorhabditis Genetics Center at the University of Minnesota.

http://embryology.med.unsw.edu.au/OtherEmb/mouse4.htm
    A full description of Theiler stages of mouse embryonic development, hosted by the University of New South Wales, Australia.

http://www.emouseatlas.org/emap/home.html
    Online atlas of the Theiler stages of mouse embryonic development, provided by the e-Map Atlas Project.

http://www.nematodes.org/
    Web site of the Blaxter Lab, Institute of Evolutionary Biology, School of Biological Sciences, The University of Edinburgh. A good portal for Nembase 4, a nematode transcriptome database.

http://www.nobelprize.org/
    An online museum that provides detailed scientific and historical information on all Nobel laureates.

http://www.sdbonline.org/fly/aimain/1aahome.htm
    The Interactive Fly, a detailed online resource for development of *Drosophila melanogaster*. An excellent catalog of genes associated with embryonic development and metamorphosis.

http://www.wormatlas.org/
    Web site of the Worm Atlas.

http://www.wormbase.org
    Web site of the Wormbase, a portal for access nematode genomic information.

http://www.wormbook.org/
    Web site of WormBook, an open-access collection of literature related to the biology of C. elegans.

http://zfin.org/
    Web site of the Zebrafish Model Organism Database.

Chapter 4 Early Events

http://www.cdc.gov/ncbddd/pregnancy_gateway/meds/
    Web site maintained by the U.S. Centers for Disease Control and Prevention that discusses the use of medication during pregnancy.

http://www.hhmi.ucla.edu/derobertis/EDR_MS/chd_page/chordin.html
    The Chordin Page, chordin-related publications and figures provided by a web site of the DeRobertis Laboratory at the University of California, Los Angeles.

http://www.ninds.nih.gov/disorders/cephalic_disorders/detail_cephalic_
disorders.htm
> A fact sheet on cephalic disorders prepared for laypersons by the National
> Institute of Neurological Disorders and Stroke of the National Institutes of
> Health. Contains a clear description of holoprosencephaly in humans.

http://www.nobelprize.org/nobel_prizes/medicine/laureates/1935/
> Interesting information about Hans Spemann and his research, much of it in
> his own words, available at the Nobel Prize web site.

http://www.stanford.edu/group/nusselab/cgi-bin/wnt/
> Visit the Wnt homepage, maintained by the Nusse laboratory at Stanford
> University, for access to further information about Wnt signaling.

Chapter 5   *Neurogenesis*

http://rarediseases.info.nih.gov/
> Information about double cortex syndrome from the National Institutes of
> Health Office of Rare Diseases Research.

http://thomas.loc.gov/home/thomas.php
> The THOMAS page of the U.S. Library of Congress, via which the full text of
> the NIH Revitalization Act of 1993 (or any other piece of U.S. federal legisla-
> tion) can be obtained.

http://www.wormbook.org/
> Online source of information on C. elegans cell lineages.

Chapter 6   *Later Events*

http://www.Nobelprize.org
> Search for Lewis, Nüsslein-Volhard, and Wieschaus to learn more about their
> Nobel Prize–winning studies on development in *Drosophila.*

http://www.sdbonline.org/fly/aimain/1aahome.htm
> Use the link titled "Effects of mutation" on the individual pages for zygotically
> transcribed genes to learn the mutant phenotypes for *Drosophila* genes related
> to segmentation.

Chapter 7   *Becoming a Neuron*

http://www.myasthenia.org/
> This web site covers the basics of the autoimmune disorder myasthenia gravis.

http://www.nature.com/scitable/topicpage/microtubules-and-filaments-14052932
> Need a refresher on the basic components of the eukaryotic cytoskeleton? Visit
> the relevant Scitable education page maintained by the Nature Publishing
> Group.

http://www.nobelprize.org/nobel_prizes/chemistry/laureates/2004/
> This web site provides additional information on the discovery of ubiquitin.

http://www.yale.edu/forschlab/
> Visit web site to view beautiful videos of axon growth from the Forscher
> laboratory at Yale University.

Chapter 8   Glia

http://www.cancer.gov/cancertopics/factsheet/detection/tumor-grade
    Fact sheets on tumors and tumor grades provided by the National Cancer
    Institute at the National Institutes of Health that will help you understand
    glial cell–derived brain tumors.

http://insitu.fruitfly.org/cgi-bin/ex/insitu.pl
    Berkeley Drosophila Genome Project site, which contains a library of in situ
    hybridization studies of embryonic development.

http://www.cell.com/neuron/supplemental/S0896-6273%2802%2900683-9
    Archives of the journal *Neuron,* which provides time-lapse videos of lateral
    line formation in the zebrafish published as supplemental data for Gilmour et
    al. (2002).

http://www.ninds.nih.gov/disorders/alexander_disease/alexander_disease.htm
    National Institute of Neurological Disorders and Stroke Alexander Disease
    Information Page which describes symptoms and contains links to informa-
    tion on clinical trials.

Chapter 9   Maturation

http://www.dana.org
    Web site of the Dana Foundation, a private philanthropic organization that
    supports brain research and neuroscience education. Use "teen brain" as a
    search term to locate numerous articles discussing the strengths and vulnera-
    bilities of the adolescent brain; other pages provide extensive resources for
    neuroscience students.

http://www.developmentalbiology.net
    Web site dedicated to providing resources for use by developmental biology
    students. Offers detailed information on postembryonic development in
    Drosophila, including detailed photographs covering all pupal stages.

http://www.loni.ucla.edu/~thompson/DEVEL/dynamic.html
    Web site containing time-lapse movies of human brain maturation based on
    MRI scans.

Chapter 10   Thinking about Intellectual Disability in the Context of Development

http://www.aaidd.org/
    Web site of the American Association of Intellectual and Developmental
    Disabilities (formerly the American Association of Mental Retardation), which
    provides clear definitions and other information on intellectual and develop-
    mental disabilities.

http://www.cdc.gov/ncbddd/fasd/
    Web site of the U.S. Centers for Disease Control and Prevention, which
    provides information on a common environmentally induced form of
    intellectual disability, fetal alcohol syndrome; other pages provide resources
    on specific developmental disabilities.

http://www.fragilex.org/
   Web site of the National Fragile X Foundation, which provides information on fragile X and related syndromes.

http://www.ncbi.nlm.nih.gov/pubmedhealth/s/diseases_and_conditions/
   Authoritative descriptions of many conditions that cause developmental disability, including neonatal hypothyroidism, fragile X syndrome, and Rett syndrome, provided by the web site of the U.S. National Library of Medicine.

http://www.ncbi.nlm.nih.gov/omim
   Web site of Online Mendelian Inheritance in Man (commonly referred to by its acronym, OMIM), a comprehensive resource providing information on human genes and genetic phenotypes; search "mental retardation" for information on inherited intellectual and developmental disabilities.

http://www.nobelprize.org/nobel_prizes/medicine/laureates/1906/
   The Nobel Prize web site, which describes the achievements of Camillo Golgi and Santiago Ramón y Cajal. Read the brief Award Ceremony Speech to be transported back a century to a time when "the central nervous system appeared as a confused mass of filaments, each as fine as the thread of a spider's web, and of microscopic cells armed with cellular processes" and "it was impossible to isolate the individual components of tissue specimens."

http://www.psych.org/
   Web site of the American Psychiatric Association, which provides information on its publication *Diagnostic and Statistical Manual of Mental Disorders*.

http://users.tinyonline.co.uk/gswithenbank/welcome.htm
   A source for the origins (or suspected origins) of English-language expressions and sayings; provides interesting reading for a study break.

FULL CITATIONS FOR INVESTIGATIVE READING EXERCISES

Chapter 1   Introduction

1. Leuner, Benedetta, Erica R. Glasper, and Elizabeth Gould. 2009. "Thymidine analog methods for studies of adult neurogenesis are not equally sensitive." *Journal of Comparative Neurology* 517: 123–33.

2. Zhang, Sheng, Mel B. Feany, Sudipta Saraswati, J. Troy Littleton, and Norbert Perrimon. 2009. "Inactivation of Drosophila Huntingtin affects long-term functioning and the pathogenesis of a Huntington's disease model." *Disease Models and Mechanisms* 2: 247–66.

3. Cheever, Thomas R., Emily A. Olson, and James M. Ervasti. 2011. "Axonal regeneration and neuronal function are preserved in motor neurons lacking beta-actin *in vivo*." *PLoS ONE* 6: e17768.

Chapter 2   Overview of Nervous System Development in Humans

1. Garel, Catherine, Emmanuel Chantrel, Hervé Brisse, Monique Elmaleh, Dominique Luton, Jean-Françoise Oury, Guy Sebag, and Max Hassan. 2001. "Fetal cerebral cortex: Normal gestational landmarks identified using prenatal MR imaging." *American Journal of Neuroradiology* 22: 184–89.

2. Yi, Hong, Lu Xue, Ming-Xiong Guo, Jian Ma, Yan Zeng, Wei Wang, Jin-Yang Cai, Hai-Ming Hu, Hong-Bing Shu, Yun-Bo Shi, and Wen-Xin Li. 2010. "Gene expression atlas for human embryogenesis." *FASEB Journal* 24: 3341–50.

3. Greely, Henry T., Mildred K. Cho, Linda F. Hogle, and Debra M. Satz. 2007. "Thinking about the human neuron mouse." *American Journal of Bioethics* 7: 27–40.

Chapter 3   Animal Models

1. Yang, Xiaojun, Jian Zou, David R. Hyde, Lance A. Davidson, and Xiangyun Wei. 2009. "Stepwise maturation of apicobasal polarity of the neuroepithelium is essential for vertebrate neurulation." *Journal of Neuroscience* 29: 11426–40.

2. Frand, Alison R., Sascha Russel, and Gary Ruvkun. 2005. "Functional genomic analysis of *C. elegans* molting." *PLoS Biology* 3: e312.

3. Nadler, Jessica J., Fei Zou, Hanwen Huang, Sheryl S. Moy, Jean Lauder, Jacqueline N. Crawley, David W. Threadgill, Fred A. Wright, and Terry R. Magnuson. 2006. "Large-scale gene expression differences across brain regions and inbred strains correlate with a behavioral phenotype." *Genetics* 174: 1229–36.

Chapter 4   Early Events

1. Kelly, Christina, Alvin J. Chin, Judith L. Leatherman, David J. Kozlowski, and Eric S. Weinberg. 2000. "Maternally controlled β-catenin-mediated signaling is required for organizer formation in the zebrafish." *Development* 127: 3899–911.

2. Pyati, Ujwal J., Ashley E. Webb, and David Kimelman. 2005. "Transgenic zebrafish reveal stage-specific roles for Bmp signaling in ventral and posterior mesoderm development." *Development* 132: 2333–43.

3. Dhara, Sujoy K. and Steven L. Stice. 2008. "Neural differentiation of human embryonic stem cells." *Journal of Cellular Biochemistry* 105: 633–40.

Chapter 5   Neurogenesis

1. Corbo, Joseph C., Thomas A. Deuel, Jeffrey M. Long, Patricia LaPorte, Elena Tsai, Anthony Wynshaw-Boris, and Christopher A. Walsh. 2002. "Doublecortin is required in mice for lamination of the hippocampus but not the neocortex." *Journal of Neuroscience* 22: 7548–57.

2. Nikolaou, Nikolas, Tomomi Watanabe-Asaka, Sebastian Gerety, Martin Distel, Reinhard W. Köster, and David G. Wilkinson. 2009. Lunatic fringe promotes lateral inhibition of neurogenesis. *Development* 136: 2523–33.

3. Malik, Saafan Z., Melissa Lewis, Alison Isaacs, Mark Haskins, Thomas Van Winkle, Charles H. Vite, and Deborah J. Watson. 2012. "Identification of the rostral migratory stream in the canine and feline brain." *PLoS ONE* 7 (): e36016.

4. Babaoglan, A. Burcu, Kate M. O'Connor-Giles, Hemlata Mistry, Adam Schickedanz, Beth A. Wilson, and James B. Skeath. 2009. "Sanpodo: a context-dependent activator and inhibitor of Notch signaling during asymmetric division." *Development* 136: 4089–96.

Chapter 6   Later Events

1. Keilani, Serene and Kiminobu Sugaya. 2008. "Reelin induces a radial glial phenotype in human neural progenitor cells by activation of Notch-1." *BMC Developmental Biology* 8: 69.

2. Janssen, Ralf, Graham E. Budd, Nikola-Michael Prpic, and Wim G.M. Damen. 2011. "Expression of myriapod pair rule gene orthologs." *EvoDevo* 2: 5.

3. Solano, Pascal Jean, Bruno Magat, David Martin, Franck Girard, Jean-Marc Huibant, Conchita Ferraz, Bernard Jacq, Jacques Demaille, and Florence Machat. 2003. "Genome-wide identification of in vivo *Drosophila* Engrailed-binding DNA fragments and related target genes." *Development* 130: 1243–54.

Chapter 7   Becoming a Neuron

1. Brown, Jacquelyn A., and Paul C. Bridgman. 2009. "The disruption of the cytoskeleton during Semaphorin 3A induced growth cone collapse correlates with differences in actin organization and associated binding proteins." *Developmental Neurobiology* 69: 633–46.

2. An, Mahru C., Wiechun Lin, Jiefei Yang, Bertha Dominguez, Daniel Padgett, Yoshie Sugiura, Prafulla Aryal, Thomas W. Gould, Ronald W. Oppenheim, Mark E. Hester, Brian K. Kaspar, Chien-Ping Ko, and Kuo-Fen Lee. 2010. "Acetylcholine nega-

tively regulates development of the neuromuscular junction through distinct cellular mechanisms." *Proceedings of the National Academy of Sciences, U.S.A.* 107: 10702–7.

3. Chin, Lih-Shen, Lian Li, Adriana Ferreira, Kenneth S. Kosik, and Paul Greengard. 1995. "Impairment of axonal development and of synaptogenesis in hippocampal neurons of synapsin I-deficient mice." *Proceedings of the National Academy of Sciences, U.S.A.* 92: 9230–34.

4. Park, Kye Won, Lisa D. Urness, Megan M. Senchuk, Carrie J. Colvin, Joshua D. Wythe, Chi-Bin Chien, and Dean Y. Li. 2005. "Identification of new netrin family members in zebrafish: Developmental expression of netrin2 and netrin4." *Developmental Dynamics* 234: 726–31.

Chapter 8   Glia

1. Soustelle, Laurent, Nivedita Roy, Gianluca Ragone, and Angela Giangrande. 2008. "Control of *gcm* stability is necessary for proper glial cell fate acquisition." *Molecular and Cellular Neuroscience* 37: 657–62.

2. Titova, Elena, Robert P. Ostrowski, Arash Adami, Jerome Badaut, Serafin Lalas, Nirmalya Ghosh, Roman Vlkolinsky, John H. Zhang, and Abdre Obenaus. 2011. "Brain irradiation improves focal cerebral ischemia recovery in aged rats." *Journal of the Neurological Sciences* 306: 143–53.

3. Lowry, Natalia, Susan K. Gocerie, Patricia Lederman, Carol Charniga, Michael R. Gooch, Kristina D. Gracey, Akhilesh Banerjee, Supriya Punyani, Jerry Silver, Ravi S. Kane, Jeffrey H. Sterna, and Sally Temple. 2012. "The effect of long-term release of Shh from implanted biodegradable microspheres on recovery from spinal cord injury in mice." *Biomaterials* 33: 2892–901.

Chapter 9   Maturation

1. Lee, Hsiu-Hsiang, Lily Yeh Jan, and Yuh-Nung Jan. 2009. "*Drosophila* IKK-related kinase IK2 and Katanin p60-like 1 regulate dendrite pruning of sensory neuron during metamorphosis." *Proceedings of the National Academy of Sciences U.S.A.* 106: 6363–68.

2. Cyrenne, De-Laine M., and Gillian R. Brown. 2011. "Ontogeny of sex differences in response to novel objects from adolescence to adulthood in Lister-hooded rats." *Developmental Psychobiology* 52: 670–76.

3. Gogtay, Nitin, Jay N. Giedd, Leslie Lusk, Kiralee M. Hayashi, Deanna Greenstein, A. Catherine Vaituzis, Tom F. Nugent III, David H. Herman, Liv, S. Clasen, Arthur W. Toga, Judith L. Rapoport, and Paul M. Thompson. 2004. "Dynamic mapping of human cortical development during childhood through early adulthood." *Proceedings of the National Academy of Sciences U.S.A.* 101: 8174–79.

4. Yildirim, Murat, Oni M. Mapp, William G. M. Janssen, Weiling Yin, John H. Morrison, and Andrea C. Gore. 2008. "Post-pubertal decrease in hippocampal dendritic spines of female rats." *Experimental Neurology* 210: 339–48.

Chapter 10   Thinking about Intellectual Disability in the Context of Development

1. den Broeder, Marjo J., Herma van der Linde, Judith R. Bouwer, Bem A. Oostra, Rob Willemsen, and René F. Ketting. 2009. "Generation and characterization of *Fmr1* knockout zebrafish." *PLoS ONE* 4: e7910.

2. Raymond, F. L. 2006. "X linked mental retardation: A clinical guide." *Journal of Medical Genetics* 43: 193–200.

3. Gibson, Joanne H., Barry Slobedman, Harikrishnan KN, Sarah L. Williamson, Dimitri Minchenko, Assam El-Osta, Joshua L. Stern, and John Christodoulou. 2010. "Downstream targets of methyl CpG binding protein 2 and their abnormal expression in the frontal cortex of the human Rett syndrome brain." *BMC Neuroscience* 11: 53.

4. Hagerman, Randy, Jacku Au, and Paul Hagerman. 2011. "FMR1 premutation and full mutation molecular mechanisms related to autism." *Journal of Neurodevelopmental Disorders* 3: 211–24.

INDEX

Page numbers for entries occurring in figures are followed by an *f,* those for entries in investigative reading questions by an *i,* and those for entries in notes by an *n.*

autophagy, 201
autoradiography, 4
axin, 146f
axis determination, defined, 63
axons, 143

BALB/c mouse strain, 41
Banker, G., 144
basic helix-loop-helix (bHLH) proteins),
    83–86, 85f, 89–93, 92f, 98, 98f, 100,
    109, 138, 181, 186–188, 187f
bazooka (baz) gene, 64
benzaldehyde, 194n
β-catenin protein, 67–68, 69f, 70, 117n,
    125, 145–147, 146f, 164n
β-galactosidase, 7
Betz cells, 137f
bHLH (basic helix-loop-helix) proteins,
    83–86, 85f, 89–93, 92f, 98, 98f, 100,
    109, 138, 181, 186–188, 187f
bicoid (bcd) gene, 64–66, 65f, 74, 122,
    124
bilaterians (Bilateria), 58, 59f, 77
Biology of the Laboratory Mouse, 60n
birthdating, 3–6, 137
bithorax gene complex (BX-C), 127–128,
    127f
blastocoel, 27, 27f
blastocyst, 26–27, 27f, 42, 43f, 58, 72
blastoderm, 46, 49–50
blastodisc, 47f
blastomeres, 26, 54
blood–brain barrier, 173, 176–177
BOLD fMRI, 18–19, 215
bone morphogenetic protein (Bmp),
    68–70, 69f, 72–73, 75, 76f, 79i–80i,
    131f, 180, 181f, 186
Book of Daniel (Bible), 238n
Bownes, M., 49
Bownes stages, 49–52
brachiopods (Phylum Brachipoda), 59f
brain imaging methods, 17–19
brain stem, 216f
BrdU (bromodeoxyuridine), 4, 5f, 20i,
    81, 106, 114, 199, 204, 236
Brenner, S., 52, 60n, 165n
broken gene complex, 135
bright-field microscopy, 9

bristles, 89–93, 91f, 96f
bromodeoxyuridine (BrdU), 4, 5f, 20i,
    81, 106, 114, 199, 204, 236
bryozoans (Phylum Bryozoa), 59f
Bungarus multicinctus, 160

Caenorhabditis elegans, 6, 52–55, 74, 75,
    78, 78n, 83–88, 84f, 87f, 89, 110,
    111, 121, 134, 135, 153, 154f, 155,
    165n, 170–171, 171f, 172f
Caenorhabditis Genetics Center (CGC),
    53
Cajal-Retzius cells, 103, 103f, 137, 138,
    139, 141n
calcipressin protein, 234–235
calcium ions, 169, 171, 172f, 185, 186,
    194n, 214, 233
CAMs (cell adhesion molecules),
    156–157
Capecchi, M. R., 60n
Carnegie Institution of Washington,
    D.C., 23
Carnegie stages, 23, 26–29
Casablanca (film), 116n
caspase enzymes, 201
CAT (CT) scans, 18
caudal (cad) gene, 122
CDKs (cyclin-dependent kinases), 81
cell adhesion molecules (CAMs),
    156–157
cell cycle, 81–82, 82f
cellular blastoderm of *Drosophila,* 50
central nervous system (CNS), definition,
    24–25, 25f
centrosomes, 65f, 110, 148
cephalization, 63
cephalochordates, 59f
Cerberus protein, 72
cerebellar granule cells, 218f, 230, 233
cerebellum, 25f, 215–216, 218f, 230, 232
cerebral cortex, 25f, 101–110, 103f, 211i,
    216f
CGG repeats, 225–226
chaete, 90, 91f
Chalfie, M. L., 10
channelrhodopsin-2, 19–20
chicken breeds, 38
chimeras, 36i

extracellular signal regulated kinase (ERK), 79
extraembryonic ectoderm, 43f

F-actin, 149–151
fasces, 165n–166n
fascicles, 156
fasciclins, 156–157, 157f
fertilization, 26, 45, 53
fetal alcohol spectrum disorder, 232
fetal alcohol syndrome, 231–234, 239n
fetal MRI, 24
fetal tissue, 117n
Feulgen stain, 3–4
fibroblastic growth factor (FGF), 72, 77, 79n
filopodia, 150, 163f
Fire, A. Z., 17
floor plate, 129, 133f, 154, 155f, 156f, 180, 181f
FLP-*FRT* system, 7–8, 203
fluorescein, 165n
fluorescence microscopy, 9–10
FMR1 gene, 225
FMRP (fragile X mental retardation protein), 225
folic acid, 33–34
follistatin, 76, 76f
forward genetics, 39–40, 40f, 45, 49, 61i
fragile X mental retardation protein (FMRP), 225
fragile X syndrome, 225–227, 226f, 240i
frazzled gene (*fra*), 165n
Frizzled family of proteins, 70
fungi, 59f

GABA, 116n, 231
GABA$_A$ receptors, 20
GABAergic interneurons, 97, 137–138
G-actin, 149
gamma neurons, 205–206, 205f
ganglion mother cells, 5f, 94–96, 199
gap genes, 121–122, 123f
gastrotrichs (Phylum Gastrotricha), 59f
gastrulation, 27–28, 50, 55, 58, 72, 75–76, 79i–80i, 129

gel electrophoresis, 11–13
gel-shift assays, 101
gene dosage, 228
gene expression methods, 10–15
gene homologs, 78
Gene Ontology Consortium, 35i–36i
gene orthologs, 78
genetic robustness, 101
genomic library, 142i
germ band, 50
GFAP (glial fibrillary acidic protein), 117n, 186, 188, 190, 191f
GFP (green fluorescent protein), 8f, 10, 52, 106, 175, 186, 202–203, 223, 238n–239n
giant (*gt*) gene, 122
glial cells missing (*gcm*) gene, 174, 195i
glial scar, 170, 191
glial fibrillary acid protein (GFAP), 117n, 186, 188, 190, 191f
glioblastoma, 192–193
glioblastoma multifore, 192
glioblasts, 174, 176
gliogenesis, 181, 184–190
gliomas, 192
glutamate, 113
glycogen synthase kinase (GSK3), 70, 146, 146f
gnathostomulids (Phylum Gnathostomulida), 59f
Golgi, C., 238n
Golgi method, 214, 219–220, 223, 226–227, 232, 239n
gonadal steroids, 206
"grain of salt," 211n
granule cells of the cerebellum, 215–219, 216f
grasshopper embryos, 94–95, 94f
gray matter, 57, 190, 208, 211i
green fluorescent protein (GFP), 8f, 10, 52, 106, 175, 186, 202–203, 223, 238n–239n
Greenough, W. T., 2
growth cones, 21i, 147–151, 152, 153, 157, 165n, 166i, 178, 205
Grunwald, D. J., 60n
gyrencephalic brains, 117n
gyri, 32–33, 35i